Advanced Structured Materials

Volume 239

Common engineering materials are reaching their limits in many applications, and new developments are required to meet the increasing demands on engineering materials. The performance of materials can be improved by combining different materials to achieve better properties than with a single constituent, or by shaping the material or constituents into a specific structure. The interaction between material and structure can occur at different length scales, such as the micro, meso, or macro scale, and offers potential applications in very different fields.

This book series addresses the fundamental relationships between materials and their structure on overall properties (e.g., mechanical, thermal, chemical, electrical, or magnetic properties, etc.). Experimental data and procedures are presented, as well as methods for modeling structures and materials using numerical and analytical approaches. In addition, the series shows how these materials engineering and design processes are implemented and how new technologies can be used to optimize materials and processes.

Advanced Structured Materials is indexed in Google Scholar and Scopus.

Raneesh Balakrishnan · Rubiya Mohammed ·
Nebu George Thomas · Yogesh Bharat Dalvi
Editors

Biocompatible Nanocomposites

From Synthesis to Applications

Springer

Editors
Raneesh Balakrishnan
Department of Physics
Catholicate College
Pathanamthitta, Kerala, India

Nebu George Thomas
Pushpagiri College of Dental Sciences
Pushpagiri Institute of Medical Sciences
and Research Centre
Thiruvalla, Kerala, India

Rubiya Mohammed
Lakehead University
Thunder Bay, ON, Canada

Yogesh Bharat Dalvi
Pushpagiri Research Centre
Pushpagiri Institute of Medical Sciences
and Research Centre
Thiruvalla, Kerala, India

ISSN 1869-8433 ISSN 1869-8441 (electronic)
Advanced Structured Materials
ISBN 978-981-96-7449-7 ISBN 978-981-96-7450-3 (eBook)
https://doi.org/10.1007/978-981-96-7450-3

This Springer imprint is published by the registered company Springer Nature Singapore Pte Ltd.
The registered company address is: 152 Beach Road, #21-01/04 Gateway East, Singapore 189721, Singapore

Preface

The field of biocompatible nanocomposites has witnessed remarkable advancements in recent years, offering immense potential for transforming healthcare through innovative applications in tissue engineering, regenerative medicine, drug delivery, and implantable medical devices. As nanotechnology continues to evolve, it plays a crucial role in developing novel biomaterials by combining the unique properties of nanomaterials with the biocompatibility required for medical applications. The integration of these advanced materials into medical devices and therapies has the potential to improve patient outcomes, personalize treatment, and enhance the overall quality of healthcare.

This book, *Biocompatible Nanocomposites: From Synthesis to Applications*, is a comprehensive resource designed to provide an in-depth overview of the latest research, trends, and advancements in the field of biocompatible nanocomposites. The book is aimed at a wide audience, including graduate and doctoral students, research scientists, engineers, and professionals working in material science, biomedical engineering, nanotechnology, biotechnology, and medical physics. It will also serve as a valuable reference for industrial practitioners involved in the design and development of medical implants, drug delivery systems, and diagnostic tools.

The book begins with an introduction and discussion of the synthesis of biocompatible nanocomposites. Each chapter explores the diverse applications of these materials in the medical field, including tissue engineering, orthopedics, cardiovascular, dental, ophthalmic, and maxillofacial applications, and provides case studies that demonstrate their real-world impact on patient care and treatment outcomes.

The multidisciplinary approach employed in this book, with contributions from experts in materials science, engineering, biology, and medicine, ensures a comprehensive exploration of the subject. By bringing together leading researchers from academia, industry, and research institutions worldwide, this book not only compiles key discoveries but also offers practical insights into the development of next-generation nanocomposite materials for medical use.

Pathanamthitta, India	Raneesh Balakrishnan
Thunder Bay, Canada	Rubiya Mohammed
Thiruvalla, India	Nebu George Thomas
Thiruvalla, India	Yogesh Bharat Dalvi

Contents

Chapter 1
Biocompatible Nanocomposites: Synthesis, Challenges, and Future Prospects in Biomedical Applications

Rubiya Mohammed, Raneesh Balakrishnan, Nebu George Thomas, and Yogesh Bharat Dalvi

Abstract Biocompatible nanocomposites are the most promising types of nanocomposites, which combine biocompatibility with mechanical, structural, and functional properties. In this chapter, we focus on a detailed analysis of the synthesis techniques, potential applications, and drawbacks associated with these novel materials. The ability to fine-tune their structure and composition enables biocompatible nanocomposites to be specifically tailored for various biomedical applications. However, certain issues, including nanoparticle dispersion, toxicity, scalability, and regulatory challenges, pose significant obstacles to their use in clinical practice. This chapter also discusses the potential future advancements in fabricating multi-functional systems and explores large-scale manufacturing methods to enhance the transition from laboratory research to practical applications. By addressing these challenges, biocompatible nanocomposites have the potential to revolutionize healthcare, offering sustainable and efficient solutions to critical medical needs.

Keywords Biocompatible · Nanocomposite · Biodegradable · Tissue engineering · Drug delivery

R. Mohammed · R. Balakrishnan (✉)
Department of Physics, Catholicate College, Pathanamthitta, Kerala, India
e-mail: raneesh.b@catholicatecollege.ac.in

R. Mohammed
e-mail: rmoham13@lakeheadu.ca

R. Mohammed
Chemistry and Materials Science Program, Lakehead University, Thunder Bay, ON, Canada

N. G. Thomas
Pushpagiri College of Dental Sciences, Pushpagiri Institute of Medical Sciences and Research, Tiruvalla, Kerala, India

Y. B. Dalvi
Pushpagiri Research Centre, Pushpagiri Institute of Medical Sciences and Research, Tiruvalla, Kerala, India

R. Balakrishnan et al. (eds.), *Biocompatible Nanocomposites*, Advanced Structured Materials 239, https://doi.org/10.1007/978-981-96-7450-3_1

1.1 Introduction

Biocompatibility refers to the ability of a material to interact with biological systems without inducing adverse reactions or toxicity, thereby ensuring safety and efficacy in biomedical applications. The development of nanocomposites has led to biocompatible variants specifically designed to interact with biological systems and enable innovative solutions (Ramakrishna et al. 2003; Platel et al. 2008; Comes Franchini et al. 2012). A nanocomposite is a multi-phase material with at least one phase at the nanoscale. Nanoscale materials have a higher percentage of surface atoms, and the interaction between these nanoscale components leads to enhanced properties beyond those of the individual constituents. As research advances, biocompatible nanocomposites are poised to revolutionize healthcare by enabling novel diagnostic tools, personalized medicine, and targeted therapies (Yadav 2015; Manoukian et al. 2018; Mondal et al. 2016). Researchers have developed various biocompatible nanocomposite architectures, including ceramic–polymer hybrids, carbon nanotube-based composites and metal–organic frameworks, which exhibit tailored surface chemistry, microstructure, and degradation profiles. Rapid progress is being made in studying biodegradable polymers for use in biocompatible nanocomposites. Novel biodegradable synthetic polymers include polylactic acid (PLA), polycaprolactone (PCL), polybutylene succinate (PBS), etc. These polymers are gaining attention for their potential applications in healthcare and in developing sustainable biomedical products (Jin et al. 2013; El-Aassar et al. 2021).

The ability to fine-tune their structure and composition enables biocompatible nanocomposites to be specifically tailored for various biomedical applications. In tissue engineering, nanocomposite scaffolds including PCL/hydroxyapatite (HA) play a crucial role in enhancing bone formation and healing. In wound healing, the nanocomposite dressings containing silver nanoparticles exhibit antimicrobial functionality, thereby minimizing the risk of infection and increasing the rate of tissue healing (Gopinath et al. 2020). Also, the nanocomposite-based biosensors have numerous benefits for the point-of-care diagnostics of diseases such as diabetes and cardiovascular disorders (Koyyada and Orsu 2021; Rezvova et al. 2020). Their improved mechanical properties and biostability also make them suitable for implants and artificial organs. Nanotechnology combined with the principles of biocompatibility has a large scope of application in the improvement of health and well-being, which creates new opportunities in the field of biomedical sciences (Ziąbka et al. 2018; Al Kasar et al. 2024).

1.2 Properties of Biocompatible Nanocomposites

Biocompatibility is one of the most important criteria in the process of selecting materials which are to be implanted into the biological systems as it shows the compatibility of the material with the living organism. In order to assess these important characteristics, a set of in vivo and in vitro tests are carried out to determine cytotoxicity, inflammation, and the immune response. Zheng and his colleagues conducted a series of in vivo and in vitro experiments to assess the biocompatibility of ZnO nanoparticles. Based on their results, they were able to show that ZnO nanoparticles exhibit both cytotoxicity and organ-damaging effect, which should be of great concern if these nanoparticles are to be used in biological systems. The findings from both the in vitro and in vivo experiments showed some adverse effects, thus underlining the importance of monitoring the levels of ZnO nanoparticles in medical and environmental fields (Zheng et al. 2009). Such studies are important since it will enable one to determine whether the materials that are being suggested to be incorporated into biological systems are safe and suitable for the intended use so as to avoid or minimize potential risks and effects. On the other hand, some materials are safe and have excellent compatibility with the body and have been investigated for different biomedical applications. For instance, Dias et al. prepared a porous, biodegradable polyurethane nanocomposite based on PCL. The results of the biocompatibility evaluation of this composite material pointed to the possibility of its application in a number of tissue engineering applications, including cartilage, bone, heart valves, nerves, muscle, bladder, liver, and other tissues. This shows how biodegradable materials can be applied in the medical field, thus underlining the possibilities of their application in regenerative medicine (Dias et al. 2010).

There are some materials that can be classified as challenging due to their hydrophobic character and potential toxicity, such as carbon nanotubes (CNTs). In their pristine form, CNTs have very limited applications in the biological field. To overcome this, scientists have opted to functionalize CNTs, which involves the addition of chemical moieties to the surface of CNTs in order to increase their solubility, dispersion, and interactions with biological systems for the purpose of utilizing them in various applications. Zul et al. developed a nanocomposite by covalently functionalizing CNTs with a copolymer of polycaprolactone (PCL) and polyethylene glycol (PEG). This functionalization improved the mechanical properties of the CNT–polymer hybrid, making it more biocompatible and suitable for biological applications, thereby expanding its potential use in the biomedical field (Cajero-Zul et al. 2019). The use of nanocomposites has continued to evolve and has led to the creation of highly compatible and bioactive materials. For instance, the study by Joy et al. focused on developing a polycaprolactone/ graphene oxide/ strontium (PCL–GO–Sr) nanocomposite. The results of the hemolytic assays proved that the nanocomposite is non-toxic to the erythrocytes and the addition of GO-Sr nanohybrids to the PCL enhanced the bioresorption of the PCL matrix. Due to its bioactivity, the PCL–GO–Sr nanocomposite has the potential to be a good scaffold material for bone tissue engineering, thus highlighting the potential of PCL-based composites (Joy et al. 2023).

Furthermore, a bilayer nanofiber scaffold was developed using biocompatible and mechanically stable polycaprolactone (PCL)/zinc oxide and Capparis spinosa L. ethyl acetate extract (CSLE)/polylactic acid (PLA) layers by the electrospinning method. The scaffold was subjected to a Cell Counting Kit-8 (CCK-8) assay, and the results, along with antibacterial tests, confirmed the scaffolds' biocompatibility and antibacterial properties. This work opens new horizons for the development of functional scaffolds for different tissue engineering applications where biocompatibility and antibacterial potential are two main concerns (Zhu et al. 2023). These studies demonstrate the ongoing efforts to create biocompatible nanomaterials with improved functionality and safety profiles, underscoring the importance of continuous evaluation to ensure their effectiveness in medical and environmental applications.

The addition of nanofillers into polymer matrix greatly improves the degradability as well as the properties of the nanocomposite. This approach has garnered considerable attention in various fields, especially in tissue engineering applications, where the combination of biodegradability and functional properties is crucial. A lot of research has been done on biodegradable polymeric nanocomposites for tissue engineering applications. The study conducted by Armentano et al. aimed at discussing some important issues concerning the sterilization and modification of nanocomposite surfaces, which are paramount to enhancing the compatibility of nanocomposites with biological tissues. These modifications are quite crucial in order to enhance the efficacy of nanocomposites when used in vivo (Armentano et al. 2010). Another development in nanocomposite materials has been studied by Abad and co-workers where they developed a new nanocarrier in the form of a hydrogel nanocomposite made of chitosan (CS), carboxymethyl cellulose (CMC), and carbon nanotubes (CNTs). This nanocarrier was designed to load the drug quercetin (QC), which is a flavonoid with antioxidant, antiviral, and anticancer effects. The findings of their research showed that the biodegradability of the nanocomposite made it a potential drug carrier for the delivery of drugs. The degradation characteristics of the nanocomposite delivery system and the safety of degradation make it highly appealing for use in medical treatments (Rezaei Abbas Abad et al. 2023). In another study, Pasha et al. have used silicon nitride (Si_3N_4) nanoparticles as the reinforcing phase in magnesium-based nanocomposites, which is fabricated via ultrasonic stir casting method. This method helped in the proper dispersion of the nanoparticles in the matrix. In vitro immersion tests in simulated physiological fluids revealed that the nanocomposites exhibited a 2.2-fold reduction in the degradation rate compared to pure magnesium. This reduction in degradation rate is highly significant, as it can lead to improved longevity and performance of the material in biomedical applications, especially where controlled degradation is necessary (Pasha et al. 2022). The mechanical properties of biomaterials have a significant impact on their therapeutic uses. Carbonated hydroxyapatite/silicon dioxide (CHA/SiO_2) nanocomposites were fabricated by Taha et al. This study evaluated their mechanical properties, including compressive strength, microhardness, and various moduli. The results revealed that incorporating SiO_2 at 900 °C significantly improved all mechanical properties compared to the base sample. However, increasing the sintering temperature from 900 to 1300 °C led to a decline in the mechanical properties of all samples (Taha et al. 2020). In a study by Khalajabadi et al., Mg/HA/MgO nanocomposites were fabricated using hydroxyapatite (HA) and periclase (MgO) nanopowders as reinforcements for pure magnesium (Mg). The milled powder was pressed into pellets using a cold isostatic press. The strength and strain of these pellets were evaluated through compression testing, while their density was measured using the Archimedes principle. The study revealed that among the nanocomposites, the 72.5 Mg–27.5HA sample exhibited the highest ultimate strength. However, reducing the hydroxyapatite (HA) content from 20 to 5 wt% and adding 15 wt% magnesium oxide (MgO) did not result in a significant improvement in ultimate strength (Khalajabadi et al. 2016).

Thermal stability is crucial for ensuring the durability and performance of materials in various applications. These materials have shown significant improvements in their thermal properties, including enhanced dimensional stability, reduced thermal expansion, and increased heat resistance. Compared to pure polymers, polymer nanocomposites often exhibit higher glass transition temperatures and better heat stability, indicating a substantial improvement in their overall thermal performance (Gacitua et al. 2005; Wang et al. 2021).

1.3 Fabrication Methods

The development of biocompatible nanocomposites is largely dependent on the advancement of fabrication techniques which are very crucial for these materials for their applications. The need for improved materials in the medical, pharmaceutical, and environmental sectors has never been more compelling, and thus the need to develop and enhance fabrication methods is even more critical. The improvement of their performance is a function of the control of structure, properties, and functionality of these products. The development of these techniques allows to control the properties of nanocomposites to ensure that they are suitable for various applications in areas like drug delivery, tissue engineering, and implant development. All these techniques have advantages that affect the microstructure, mechanical properties, and the biocompatibility of the nanocomposite. The method to be used should be chosen carefully, as it determines the compatibility of the composite material with biological systems. This section will focus on some of the most common techniques used in the fabrication of biocompatible nanocomposites and discuss the mechanisms involved and the advantages of each technique.

1.3.1 Sol–Gel Process

The sol–gel process is one of the widely used and effective methods for synthesizing nanocomposites. The precision and the control over the process make it simple and adaptable. This method enables the synthesis of diverse nanostructures with desired characteristics for certain applications. The process starts with the choice of precursors which can be organic compounds or metal alkoxides. These

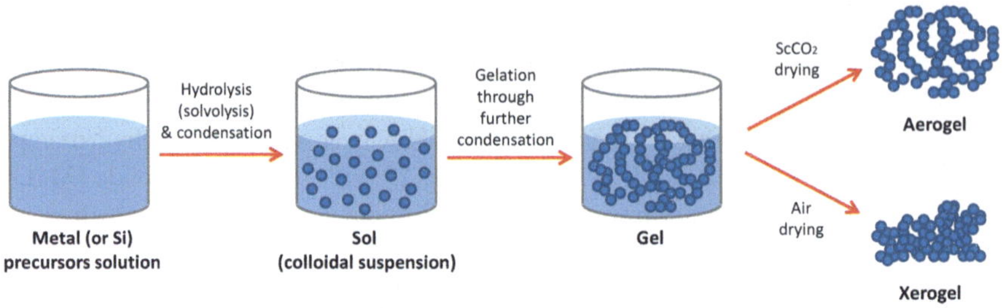

Fig. 1.1 General scheme of a sol–gel synthesis method. Reprinted from the reference Tao et al. (2018) (CC BY)

precursors undergo hydrolysis in which they form M–OH groups and then condensation in which the formation of M–O–M bonds occurs (Schubert 2015; Hench and West 1990; Orgaz and Rawson 1986). Using the sol–gel technique, researchers have successfully developed biocatalysts, such as those using methylotrophic yeast *Ogataea polymorpha*, immobilized in polymer-based nanocomposites, for the treatment of methanol-contaminated wastewater. These biocatalysts demonstrated high sensitivity, long-term stability (lasting 10–15 days), and protection against UV radiation and heavy metal toxicity (Lavrova et al. 2023). In another study, Taeby et al., preferred the sol–gel method over solid-state techniques for synthesizing strontium ferrite ($SrFeO_3$) nanoparticles (Taeby et al. 2024). For low-temperature synthesis of perovskite materials with large surface areas, high porosity, and fine-grained powders, the sol–gel process has proven to be the most suitable. Figure 1.1 illustrates the sol–gel process.

1.3.2 Electrospinning

Electrospinning is another widely used fabrication technique for nanocomposites, capable of producing continuous fibers with diameters on the nanoscale. The process employs electrical forces to draw fibers from a polymer solution. A high-voltage electric field (typically 10–30 kV) is applied between a needle and a collector, causing the polymer solution to be pumped through the needle and form a Taylor cone. As the solution exits the needle, the electric field stretches and thins the jet, while the solvent evaporates, leaving behind solid fibers. These fibers are collected on a substrate, forming a nonwoven mat (Bhardwaj and Kundu 2010; Rutledge and Fridrikh 2007). Electrospinning provides precise control over fiber diameter, alignment, and composition. However, one drawback of needle-based electrospinning systems is the challenge of maintaining a constant needle flow rate, which is necessary to form a stable and uniform Taylor cone (Li et al. 2021). Nanoparticles or biomolecules can be incorporated into the polymer solution to enhance the functionality of the resulting fibers. Electrospun nanocomposites have a broad range

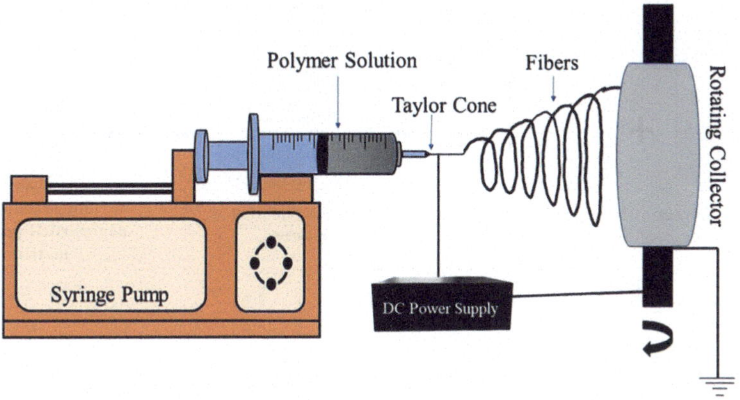

Fig. 1.2 Illustration of the electrospinning process. Reprinted with permission from reference Mohammed et al. (2024). Copyright 2024 Wiley

of applications, including tissue engineering, filtration, and biomedical devices. For instance, Canales et al. utilized electrospinning to incorporate calcium oxide (n-CaO) into a polylactic acid (PLA) matrix at filler contents of 10 wt% and 20 wt%, producing PLA/n-CaO fibers with uniform shape and diameter, which could potentially serve as scaffolds for bone tissue regeneration (Canales et al. 2023). Additionally, integrating cerium nanoparticles into chitosan–polyethylene oxide electrospun mats, as demonstrated by Petrova et al., led to enhanced material characteristics, such as reduced fiber diameter, increased swelling capacity, and reinforced mechanical properties, evidenced by a higher Young's modulus (Petrova et al. 2024). Figure 1.2 illustrates the electrospinning process.

1.3.3 Melt Intercalation Process

The melt intercalation process is a widely used, eco-friendly method for producing thermoplastic polymer nanocomposites. The process involves heating the polymer to high temperatures, adding fillers such as ceramic particles or nanoparticles, and then mixing thoroughly to ensure even dispersion. It is particularly suitable for large-scale manufacturing, making it a cost-effective choice. However, careful control of processing conditions is necessary to prevent damage to the surface modifications of the fillers, especially at elevated temperatures. This method enables the formation of intercalated or exfoliated nanocomposites, utilizing weak electrostatic interactions between the fillers and the polymer matrix for enhanced integration (DineshKumar 2016; Xu et al. 2005). Melt intercalation was employed by Huang et al. to incorporate nanoparticle hydroxyapatite (np-HAp) into polylactic acid (PLA), allowing PLA chains to penetrate the np-HAp layers without the use of solvents (Huang et al. 2020). A notable limitation of this process is the tendency for particles to aggregate within

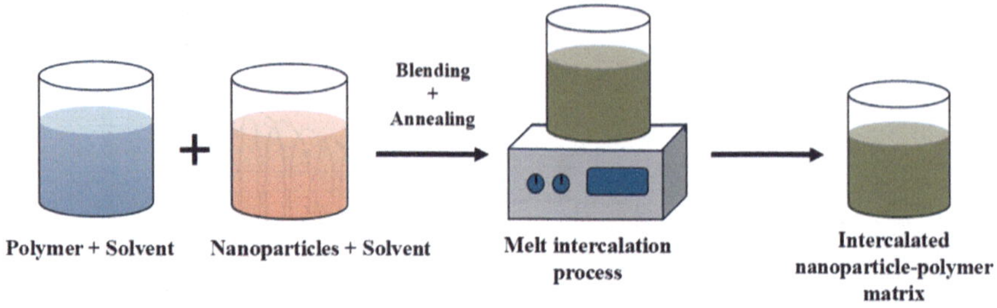

Fig. 1.3 Illustration of the melt intercalation process

the polymer matrix, especially at higher filler loadings, due to increased viscosity (Verma and Goh 2019). Figure 1.3 illustrates the melt intercalation process.

1.3.4 Solvent Casting

Solvent casting is one of the most common methods of preparing nanocomposites, wherein polymers and nanoparticles are dissolved in a solvent to form a solution. This solution is then cast onto a substrate like glass or silicone and when the solvent is allowed to evaporate, a thin film of nanocomposite remains. The thickness of the film can be regulated through the concentration of the solution used. The proper dispersion of nanoparticles in the matrix is very important to acquire the target properties. Solvent casting offers a high degree of accuracy in the control of the composition and the structure of nanocomposites and is therefore suitable for the fabrication of thin films, membranes, and coatings. This is a very simple and cost-effective method which is used extensively in various applications including biomedical instruments, sensors, and energy storage systems (Prasad et al. 2017; Borbolla-Jiménez et al. 2023). For example, Wang et al. have used solvent casting to produce the blend films of poly(lactide) (PLA) and poly(butylene adipate-co-terephthalate) (PBAT) to improve the properties of PLA. The resulting films showed higher flexibility, UV blocking capacity, and anti-fogging capabilities and at the same time had good mechanical properties and transparency (Wang et al. 2016). In another work, Govindasamy et al., improved the properties of chitosan by incorporating carboxymethyl cellulose, microcrystalline cellulose, and halloysite nanotubes using solvent casting method. The mechanical, chemical, and morphological properties of the resulting composites were evaluated, with the CS/MCC membranes showing the best mechanical performance (Govindasamy et al. 2015). A limitation of this method is that the polymer must be soluble in a volatile solvent or water, and the solution must maintain optimal viscosity and solid content for successful film formation. Figure 1.4 illustrates the solvent casting process.

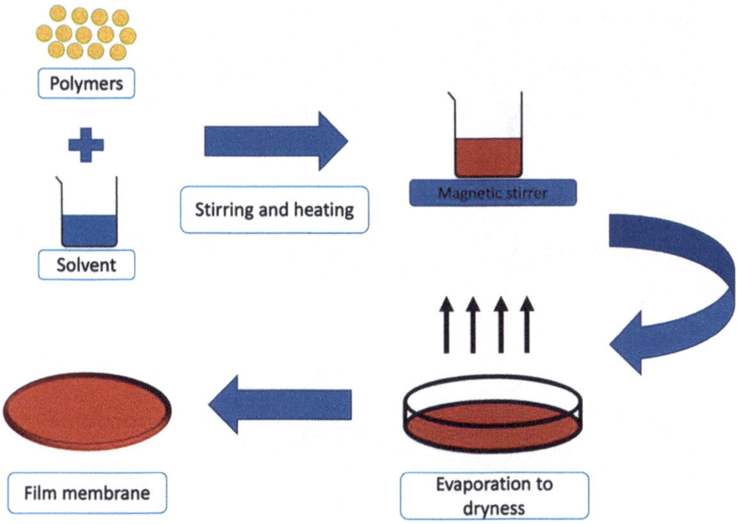

Fig. 1.4 Illustration of the solvent casting process. Reprinted from reference Prete et al. (2023). (CC BY)

1.3.5 *Spray Pyrolysis*

Another simple and affordable method for fabricating a range of materials, such as powders, ceramic coatings, thin and thick films, is spray pyrolysis. It fabricates thin films by spraying a solution onto a heated surface, where the ingredients react to form a compound (Salehabadi et al. 2023). Compared to other deposition techniques, spray pyrolysis has several advantages, such as high purity, excellent control over chemical homogeneity, and stoichiometry in multi-component systems (Panda and Palit 2024). The spray pyrolysis system comprises four key components: an atomizer, precursor solution, substrate heater, and temperature controller. The atomizer generates a liquid stream from the precursor solution. Depending on the atomization technique employed, droplets of varying sizes are generated from the precursor solution. The surface is preheated on a hot plate before applying the coating. The coating temperature and droplet size greatly influence the morphology and thickness of the final coating (Salehabadi et al. 2023). Bargavi et al. reported the use of atomized spray pyrolysis to coat commercially pure titanium (Cp-Ti) substrates with bioglass-zirconia (BG-Zr) composites. The coatings had an average thickness of 5–10 μm and demonstrated excellent bioactivity, cytocompatibility, and antibacterial properties (Bargavi et al. 2020). Figure 1.5 illustrates the spray pyrolysis process.

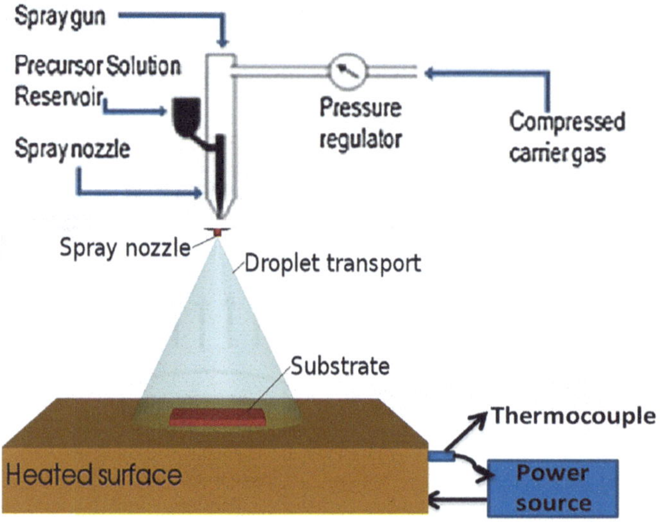

Fig. 1.5 Illustration of the spray pyrolysis process. Reproduced with permission from (Slimani et al. 2024). Copyright 2024 Elsevier

1.4 Overview of Applications

Biocompatible nanocomposites have revolutionized various biomedical fields, offering a diverse range of applications including regenerative medicine, controlled drug delivery, advanced implantable devices, and cancer therapy. Their unique combination of properties—biocompatibility, tunable mechanical strength, and responsiveness to external stimuli—has made them invaluable for addressing complex challenges in orthopedics, cardiovascular treatments, dental care, and beyond (Michael et al. 2018; Kiradzhiyska and Mantcheva 2019).

One notable example of these advanced materials is the nano-hydroxyapatite (nHA) embedded composite magnesium foam, fabricated via the powder metallurgy route by Parai et al. This material exhibits a bio-mimetic structure that closely resembles natural bone, effectively addressing both the structural and functional requirements of scaffolds for biomedical applications. The mechanical properties of these nanocomposites align remarkably well with those of cancellous bone, providing the necessary support for bone tissue engineering. Furthermore, their bio-mimetic design facilitates bone tissue regeneration, making them highly promising candidates for applications where mechanical integrity and biological compatibility are critical (Parai and Bandyopadhyay-Ghosh 2019). Similarly, Gelatin-PCL-nHA nanocomposites, developed through electrospinning, demonstrate exceptional potential for bone tissue engineering. These materials support proliferation and exhibit good viability in human osteoblasts, indicating their robust biocompatibility and ability to enhance cellular activities essential for bone regeneration. The alignment of their mechanical and biological properties underscores their suitability for creating a conducive environment for bone repair and growth (Gautam et al. 2021).

In addition to bone regeneration, nanocomposites like CNT/chitosan/glycerophosphate, prepared using the freeze-drying method, have been shown to exhibit enhanced electrical and mechanical properties. These features make them particularly suitable for applications requiring both structural strength and electrical conductivity. Importantly, these nanocomposites demonstrate no significant cytotoxicity, ensuring their safety and compatibility for biomedical use (Gholizadeh et al. 2017). Apart from their use in bone tissue engineering, biocompatible nanocomposites have been found to be effective in neural tissue engineering. For instance, hollow coaxial microfibers fabricated using a liquid–liquid coflowing electrospinning method with a poly-ε-caprolactone (PCL) sheath and a polyvinyl alcohol (PVA) core mixed with pheochromocytoma 12 (PC12) cells provide a 3D scaffold that guides neuronal growth. After selectively dissolving the PVA core, PC12 cells can grow, proliferate, and differentiate inside the hollow fibers, forming interconnected neuronal networks. This structural mimicry is crucial for the enhancement of cell differentiation as well as tissue formation and highlights the role of these material in the treatment of neural injuries and disorders (Wu et al. 2018).

The development of sophisticated scaffolds for tissue engineering has resulted in the generation of extremely effective structures that can mimic the characteristics of natural extracellular matrix and support the process of tissue formation. Of these, the α-Chitin hydrogel/nHA scaffolds fabricated through the freeze-drying method have the potential to be used in wound healing applications. These scaffolds have highly interconnected micropores, proper swelling characteristics, and very high protein adsorption capacity, which makes them ideal candidates for supporting cellular activities. The mineralization of the scaffolds along with their controlled degradation ensures that they have adequate mechanical strength, which gradually gets replaced by the new bone tissue. In addition, they are highly cytocompatible, which means that cells can easily adhere, proliferate, and spread throughout the scaffold in a uniform manner, thereby helping in the healing process. The α-Chitin hydrogel/nHA scaffolds have all the characteristics that are desirable for applications in wound healing which include mechanical strength and biological activity essential for the healing process (Kumar et al. 2011).

Besides wound healing, biomimetic composites have been developed by researchers that can mimic the structure, physical properties, and biochemistry of articular cartilage. Camarero-Espinosa et al. developed a multilayer polymer nanocomposite scaffold that mimics the distinct collagen orientation and extracellular matrix components of mature articular cartilage. It has mechanical properties similar to natural cartilage, and with the help of chemical cues, it stimulates the production of hyaluronic acid, which is very important for the integration of the cartilage tissue with the subchondral bone. Therefore, the biomimetic characteristics of this composite scaffold provide a suitable substrate for cartilage regeneration, thus making it a potential candidate for the treatment of articular cartilage defects (Camarero-Espinosa et al. 2016). Nanocomposites have been extensively used in regenerative medicine and drug delivery systems. These materials are developed for the enhancement of tissue healing and restoration with potential for specific drug delivery systems. Through solving problems like the specific delivery and duration of effects, biocompatible nanocomposites help to improve the results of treatments that are more likely to be less invasive.

The possibility of applying biocompatible nanocomposites in cancer therapy for targeted and controlled delivery of chemotherapeutic drugs is very promising. These nanocomposites are a combination of polymers, ceramics, and metals. They have been designed to attach specifically to tumor cells, thereby minimizing the effect on normal tissues. Hence, they reduce the side effects caused by traditional chemotherapy. By functionalizing the nanocomposites with targeting ligands, researchers aim to enhance drug specificity, address challenges such as drug resistance, and improve drug bioavailability (Shende and Shah 2021; Du et al. 2021). There is a growing interest in the use of synthetic polymers, including polycaprolactone (PCL), in the development of sutures, drug delivery systems, tissue engineering, and wound healing. Due to the biocompatibility and biodegradability, PCL has been found to be very useful in these applications. This is because it is capable of degrading into non-hazardous products when exposed to heat or chemicals, thus ensuring that it does not cause any harm to the surrounding tissues, which makes it suitable for use where the implant is required to dissolve after some time. The physicochemical and mechanical properties of pure PCL can be adjusted through different processes or the addition of different types of fillers into the basic matrix to enhance its performance (Rahimkhoei et al. 2023; Shirzaei Sani et al. 2021; Lam et al. 2007; Pedrosa et al. 2021).

1.5 Brief Overview of Upcoming Chapters

The upcoming chapters of this book explore the impact of nanocomposites in modern medicine as a means of solving the most pressing healthcare issues. It starts with state-of-the-art nano-biocomposites designed for specific medical uses, such as devices, orthopedics, cardiovascular systems, dental systems, and ophthalmic devices. These materials greatly improve the mechanical properties, biocompatibility, and stability of the devices, thus enhancing their integration with biological systems. The discussions also cover new material choices that minimize issues, enhance the patient's healing process, and lead to more favorable clinical outcomes, with examples from current practices and research.

The chapters evolve further to demonstrate the potential of nanocomposites in the fields of tissue engineering, regenerative medicine, and drug delivery. Relevant discussions include electrospinning, self-assembly, advanced 3D fabrication printing methods, etc., for the generation of scaffolds that are biomimetic in nature and the development of highly targeted drug carriers. The ability of these materials to support the microenvironment that is suitable for tissue reconstruction and also to provide the controlled release of drugs are discussed in detail. Also, there is a comprehensive analysis of the biocompatible and biodegradable material like polycaprolactone and its composites that are used in the preparation of bioabsorbable implants, sutures, and tissue engineering scaffolds.

The use of stimulus-responsive nanocomposites and their application in emerging fields such as 4D printing and advanced cancer therapies is discussed in detail. The materials that are sensitive to environmental factors, including temperature, pH, and light, present an opportunity for revolutionary advancements in the field of personalized medicine. In cancer therapy, the nanocomposites are studied with regard to their ability to release anticancer drugs with high precision, thereby reducing the side effects that are seen in conventional treatments and to target only the tumor cells.

Finally, the exploration is extended to enhance the stability of nanocomposites in vaccine delivery and diagnostic imaging. These materials significantly enhance vaccine stability, immunogenicity, and targeted delivery to cells, improving efficacy. Likewise, diagnostic imaging is also enhanced with nanocomposites that provide high image resolution and allow for the detection of diseases at earlier stages. All these sections together are indicative of the fact that nanocomposites have the possibility of transforming biomedical technologies and solving some of the existing problems in the process. This discourse is well rounded with both the current practical applications and the future research directions, thus giving a clear view about the topic.

1.6 Challenges and Opportunities

The use of nanocomposites (NC) has recently increased due to their improved mechanical properties, biodegradability and potential in bone tissue engineering and regeneration. The incorporation of bioactive nanoparticles into polymer matrices has been found to have the capability of replicating the properties of natural bone, thus making these materials suitable for development of bone scaffolds and fracture fixation. Also, there has been improvement in the functionalization of nanoparticles which has enabled control over the size and shape of the particles, thus improving the properties of polymer NC systems. Nevertheless, issues such as nanoparticle dispersion, toxicity, and biodegradability still pose a significant challenge to their adoption in clinical practice.

In the same way, 2D nanomaterials including graphene and its counterparts have been extensively studied for their potential in drug delivery and antimicrobial treatments. However, the gap between laboratory findings and industrial applications is significant due to issues such as reproducibility, viability, and safety of these materials especially in the complex environment of a biological system. Studies on the interactions of 2D materials with biological systems need to be expanded with a focus on genotoxicity, effects on the immune system and biodegradability in order to assess the safety and effectiveness of these materials for clinical use. Even though there is a huge potential of biocompatible nanocomposites in enhancing healthcare and sustainability, there are still important issues regarding the structure–property relationships, long-term biocompatibility, and the safety of these products for human use still need to be addressed in future research.

Future work on biocompatible nanocomposites needs to focus on how to address the main issues to enhance their use in the biomedical as well as industrial sectors. There should be more focus on developing manufacturing techniques that maintain product efficiency and quality consistently. In addition, toxicity, genotoxicity, and biodegradability analysis in sophisticated biological models are essential to reduce the possibilities of adverse effects and improve safety margins. The improvements in nanoparticle functionalization should remain a focus as they help in enhancing the mechanical and biological properties of the polymers. In addition, there is a need to conduct more preclinical and clinical trials to establish the safety, effectiveness, and durability of these materials when used in human applications. To enhance the application of nanocomposites, current research should be extended to develop multifunctional systems that combine drug delivery, tissue regeneration, and antimicrobial properties. By following such directions, it will be possible to further develop the application of nanocomposites in different areas with the emphasis on healthcare.

1.7 Conclusion

Biocompatible nanocomposites have the potential to revolutionize the biomedical sciences, as they provide a wide range of materials with various functionalities that can potentially replace conventional biomaterials and meet the present and future demands of the healthcare sector. The utilization of these materials in the areas of tissue engineering, drug delivery, cancer treatment, and regenerative medicine shows the flexibility and relevance of these materials in addressing medical problems. However, there are still several challenges that need to be addressed to enhance the use of these materials, including toxicity, scalability, and regulatory issues. Through the improvement of the fabrication methods, enhancement of biocompatibility, and encouragement of multidisciplinary studies, these materials will provide the platform for advanced medical treatments. Future research into the properties and applications of these materials is expected to advance the goal of developing personalized, efficient, and environmentally friendly healthcare products and services.

References

AL Kasar BCM, Asl SK, Asgharzadeh H, Peighambardoust SJ (2024) AIMS Bioeng 11:241

Armentano I, Dottori M, Fortunati E, Mattioli S, Kenny JM (2010) Polym Degrad Stab 95:2126

Bargavi P, Chitra S, Durgalakshmi D, Radha G, Balakumar S (2020) Mater Today Commun 25:101253

Bhardwaj N, Kundu SC (2010) Biotechnol Adv 28:325

Borbolla-Jiménez FV, Peña-Corona SI, Farah SJ, Jiménez-Valdés MT, Pineda-Pérez E, Romero-Montero A, Del Prado-Audelo ML, Bernal-Chávez SA, Magaña JJ, Leyva-Gómez G (2023) Pharmaceutics 15:1914

Cajero-Zul LR, López-Dellamary FA, Gómez-Salazar S, Vázquez-Lepe M, Vera-Graziano R, Torres-Vitela MR, Olea-Rodríguez MA, Nuño-Donlucas SM (2019) J Biomater Sci Polym Ed 30:163
Camarero-Espinosa S, Rothen-Rutishauser B, Weder C, Foster EJ (2016) Biomaterials 74:42
Canales D, Moyano D, Alvarez F, Grande-Tovar CD, Valencia-Llano CH, Peponi L, Boccaccini AR, Zapata PA (2023) Biomater Adv 153:213578
Comes Franchini M, Pucci A, Locatelli E, Naddaka Milani M, Passoni L, Matteoli M, Llop J, Reese T, Israel LL, Lellouche JP, Gil L, Gomez-Vallejo V (2012) Int J Nanomedicine 6021
Dias RCM, Góes AM, Serakides R, Ayres E, Oréfice RL (2010) Mater Res 13:211
DineshKumar PSS (2016) Int J Res Eng Sci (IJRES) 4:78
Du W, Zong Q, Guo R, Ling G, Zhang P (2021) Macromol Biosci 21:2100186
El-Aassar MR, Ibrahim OM, Al-Oanzi ZH (2021) Polymers (Basel) 13:3734
Gacitua W, Ballerini A, Zhang J (2005) Maderas. Ciencia y tecnología 7:159
Gautam S, Sharma C, Purohit SD, Singh H, Dinda AK, Potdar PD, Chou C-F, Mishra NC (2021) Mater Sci Eng, C 119:111588
Gholizadeh S, Moztarzadeh F, Haghighipour N, Ghazizadeh L, Baghbani F, Shokrgozar MA, Allahyari Z (2017) Int J Biol Macromol 97:365
Gopinath V, MubarakAli D, Vadivelu J, Manjunath Kamath S, Syed A, Elgorban AM (2020) Process Biochem 99(348)
Govindasamy K, Ramli MH, Pasbakhsh P, Pushpamalar V, Salamatinia B (2015) Polym Polym Compos 23:325
Hench LL, West JK (1990) Chem Rev 90:33
Huang Z, Wan Y, Peng M, Yang Z, Luo H (2020) Compos Sci Technol 185:107903
Jin G, Prabhakaran MP, Kai D, Annamalai SK, Arunachalam KD, Ramakrishna S (2013) Biomaterials 34:724
Joy A, Unnikrishnan G, Megha M, Haris M, Thomas J, Kolanthai E, Muthuswamy S (2023) Appl Nanosci 13:4471
Khalajabadi SZ, Abdul Kadir MR, Izman S, Marvibaigi M (2016) J Alloys Compd 655:266
Kiradzhiyska DD, Mantcheva RD (2019) Folia Med (Plovdiv) 61:34
Koyyada A, Orsu P (2021) Biomed Mater 16:065025
Kumar PTS, Srinivasan S, Lakshmanan V-K, Tamura H, Nair SV, Jayakumar R (2011) Int J Biol Macromol 49:20
Lam CX, Teoh SH, Hutmacher DW (2007) Polym Int 56:718
Lavrova DG, Zvonarev AN, Alferov VA, Khonina TG, Shadrina EV, Alferov SV, Ponamoreva ON (2023) Polymers (Basel) 15:458
Li Y, Zhu J, Cheng H, Li G, Cho H, Jiang M, Gao Q, Zhang X (2021) Adv Mater Technol 6
Manoukian OS, Arul MR, Sardashti N, Stedman T, James R, Rudraiah S, Kumbar SG (2018) J Appl Polym Sci 135
Michael FM, Khalid M, Walvekar R, Siddiqui H, Balaji AB (2018) Biodegradable and biocompatible polymer composites, Elsevier, p 33
Mohammed R, Chacko SK, Balakrishnan R, Thomas NG, Binsi PK, Muhamed Ashraf P, Krishnan N, Anil S (2024) J Appl Polym Sci e56308
Mondal D, Griffith M, Venkatraman SS (2016) Int J Polym Mater Polym Biomater 65:255
Orgaz F, Rawson H (1986) J Non Cryst Solids 82:57
Panda B, Palit N (2024) Comprehensive materials processing. Elsevier, p 200
Parai R, Bandyopadhyay-Ghosh S (2019) J Mech Behav Biomed Mater 96:45
Pasha MB, Rao RN, Ismail S, Özcan M, Prasad PS, Gupta M (2022) J Market Res 19:3803
Pedrosa MCG, dos Anjos SA, Mavropoulos E, Bernardo PL, Granjeiro JM, Rossi AM, Dias ML (2021) J Bioact Compat Polym 36:314
Petrova VA, Poshina DN, Golovkin AS, Mishanin AI, Zhuravskii SG, Yukina GY, Naumenko MY, Sukhorukova EG, Savin NA, Erofeev AS, Gofman IV, Ivan'kova EM, Dubashynskaya NV, Yakimansky AV, Skorik YA (2024) Polymers (Basel) 16(1787)
Platel RH, Hodgson LM, Williams CK (2008) Polym Rev 48:11

Prasad A, Sankar MR, Katiyar V (2017) Mater Today Proc 4:898
Prete S, Dattilo M, Patitucci F, Pezzi G, Parisi OI, Puoci F (2023) Natural and synthetic polymeric biomaterials for application in wound management. J Funct Biomater 14
Rahimkhoei V, Padervand M, Hedayat M, Seidi F, Dawi EA, Akbari A (2023) Int J Biol Macromol 253:126642
Ramakrishna S, Huang Z-M (2003) Comprehensive structural integrity. Elsevier, p 215
Rezaei Abbas Abad F, Pourmadadi M, Abdouss M, Behzadmehr R, Rahdar A, Pandey S (2023) Inorg Chem Commun 158(111621)
Rezvova MA, Glushkova TV, Makarevich MI, Nikishau PA, Kostjuk SV, Klyshnikov KYu, Ovcharenko EA (2020) Russ J Appl Chem 93:1412
Rutledge GC, Fridrikh SV (2007) Adv Drug Deliv Rev 59:1384
Salehabadi A, Enhessari M, Ahmad MI, Ismail N, Gupta BD (2023) Metal chalcogenide biosensors. Elsevier, p 143
Schubert U (2015) The sol-gel handbook. Wiley, p 1
Shende P, Shah P (2021) Int J Biol Macromol 175:281
Shirzaei Sani I, Rezaei M, Baradar Khoshfetrat A, Razzaghi D (2021) Int J Biol Macromol 182:1638
Slimani H, Nkele AC, Dagher S, Alshoaibi A, Bessous N, Akhozheya B, Ezema FI (2024) Opt Mater (Amst) 157:116405
Taeby M, Ashoub MH, Asghari M, Farsinejad A, Amiri M (2024) J Solgel Sci Technol 109:56
Taha MA, Youness RA, Ibrahim M (2020) Ceram Int 46:23599
Tao Y, Pescarmona PP (2018) Catalysts 8:212
Verma D, Goh KL (2019) Functionalized graphene nanocomposites and their derivatives. Elsevier, p 219
Wang LF, Rhim JW, Hong SI (2016) LWT - Food Sci Technol 68:454
Wang S, Chi H, Chen L, Li W, Li Y, Li G, Ge X (2021) Adv Polym Technol 2021:1
Wu Y, Ranjan VD, Zhang Y (2018) Adv Biosyst 2(5):e1700218
Xu J, Li RKY, Xu Y, Li L, Meng YZ (2005) Eur Polym J 41:881
Yadav P, Yadav H, Shah VG (2015) J Clin Diagn Res 9(9):ZE21–ZE25
Zheng Y, Li R, Wang Y (2009) Int J Mod Phys B 23:1566
Zhu P, Yin H, Wei J, Wu J, Ping D, Zhang X (2023) Int J Biol Macromol 242:125093
Ziąbka M, Menaszek E, Tarasiuk J, Wroński S (2018) Nanomaterials 8:764

Chapter 2
Biocompatible Nanocomposites: An Overview of Materials Used in Biomedical Applications

Sukumaran Anil, Nandita Suresh, Vishnupriya K. Sweety, Amartya Raj Suraj, and Nebu George Thomas

Abstract Biocompatible nanocomposites have emerged as a promising frontier in biomedicine, offering unprecedented opportunities for advanced healthcare solutions. This chapter provides a comprehensive overview of the materials used in biocompatible nanocomposites for various biomedical applications. It explores the unique properties of these nanocomposites stemming from the synergistic interactions between matrix materials and nanometer-sized fillers. The chapter delves into the classification of biocompatible nanocomposites, including polymer-based, ceramic-based, and metallic nanoparticle-polymer nanocomposites, discussing their synthesis methods, properties, and applications in areas such as orthopedic implants, dental materials, wound healing, and drug delivery systems. Specific nanocomposites, including hydroxyapatite-polymer, carbon nanotube-based, graphene-based, and protein-based nanocomposites, are examined in detail, highlighting their potential in addressing healthcare challenges and inspiring hope for the future of the field. The chapter also addresses the challenges facing these materials' development and clinical translation, such as long-term safety concerns, scalability issues, and evolving regulatory landscapes. Furthermore, it explores emerging trends in the field, such as

S. Anil (✉) · A. R. Suraj
Department of Dentistry, Oral Health Institute, Hamad Medical Corporation, Doha, Qatar
e-mail: drsanil@gmail.com

S. Anil
College of Dental Medicine, Qatar University, Doha, Qatar

N. Suresh
Department of Oral and Maxillofacial Diseases, Helsinki University and University Hospital, Helsinki, Finland

N. Suresh · V. K. Sweety
Pushpagiri Institute of Medical Sciences and Research Centre, Medicity, Perumthuruthy, Tiruvalla, Kerala, India

N. G. Thomas
Department of Periodontics, Pushpagiri College of Dental Sciences, Pushpagiri Institute of Medical Sciences and Research Centre, Medicity, Tiruvalla, Kerala, India
e-mail: nebugeorgethomas@pushpagiri.in

R. Balakrishnan et al. (eds.), *Biocompatible Nanocomposites*, Advanced Structured Materials 239, https://doi.org/10.1007/978-981-96-7450-3_2

smart, stimuli-responsive nanocomposites, and integrating nanocomposites with 3D printing technologies. The chapter discusses future research directions, emphasizing the need for comprehensive long-term safety studies, improved synthesis methods, and the development of multifunctional nanocomposites. By providing an in-depth analysis of biocompatible nanocomposites' current state and prospects, this chapter serves as a valuable resource for researchers, clinicians, and industry professionals in biomedical engineering and nanomedicine, inspiring them with the potential of nanocomposites in addressing healthcare challenges.

Keywords Biocompatible nanocomposites · Tissue engineering · Drug delivery systems · Nanomaterials · Biomedicine · Regenerative medicine

2.1 Introduction

Biocompatible nanocomposites, a class of materials composed of matrix material and nanometer-sized (1–100 nm) fillers, have emerged as a promising frontier in biomedicine. Their unique properties, stemming from the synergistic interactions between the matrix and nanofillers, offer unprecedented opportunities to develop advanced healthcare solutions (Alshangiti et al. 2023). These nanocomposites are designed to interact harmoniously with biological systems, making them suitable for various applications, from drug delivery and tissue engineering to medical implants and diagnostics. (Njuguna et al. 2014). Biocompatibility plays a pivotal role in designing and developing nanocomposites for biomedical applications. It refers to the material's ability to perform its intended function without causing any adverse effects on the surrounding biological environment (Kyriakides et al. 2021). Developing biocompatible nanocomposites that can be safely introduced into the human body without triggering an immune response or causing toxicity is a significant achievement.

One of the key advantages of biocompatible nanocomposites is their ability to mimic the natural extracellular matrix (ECM) of tissues. The ECM is a complex network of proteins, glycosaminoglycans, and other biomolecules that provide structural support and regulate cellular behavior (Zhao et al. 2021). By incorporating nanofibers, nanoparticles, or other nanostructures into biocompatible polymers, researchers can create scaffolds that closely resemble the native ECM. These scaffolds provide a conducive environment for cell adhesion, proliferation, and differentiation, making them ideal for tissue engineering applications. For instance, nanocomposite scaffolds composed of collagen and hydroxyapatite nanoparticles have successfully regenerated bone tissue, as they mimic bone ECM's natural composition and structure (Bauso et al. 2024).

Another application of biocompatible nanocomposites is in drug delivery systems. Conventional drug delivery methods often need help with poor bioavailability, non-specific targeting, and rapid clearance from the body. Nanocomposites offer a promising solution to these limitations by enabling controlled and targeted drug release (Sultana et al. 2022). By encapsulating drugs within nanoparticles or embedding them in biocompatible polymeric matrices, researchers can achieve sustained release profiles and enhance the therapeutic efficacy of drugs. Moreover, the surface of nanocomposites can be functionalized with targeting ligands that specifically bind to receptors overexpressed on diseased cells, allowing for selective drug delivery to the desired site of action. This targeted approach minimizes off-target effects and reduces the systemic toxicity associated with conventional drug therapies (Chehelgerdi et al. 2023).

Biocompatible nanocomposites have also found applications in diagnostic imaging and biosensing. Nanoparticles with unique optical, magnetic, or electrical properties can be incorporated into biocompatible matrices to create highly sensitive and specific imaging probes or biosensors. For example, superparamagnetic iron oxide nanoparticles (SPIONs) have been widely used as contrast agents in magnetic resonance imaging (MRI) due to their excellent biocompatibility and strong magnetic properties (Materón et al. 2021). When conjugated with targeting ligands, SPIONs can accumulate in specific tissues or cells, enabling non-invasive visualization and diagnosis of diseases like cancer. Similarly, nanocomposites containing gold nanoparticles or carbon nanotubes have been employed in developing biosensors for detecting various biomolecules, such as glucose, proteins, and nucleic acids. These nanocomposite-based biosensors offer high sensitivity, selectivity, and rapid response times, making them valuable tools for early disease detection and monitoring (Malik et al. 2023).

2.2 Classification

Biocompatible nanocomposites typically comprise a biocompatible polymer matrix reinforced with nanoscale fillers, such as nanoparticles, nanotubes, nanofibers, or nanoplatelets. These fillers can be derived from various materials, including metals, ceramics, carbon-based structures, or natural sources like biopolymers and minerals. Biocompatible nanocomposites are typically classified based on their constituent matrix and the type of nanofiller incorporated (Fig. 2.1, Table 2.1) (Rashid et al. 2024).

2.3 Biomedical Applications

Biocompatible nanocomposites have emerged as crucial materials in the biomedical field, attracting significant attention due to their exceptional properties and versatility (Khane et al. 2024). Their importance is multifaceted, encompassing enhanced mechanical properties that make them suitable for load-bearing applications like orthopedic implants and tailorable characteristics that allow fine-tuning for specific biomedical needs. These nanocomposites exhibit improved bioactivity due

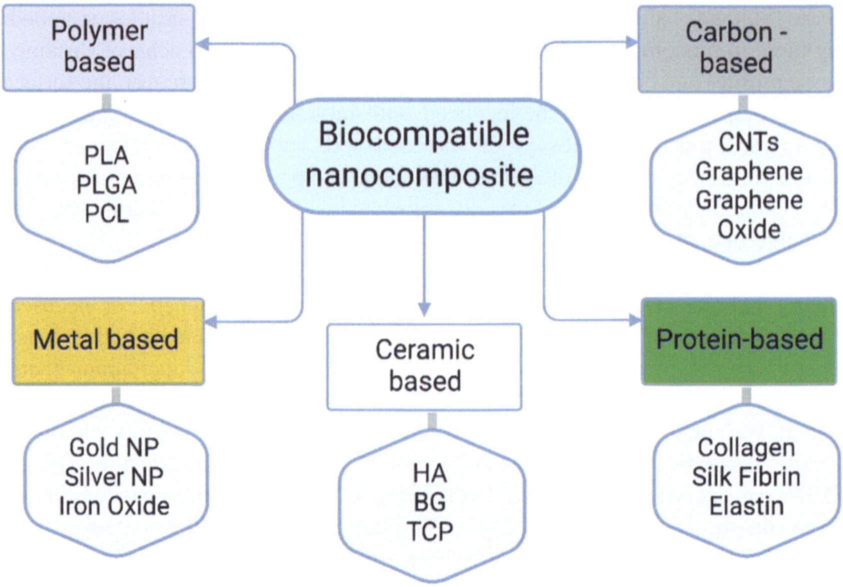

Fig. 2.1 Schematic representation of major categories and subcategories of biocompatible nanocomposites. Each category includes key examples of materials used in biomedical applications (PLA: Poly(lactic acid); PLGA: Poly(lactic-co-glycolic acid); PCL: Polycaprolactone; TCP: Tricalcium phosphate; NPs: Nanoparticles; CNTs: Carbon nanotubes; GO: Graphene oxide)

to their high surface area-to-volume ratio, promoting better interactions with biological systems (Alshangiti et al. 2023). Their multifunctionality enables the combination of various capabilities, such as antimicrobial properties and drug release, within a single material. Moreover, the hierarchical structure of many nanocomposites closely mimics natural tissues, potentially enhancing biocompatibility and functionality. These advantages have led to the widespread application of biocompatible nanocomposites across various biomedical domains, including tissue engineering, drug delivery, medical imaging, biosensors, and implantable devices (Goldberg et al. 2007) (Fig. 2.2).

2.3.1 Orthopedic Implants

One of the most significant applications of biocompatible nanocomposites is in the field of orthopedic implants. While strong and durable, traditional metallic implants often suffer from stress shielding, poor osseointegration, and the potential release of toxic ions. Nanocomposites address these challenges by offering materials with mechanical properties more closely matched to natural bone, improved bioactivity, and enhanced long-term stability. For instance, hydroxyapatite (HA)/

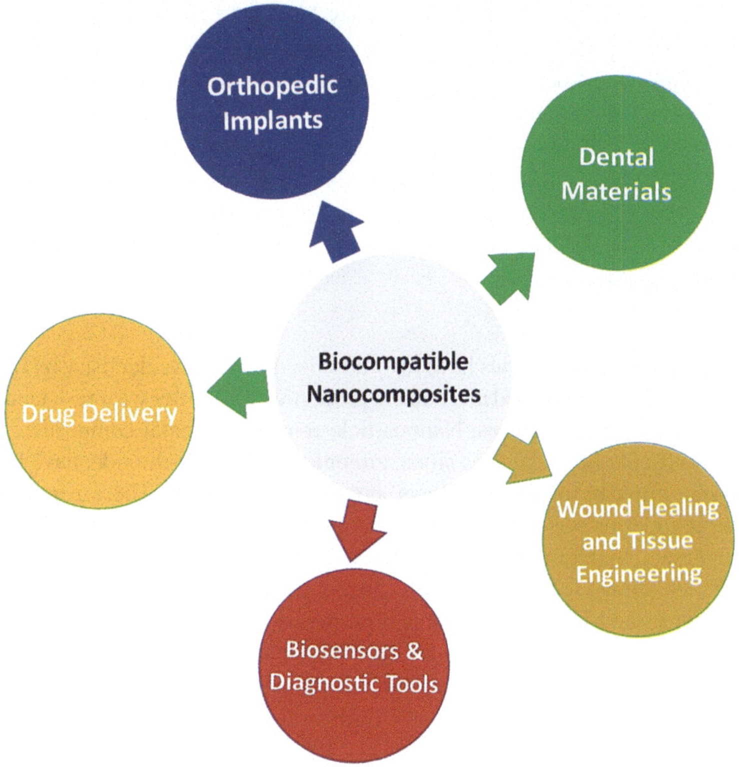

Fig. 2.2 Biomedical applications of nanocomposites–Schematic illustration depicting the major biomedical applications of nanocomposites

polymer nanocomposites have gained considerable attention for bone tissue engineering and implant applications (Abbas et al. 2023). These materials combine HA's osteoconductivity with polymers' flexibility and processability, resulting in implants that can better mimic the structure and properties of natural bone. The incorporation of HA nanoparticles into polymer matrices such as poly (lactic acid) (PLA) or poly(ε-caprolactone) (PCL) has been shown to enhance osteoblast adhesion, proliferation, and differentiation significantly (Canales et al. 2023). Moreover, the nanoscale dispersion of HA particles within the polymer matrix leads to improved mechanical properties, including increased strength and modulus, without compromising the material's ductility.

Another exciting development in orthopedic applications is using carbon nanotube (CNT)-reinforced metal matrix nanocomposites. By incorporating CNTs into titanium or titanium alloy matrices, researchers have developed implants with superior mechanical strength and reduced elastic modulus mismatch with natural bone, thereby minimizing stress shielding effects (Abazari et al. 2020). The exceptional strength-to-weight ratio of CNTs allows for the creation of lighter implants without compromising strength, which is particularly beneficial for load-bearing applications. Furthermore, the nanoscale roughness introduced by CNTs on the implant surface has promoted better cell adhesion and osseointegration. Studies have also reported on the potential antimicrobial properties of CNT-containing nanocomposites, which could help reduce the risk of implant-associated infections, a significant concern in orthopedic surgeries (Saleemi et al. 2022).

2.3.2 Dental Materials

Biocompatible nanocomposites have revolutionized restorative dentistry by offering improved aesthetics, enhanced mechanical properties, and better wear resistance than conventional dental composites. Nanoparticle-reinforced dental composites, incorporating nanoscale fillers such as silica, zirconia, or titanium dioxide, have become increasingly popular for dental fillings, crowns, and bridges. These nanocomposites exhibit superior polish retention and color stability, crucial for dental restorations' longevity, and aesthetic appeal (Azmy et al. 2022). The high surface area of nanoparticles allows for increased filler loading, resulting in reduced polymerization shrinkage and improved mechanical properties. For instance, dental nanocomposites containing silica nanoparticles have demonstrated significantly higher flexural strength and modulus than micro-filled composites while maintaining excellent polishability and optical properties that closely mimic natural teeth (Ren et al. 2024). Another notable application in dentistry is using nano-hydroxyapatite-based materials for enamel remineralization and treatment of dental hypersensitivity. Due to their small size and similarity to natural tooth minerals, nano-hydroxyapatite particles can effectively penetrate dentinal tubules and promote enamel reconstruction (Pushpalatha et al. 2023). When incorporated into toothpaste or mouthwash formulations, these nanocomposites have shown promising results in reducing dentinal hypersensitivity and enhancing the remineralization of early carious lesions. The bioactivity of nano-hydroxyapatite and its excellent biocompatibility make it an ideal candidate for preventive and restorative dental applications.

2.3.3 Wound Healing and Tissue Engineering

Biocompatible nanocomposites have also made significant strides in wound healing and tissue engineering. Advanced wound dressings incorporating nanocomposites offer enhanced healing capabilities, antimicrobial properties, and the ability to deliver therapeutic agents directly to the wound site. Chitosan-based nanocomposite scaffolds, for example, have gained attention for their ability to promote wound healing while providing antimicrobial protection. By incorporating silver or zinc oxide nanoparticles into chitosan matrices, researchers have developed wound dressings that combine chitosan's inherent wound-healing properties with metal nanoparticles' potent antimicrobial activity (Alven and Aderibigbe 2024). These nanocomposite dressings not only prevent bacterial infection but also create a moist environment conducive to healing, promote cell proliferation, and in some cases, exhibit controlled release of growth factors to accelerate the wound-healing process. In tissue engineering, nanocomposite scaffolds provide a suitable microenvironment for cell growth and tissue regeneration. Electrospun nanocomposite scaffolds, composed of biodegradable polymers reinforced with bioactive nanoparticles, have shown great promise in mimicking the natural extracellular matrix (ECM). These fibrous scaffolds offer high porosity, interconnected pore structures, and nanoscale surface features that promote cell adhesion, proliferation, and differentiation (Anjum et al. 2022). For instance, poly(lactic-co-glycolic acid) (PLGA) nanofibers incorporating hydroxyapatite nanoparticles have enhanced the osteogenic differentiation of mesenchymal stem cells, making them suitable for bone tissue engineering applications. The ability to tailor the composition and structure of nanocomposite scaffolds allows for the creation of biomimetic environments that can guide tissue regeneration in a controlled manner.

2.3.4 *Drug Delivery Systems*

The advent of biocompatible nanocomposites has significantly impacted the field of drug delivery. These materials offer unique advantages such as controlled and sustained drug release, protection of sensitive therapeutic agents, targeted delivery to specific tissues or cells, and improved drug solubility and bioavailability. Stimuli-responsive hydrogel nanocomposites represent a particularly innovative approach to drug delivery. These intelligent materials can change their properties in response to environmental stimuli such as pH, temperature, or magnetic fields, allowing for precise control over drug release kinetics. For example, nanocomposites combining temperature-sensitive polymers like poly(N-isopropylacrylamide) (PNIPAM) with magnetic nanoparticles have been developed for remotely controlled drug release (Sun et al. 2024). When exposed to an alternating magnetic field, the magnetic nanoparticles generate heat, causing the polymer network to collapse and release the encapsulated drug. This approach enables on-demand drug release and has shown promise in cancer therapy, where localized and controlled drug delivery is crucial (Farzin et al. 2020). Layer-by-layer (LbL) nanocomposite films represent another innovative approach to drug delivery. These thin films, composed of alternating layers of polymers and nanoparticles, can coat medical devices or create standalone drug delivery systems with precise control over release kinetics. The LbL assembly technique incorporates multiple therapeutic agents within a single nanocomposite film, enabling combination therapies or sequential drug release (Diez-Pascual and Rahdar 2022). For instance, antibiotic-loaded LbL nanocomposite coatings on orthopedic implants have shown the potential to prevent implant-associated infections while promoting osseointegration. The versatility of the LbL technique, combined with the unique properties of nanocomposites, opens up new possibilities for personalized medicine and targeted drug delivery.

2.3.5 *Biosensors and Diagnostic Tools*

Biocompatible nanocomposites have enabled the development of highly sensitive and specific biosensors for various applications. These nanocomposite-based biosensors offer advantages such as enhanced sensitivity, improved signal transduction, and the ability to detect multiple analytes simultaneously (Malik et al. 2023). Graphene-based nanocomposite biosensors have gained particular attention due to graphene's exceptional electronic properties and high surface area. By combining graphene with metal nanoparticles or enzymes, researchers have created electrochemical sensors capable of detecting biomarkers, pathogens, or small molecules with unprecedented sensitivity. For example, gold nanoparticle-decorated graphene oxide nanocomposites have been used to develop electrochemical immunosensors to detect cancer biomarkers, offering lower detection limits and wider linear ranges than conventional methods (Baruah et al. 2024). Quantum dot-polymer nanocomposites represent another exciting development in biosensing and bioimaging applications. Quantum dots offer significant advantages over traditional organic fluorophores with their unique optical properties and size-tunable emission. When incorporated into polymer matrices, quantum dots can be protected from the external environment while retaining fluorescent properties (Wagner et al. 2019). These nanocomposites have been used to develop fluorescence-based biosensors with improved sensitivity and multiplexing capabilities. For instance, quantum dot-embedded silica nanoparticles have been functionalized with antibodies to detect multiple cancer biomarkers in serum samples. The high quantum yield and photostability of quantum dot nanocomposites make them particularly suitable for long-term cellular imaging and in vivo diagnostic applications (Mladenovic et al. 2024).

2.4 Polymer-Based Nanocomposites

Polymers are commonly used as matrix materials due to their versatility and excellent biocompatibility. Natural polymers such as collagen, chitosan, and alginate, as well as synthetic polymers like polylactic acid (PLA), polyglycolic acid (PGA), and their copolymers (PLGA), are frequently employed (Satchanska et al. 2024). These polymers can be quickly processed and are generally biodegradable, which is beneficial for applications requiring temporary scaffolds in tissue engineering.

2.4.1 Polyethylene Glycol (PEG) Nanocomposites

Polyethylene glycol (PEG) is a biocompatible, hydrophilic polymer that uses various pharmaceutical and biomedical applications. It is frequently functionalized with reactive groups or grafted onto surfaces to minimize protein adsorption and impart stealth capabilities that evade immune recognition (Padin-Gonzalez et al. 2022). This has fueled significant interest in PEG nanocomposites fabricated by incorporation within inorganic nanoparticles, bioactive polymers, peptides, or drug conjugates. Such multifunctional integrated systems amalgamate beneficial aspects of PEG alongside tunable features tailored toward expanding biomedical uses like long-circulating drug carriers, tissue engineering scaffolds, biosensors, and anti-biofouling coatings (Omidian et al. 2023).

Polyethylene oxide polymers (PEG) are synthesized via catalytic ring opening polymerization of ethylene oxide monomers, which allows control over molecular weights. Copolymerization with other constituents like polylactic acid, polycaprolactone, etc., further enables tailoring physical properties like crystallinity and melting points. Strategies encompass grafting to preformed nanoparticles from particle surfaces outward via initiated polymerization and through PEG macromonomer insertion into propagating polymer chains. Optimizing density, conformation, and ligand lengths prevents aggregation in physiological media while benefiting circulation and biodistribution. Excess PEGylation risks blocking active sites, altering intended functions, or increasing immunogenicity conversely (Hoang Thi et al. 2020). Nano-construct design parameters like PEG configurations (linear, multi-arm branched), chain lengths, fusions with other polymers, and conjugation chemistries influence resultant properties like protein resistance, immune evasion, mucoadhesion, and even endosomal buffering capacities for gene delivery relevant for biomedical applications.

Biomedical Applications: PEG nanocomposites demonstrate expansive multifaceted utility leveraging anti-biofouling, immune evasion, and stabilization effects by tentative biointegration when configuring nanoparticle, protein, gene, and small molecule therapeutic carriers. Extensive investigations have expanded the fundamental understanding of structure–property co-dependencies, offering extensive fine-tuning flexibility toward desired biomedical outcomes spanning imaging, sensing, drug delivery, and scaffold engineering applications (Li et al. 2015).

Stealth Drug Delivery: PEGylation allows evading immune surveillance to sustain circulatory lifetimes for enhanced drug targeting via "passive" accumulation in pathological sites like tumors or inflammation loci with leaky vasculatures. Grafting alkyl chains also assists membrane fusion for intracellular delivery. Cleavable linkers permit programmed intracellular cargo release responding to definitive triggers like pH, enzymes, or externally applied alternating magnetic fields (Suk et al. 2016). Disease-site bioavailability improvements up to fivefold support reducing dosages and attendant toxicity risks.

Protein & Gene Therapy: Besides circulation longevity benefits, PEG addition preserves structural and functional protein stability against denaturation, undesirable cleavage, or enzymatic attack during storage and delivery phases. Improvements in efficacy were consistently demonstrated with PEGylated enzyme replacement formulations in treating lysosomal indications where rapid catalytic degradation otherwise limits bioavailability (Liu et al. 2022). Sustained DNA/RNA interference release for gene therapy also relies on cushioning protection, improved unpacking, and cytosolic escape afforded by PEG-polycation composite micellar carriers. However, the adverse effects of such introduced genetic manipulation cannot be ruled out. So exhaustive long-term evaluations, even in non-fatal non-communicable diseases, remain imperative.

Tissue Engineering: PEG incorporation as bioinert coatings over scaffolds comprising structural biomaterials like collagen prevents undesirable protein adsorption and cell adhesion for guided regeneration in desired cell types alone. Cell-resistant properties also minimize thrombogenic risks when blood contacts require devices like extracorporeal oxygenators (Zhu 2010). Further covalent enzyme immobilization introduces active cues promoting desired differentiation pathways.

Surface Grafting: Leveraging antifouling aspects, PEG coatings minimize biofilm formation on implanted devices like urinary catheters. They are also exploited as non-toxic dynamic surface coatings for biomedical microfluidics. When thermally or electrochemically triggered, they enable flow control, valves, and pumps via actuated transitions between non-fouling and adhesive states (Ozkan et al. 2022). Antibacterial properties can also stem from integrated silver nanoparticles or nitric oxide donors. Radiopaque liposomal contrast formulations use PEG-inserted bilayers for improved vascular imaging.

2.4.2 Poly (Lactic-Co-Glycolic Acid) (PLGA) Nanocomposites

The copolymer poly(lactic-co-glycolic acid) (PLGA) is a biodegradable and biocompatible polymer used widely for therapeutic devices and drug delivery systems. It undergoes hydrolytic degradation via ester linkages to metabolite monomers lactic acid and glycolic acid that enter metabolic pathways in the body (Makadia and Siegel 2011). By altering molecular weights and comonomer ratios, extensive tunability over degradation rates arises from days to months. This assists in the programmable release of encapsulated contents like small molecule drugs, proteins, cells, and genes (Kamaly et al. 2016). However, some areas for improvement around low drug loading capacities, incomplete release, the tendency for denaturing sensitive payloads, and limited functional capabilities motivate engineering advanced materials. PLGA nanocomposites integrate nanoparticles, structured matrices, or bioactive components alongside PLGA to overcome certain restrictions. Carbon nanotubes, magnetic particles, hydroxyapatite, chitosan, and gold nanoparticles constitute some commonly paired moieties (Gao et al. 2016).

Strategic pairings of nanoscale constituents with PLGA matrices give rise to multifunctional composites with synergistically tunable characteristics. Incorporating rigid nanofillers like organically modified montmorillonite clay, graphene derivatives, and nanodiamonds demonstrably enhances mechanical stiffness, hardness, tensile strength, and glass transition temperatures from additional crosslinking and filler interactions (Rocha et al. 2022). This counters otherwise rapid strength losses during degradation. Substituting fast-eroding glycolide-rich PLGA for slower lactic acid-releasing formulations creates multilayered constructs with sequential dismantling. Coupling hydrophobic polycaprolactone layers further influences moisture permeability and hydrolytic erosion gradients for sustained multiphasic release profiles (Makadia and Siegel 2011). Magnetic nanoparticles like iron oxide introduced within polymeric matrices render composites susceptible to external magnetic field control over locations. This benefits diagnostic agents, implant guidance, tissue engineering scaffolds, implantable biosensors, etc., requiring careful targeting or positioning flexibility (Mourdikoudis et al. 2021). Thermal or pH sensitivity arises through smart polymer integuments, while graphene oxides bestow photo-responsivity. This assists in triggering drug release activated by spatial tissue microenvironment signals or external heating/ irradiation cues. Functional ligands like antibodies grafted on surfaces or bioactive peptides/growth factors loaded in matrices render composites "smart." This imparts selective binding, targeting capabilities, improved biocompatibility, and recruitment/regulation of endogenous regenerative cell populations (El-Hammadi and Arias 2022).

Biomedical Applications: PLGA nanocomposites manifest promising multifunctionality for pharmaceutical, musculoskeletal, and diagnostic applications by amalgamating favorable attributes from structural matrices and functional nanoscale agents. Specifically, four areas witness extensive burgeoning innovation—(1) cancer therapeutics via improved targeting, controlled intratumoral activation, and combined therapy-imaging vehicles (2) immunomodulation using sequentially releasing films to limit fibrosis upon implantation (3) bone regeneration by recapitulating the microenvironmental stimuli guiding new tissue growth, and (4) bioimaging leveraging inherent material signatures.

Cancer Therapeutics: Chemotherapeutics loaded PLGA nanoparticles surface-conjugated to targeting moieties like folic acid and cell-penetrating peptides enhance permeability and retention while minimizing efflux in notoriously resistant neoplastic cells. Magnetic or fluorescent label tracking locations assist diagnostics, too. Sustained gene silencing is also attained through small interfering RNA payload homing using cationic polymer functionalization over PLGA cores (Zhang et al. 2022a). External stimuli-activated clever cargo releases from implanted composite depots further improve spatiotemporal control, which is otherwise challenging with systemic injections. Entrapped gold nanorod clusters generate photothermal ablation effects upon near-infrared (NIR) excitation. Thus, payload protection in circulation, programmed dissemination at malignant sites, reversal of chemoresistance, remotely triggered activation, and imaging/thermal duality underpin improved anticancer efficacy (Zhou et al. 2022). However, further evaluations against aggressive and metastatic varieties plus investigations ensuring no toxic off-target accumulations remain vital prerequisites before clinical adoption.

Immunomodulation: As foreign bodies, biomaterial implants often trigger inflammatory tissue reactions, prompting fibrotic encapsulation and device failure. So, immunosuppressants/anti-inflammatories require targeted delivery for extended functional lifetimes. Layer-by-layer configured PLGA composite films enable the sequential release of dexamethasone and antifouling polyethylene glycol to limit initial inflammatory and late fibrotic responses, respectively (Milosevic et al. 2020). Preliminary in vivo validations portray reduced foreign body reactions. If corroborated by expanded trials, such multilayered films can enhance biocompatible integration.

Bone Regeneration: Bone graft substitutes utilize osteoinductive PLGA composites to repair tissue deficits from trauma, infections, or skeletal abnormalities. Adding osteoconductive bone-like minerals (hydroxyapatite) and bioactive glass alongside osteogenic inductors (BMP peptides) creates multifunctional scaffolds mimicking native bone extracellular matrix (Ma et al. 2023). High porosity assists cellular infiltration while still providing mechanical integrity when load bearing. Degradation rates can be tuned to match tissue ingrowth to avoid collapsing scaffolds before maturation. Ingrowth quantification via longitudinal in vivo imaging in clinically relevant animal models constitutes the crucial next step for product development.

Bioimaging and Diagnostics: Introducing imaging-capable inorganic phases like luminescent quantum dots, gold nanoparticles, carbon nanotubes, and iron oxide nanoparticles makes PLGA composites inherently traceable via external optical, photoacoustic, or magnetic resonance scans. Conjugating targeting moieties like folic acid additionally yields theranostic "seek and detect" capabilities for malignant tissues (Jokerst et al. 2011). Chemilumniscence and fluorescence resonance energy transfer (FRET) techniques also facilitate highly sensitive diagnostics by incorporating efficient indicator dyes with stabilized enzyme systems. Spatiotemporal emission patterns shed light on localized biochemical events. Photoacoustic contrast improvements benefit surgical planning. Overall, such theranostic platforms bridge therapy and diagnosis for personalized treatments (Bhuckory et al. 2019).

2.5 Ceramic-Based Nanocomposites

Hydroxyapatite-polymer nanocomposites.

- Hydroxyapatite-polyethylene (HA-PE)
- Hydroxyapatite-polylactic acid (HA-PLA)
- Hydroxyapatite-polyglycolic acid (HA-PGA)
- Hydroxyapatite-polycaprolactone (HA-PCL)

Hydroxyapatite (HA) is a naturally occurring mineral form of calcium phosphate with the chemical formula $Ca_{10}(PO_4)_6(OH)_2$. It is the primary inorganic component of bone and teeth in vertebrates, making up approximately 60–70% of bone by weight. Due to its excellent biocompatibility, osteoconductivity, and similarity to the mineral phase of natural bone, hydroxyapatite has garnered significant interest in biomedical engineering for various applications, such as bone tissue engineering, drug delivery, and dental implants (Zastulka et al. 2023). However, the mechanical properties of pure hydroxyapatite, such as low fracture toughness and brittleness, limit its use in load-bearing applications. To overcome these limitations, researchers have focused on developing hydroxyapatite-polymer nanocomposites that combine the bioactive properties of hydroxyapatite with the mechanical strength and flexibility of polymers. These nanocomposites have shown great promise in improving the overall performance of hydroxyapatite-based biomaterials (Pielichowska and Blazewicz 2010).

Hydroxyapatite-polymer nanocomposites have emerged as promising materials for biomedical applications, particularly bone tissue engineering. Combining hydroxyapatite's bioactive properties with polymers' mechanical strength and flexibility, these nanocomposites offer enhanced performance compared to pure hydroxyapatite or polymer materials. The four commonly used hydroxyapatite-polymer nanocomposites discussed in this article, namely HA-PE, HA-PLA, HA-PGA, and HA-PCL, have shown great potential in promoting bone regeneration, improving the bioactivity and osseointegration of orthopedic implants, and serving as drug delivery systems (Mo et al. 2023). The synthesis methods for these nanocomposites, such as solution casting, melt compounding, and electrospinning, allow for the fabrication of various structures, including porous scaffolds and nanofibers, which can mimic the natural extracellular matrix of bone. The properties of these nanocomposites, such as mechanical strength, biodegradability, and bioactivity, can be tailored by adjusting the composition and processing conditions to suit specific biomedical applications.

HAp nanocomposite is an up-and-coming next-generation advanced material platform displaying multifaceted biofunctional capacities leveraging combined, augmented, and emergent properties from rationally pairing synergistically with polymeric or bio-macromolecular matrices (Fang et al. 2019). Calcium phosphate ceramics' history in medicine traces back over four decades, pioneered as bone graft substitute materials given innate capacities encouraging bone regeneration. Exciting advances since combining HAp with assorted matrices has expanded prospective uses beyond merely inert defect fillers toward bioactive platforms triggering stem cell recruitment for inductive regeneration, dynamic implants with drug delivery multitasking functionality, and injectable cement polymerizing in-situ for tissue-specific augmented healing responses (Yu et al. 2015).

2.5.1 Hydroxyapatite-Polyethylene (HA-PE) Nanocomposites

Polyethylene (PE) is a synthetic polymer known for its excellent mechanical properties, chemical resistance, and biocompatibility. It is widely used in biomedical applications, such as joint replacements and surgical sutures. Incorporating hydroxyapatite nanoparticles into a polyethylene matrix can enhance the bioactivity and osteoconductivity of the composite material while maintaining the mechanical strength and durability of polyethylene. HA-PE nanocomposites exhibit improved mechanical properties compared to pure polyethylene, with increased stiffness and strength (Ornaghi et al. 2023). Hydroxyapatite nanoparticles also enhance the composite's bioactivity, promoting bone cell adhesion, proliferation, and differentiation. These properties make HA-PE nanocomposites suitable for biomedical applications, such as bone tissue engineering scaffolds, orthopedic implants, and dental fillers. In bone tissue engineering, HA-PE nanocomposites can be fabricated into porous scaffolds that mimic the structure and composition of natural bone. These scaffolds provide a conducive environment for bone cell growth and facilitate the regeneration of damaged or diseased bone tissue (Saurav et al. 2024). HA-PE nanocomposites have also been explored as coatings for metallic implants to improve their osseointegration and long-term stability.

2.5.2 Hydroxyapatite-Polylactic Acid (HA-PLA) Nanocomposites

Polylactic acid (PLA) is a biodegradable and biocompatible polymer derived from renewable resources, such as corn starch or sugarcane. It has gained significant attention in the biomedical field due to its excellent mechanical properties, processability, and ability to degrade into non-toxic byproducts (Wu et al. 2023). Combining PLA with hydroxyapatite nanoparticles results in a nanocomposite material with enhanced bioactivity and osteoconductivity. HA-PLA nanocomposites can be synthesized using various methods, including solution casting, melt compounding, and electrospinning. Solution casting involves dissolving PLA in a suitable solvent, such as chloroform or dichloromethane, and dispersing hydroxyapatite nanoparticles in the solution. The mixture is then cast into a mold or on a substrate, followed by solvent evaporation. HA-PLA nanocomposites exhibit improved mechanical properties compared to pure PLA, with increased stiffness and strength. Incorporating hydroxyapatite nanoparticles also enhances the composite's bioactivity and osteoconductivity, promoting bone cell adhesion and differentiation (Bikiaris et al. 2023). Moreover, PLA's biodegradability allows for the controlled release of hydroxyapatite nanoparticles as the polymer matrix degrades, which can stimulate bone regeneration. HA-PLA nanocomposites have been extensively studied for bone tissue engineering applications. They can be fabricated into porous scaffolds with interconnected pores that facilitate cell infiltration, nutrient transport, and vascularization. These scaffolds have shown promising results in promoting bone regeneration in various animal models. HA-PLA nanocomposites have also been explored as coatings for orthopedic implants to improve their bioactivity and osseointegration.

2.5.3 Hydroxyapatite-Polyglycolic Acid (HA-PGA) Nanocomposites

Polyglycolic acid (PGA) is a biodegradable polymer with excellent mechanical strength and rapid degradation kinetics. It is widely used in biomedical applications, such as surgical sutures and tissue engineering scaffolds. Combining PGA with hydroxyapatite nanoparticles results in a nanocomposite material with enhanced bioactivity and osteoconductivity (Rocha et al. 2022). HA-PGA nanocomposites exhibit excellent mechanical strength and rapid degradation kinetics. The incorporation of hydroxyapatite nanoparticles enhances the composite's bioactivity and osteoconductivity, promoting bone cell adhesion and differentiation. The rapid degradation of PGA allows for the controlled release of hydroxyapatite nanoparticles, which can stimulate bone regeneration. HA-PGA nanocomposites have been explored for biomedical applications, particularly in bone tissue engineering. They can be fabricated into porous scaffolds with interconnected pores facilitating cell infiltration and nutrient transport (Eshgh et al. 2022). These scaffolds have shown promising results in promoting bone regeneration in vitro and in vivo. HA-PGA nanocomposites have also been investigated as bone fillers and dental implant coatings.

2.5.4 Hydroxyapatite-Polycaprolactone (HA-PCL) Nanocomposites

Polycaprolactone (PCL) is a biodegradable and biocompatible polymer with excellent mechanical properties and slow degradation kinetics. It is widely used in biomedical applications, such as drug delivery systems, tissue engineering scaffolds, and medical devices. Incorporating hydroxyapatite nanoparticles into a PCL matrix results in a nanocomposite material with enhanced bioactivity and osteoconductivity (Kayan and Kayan 2023). HA-PCL nanocomposites exhibit improved mechanical properties compared to pure PCL, with increased stiffness and strength. The incorporation of hydroxyapatite nanoparticles enhances the composite's bioactivity and osteoconductivity, promoting bone cell adhesion, proliferation, and differentiation. The slow degradation kinetics of PCL allow for the sustained release of hydroxyapatite nanoparticles, which can stimulate long-term bone regeneration. HA-PCL nanocomposites have been extensively studied for bone tissue engineering applications. They can be fabricated into porous scaffolds with interconnected pores that facilitate cell infiltration, nutrient transport, and vascularization (Wang et al. 2024).

These scaffolds have shown promising results in promoting bone regeneration in various animal models. HA-PCL nanocomposites have also been explored as coatings for orthopedic implants to improve their bioactivity and osseointegration. In addition, HA-PCL nanocomposites have been investigated as drug delivery systems for the controlled release of therapeutic agents, such as antibiotics and growth factors, to enhance bone healing and prevent infections (Chen et al. 2020).

Despite the significant progress in developing hydroxyapatite-polymer nanocomposites, their clinical translation challenges still need to be addressed. Further research is needed to optimize the synthesis methods, characterize these materials' long-term stability and degradation behavior in vivo, and evaluate their safety and efficacy in large animal models and clinical trials. Additionally, the scalability and reproducibility of the manufacturing processes need to be addressed to enable the commercialization of these nanocomposites. Hydroxyapatite-polymer nanocomposites are promising to advance biomedical engineering, particularly in bone tissue engineering and regenerative medicine (Ielo et al. 2022). With continued research and development, these materials have the potential to revolutionize the treatment of bone defects, fractures, and diseases, improving patient outcomes and quality of life.

2.6 Metallic Nanoparticle-Polymer Nanocomposites

Metallic nanoparticles, such as gold, silver, and iron oxide nanoparticles, have attracted significant attention in biomedical engineering due to their unique optical, electrical, and magnetic properties. These nanoparticles have been explored for various biomedical applications, including drug delivery, cancer therapy, bioimaging, and biosensing. However, the stability and biocompatibility of metallic nanoparticles in physiological environments remain challenging (Mody et al. 2010). To overcome these limitations, researchers have focused on developing metallic nanoparticle-polymer nanocomposites that combine the unique properties of metallic nanoparticles with the biocompatibility and functionality of polymers. The three commonly used metallic nanoparticle-polymer nanocomposites: gold nanoparticle-polyethylene glycol (AuNP-PEG), silver nanoparticle-chitosan (AgNP-chitosan), and iron oxide nanoparticle-dextran (IONP-dextran) (Falsafi et al. 2023).

1. Gold nanoparticle-polyethylene glycol (AuNP-PEG)
2. Silver nanoparticle-chitosan (AgNP-chitosan)
3. Iron oxide nanoparticle-dextran (IONP-dextran).

2.6.1 Gold Nanocomposites

Gold nanoparticles (AuNPs), nanoclusters, and nanostructures display unique optoelectronic properties attributable to quantum confinement and surface plasmon resonance. Attributes like biocompatibility, facile surface bioconjugation, and photothermal and photodynamic capabilities propelled gold nanostructures into diverse sensing and therapeutic realms. However, applications for standalone AuNPs remain restricted by aggregation risks, poor load capacities for drug delivery, and lack of stability in physiological contexts over long periods (Sarfraz and Khan 2021). This drove significant efforts toward fabricating gold nanocomposites by strategically integrating AuNPs within functional matrices like polymers, ceramics, or bioactive phases. Resultant multifunctional systems amalgamate beneficial characteristics from both constituents for expanding biomedicine utilities encompassing imaging, gene/drug delivery, tissue engineering, antimicrobials, biosensing, and cancer therapy (Vial et al. 2017).

AuNP synthesis frequently relies on the chemical reduction of gold salts (like HAuCl4) dissolved in aqueous or organic solutions using agents like citrate and borohydrides. This facilitates exquisite morphology selections like spheres, rods, and cages by tuning parameters like temperature, capping agents, and ratios. Capping molecules such as mercaptoundecanoic acid enable dispersion stability in colloidal suspensions for further fabrication through electrostatic or steric effects. Polymer phases aid processability and allow the coating to outperform AuNPs using layers-by-layer assembly, in-situ reduction, surface grafting, and entrapment. Biopolymers also bestow inherent biocompatibility alongside biofunctionality via integrated cell signaling peptides. Conversely, AuNPs grant optical tunability, photothermal conversions, electromagnetic field enhancements, and high X-ray absorption (Xue et al. 2017). So, emergent capabilities become realizable through such strategic composite building. Characterization techniques like UV–Vis spectroscopy, X-ray diffraction, electron microscopy, etc., help optimize configurations for intended functions by unraveling optical resonances, crystallinity, and composite homogeneity.

Photothermal Cancer Therapies: Visible/NIR resonance of AuNPs produces heat upon irradiation. So, targeted delivery to malignant tissues allows selective photothermal ablation by molecular or magnetic guidance. Co-delivery alongside chemotherapy/immunotherapy agents exerts synergistic effects (Alle et al. 2022). Moreover, AuNPs affect DNA replication and cell cycles to suppress growth further non-thermally. Multifunctionality arises from integrating imaging, thermotherapy, and drug action concertedly. Au-chitosan nanocomposites mixed with methotrexate demonstrated 50-fold higher uptake in HeLa cells via folate receptor interactions. This permitted combined photothermal and chemotherapeutic action using minute near-infrared doses with high biocompatibility. Similarly, Au-grafted graphene oxide complexes stabilize antitumor siRNA payloads for photothermal-gene therapy through extensive hydrogen bonding while enabling photothermal capacities. Au nanocomposites manifest versatility as "all-in-one" agents merging diagnostics, transport, and multimodal targeted treatment modalities (Dykman and Khlebtsov 2019). However, questions around long-term fate, metabolism, and toxicity from chronic low-level exposures remain largely uninvestigated.

Biosensing: AuNPs elicit quantifiable optical or electrochemical signals upon target analyte binding events. This assists ultrasensitive detection down to femto/picomolar levels for applications like medical diagnostics, food safety screening, etc. Conjugation chemistries and stabilizing matrices drive specificity. However, most efforts are limited to proof-of-concept in vitro demonstrations rather than testing in clinical samples at-site (Zhou et al. 2015). So, significant optimization and product development efforts remain before enabling point-of-care adoption.

Antimicrobial Action: Oligopeptides attached to AuNPs penetrate bacterial cell walls without engendering resistance. Photothermal excitation additionally assists localized synergy (Okkeh et al. 2021). Composite formulations can be coated on wound dressings or surgical tools like scalpels and sutures to prevent infectious site colonization.

Tissue Engineering: AuNP doping within scaffolds like collagen/hydroxyapatite sometimes boosts osteogenic differentiation of seeded mesenchymal cells and upregulates bone markers (Shen et al. 2021). Photo-crosslinking capabilities also assist in shaping flexibility. Evidence on short/long-term in vivo functionality remains to be sparse.

2.6.2 Silver Nanoparticle-Chitosan (AgNP-Chitosan) Nanocomposites

Silver nanoparticles (AgNPs) have gained significant attention in biomedical applications due to their broad-spectrum antimicrobial activity, attributed to the release of silver ions and their interaction with bacterial cell membranes. However, the stability and biocompatibility of AgNPs in physiological environments are a concern (Bruna et al. 2021). Chitosan, a natural polysaccharide derived from the deacetylation of chitin, has been used to functionalize AgNPs to improve their stability and biocompatibility while retaining their antimicrobial properties. AgNP-chitosan nanocomposites can be synthesized using various methods, including chemical reduction, microwave-assisted synthesis, and green synthesis approaches (Darakai et al. 2024).

AgNP-chitosan nanocomposites exhibit enhanced stability and biocompatibility compared to bare AgNPs. The chitosan functionalization prevents the aggregation of AgNPs and provides a protective layer that reduces their toxicity to mammalian cells while maintaining their antimicrobial activity. AgNP-chitosan nanocomposites have been explored for various biomedical applications, such as wound healing, tissue engineering, and drug delivery (Cinteza et al. 2018). In wound-healing applications, AgNP-chitosan nanocomposites can be incorporated into wound dressings or hydrogels to sustain the release of silver ions, promoting wound healing by reducing bacterial infection and inflammation. The antimicrobial properties of AgNPs, combined with the biocompatibility and wound-healing properties of chitosan, make AgNP-chitosan nanocomposites attractive for treating chronic wounds and burns. AgNP-chitosan nanocomposites can fabricate antimicrobial scaffolds that prevent bacterial colonization while promoting cell adhesion and proliferation in tissue engineering applications (Alven and Aderibigbe 2024). These scaffolds can regenerate tissues such as skin, bone, and cartilage.

2.6.3 *Iron Oxide Nanoparticle-Dextran (IONP-Dextran) Nanocomposites*

Iron oxide nanoparticles (IONPs), such as magnetite (Fe_3O_4) and maghemite (γ-Fe_2O_3), have been widely studied for biomedical applications due to their unique magnetic properties, biocompatibility, and biodegradability (Nguyen et al. 2021). IONPs exhibit superparamagnetic behavior, which allows them to be easily manipulated by an external magnetic field, making them attractive for magnetic resonance imaging (MRI), magnetic hyperthermia, and targeted drug delivery applications. Dextran, a natural polysaccharide composed of glucose units, has functionalized IONPs to improve their stability and biocompatibility in physiological environments. IONP-dextran nanocomposites can be synthesized using various methods, including co-precipitation, thermal decomposition, and microwave-assisted synthesis. Co-precipitation involves the simultaneous precipitation of iron precursors (Fe2+ and Fe3+) in the presence of dextran, which acts as a stabilizing ligand. This method allows for the one-step synthesis of IONP-dextran nanocomposites with controllable size and shape (Hauser et al. 2015). Thermal decomposition involves the high-temperature decomposition of iron precursors in the presence of dextran, resulting in the formation of monodisperse IONP-dextran nanocomposites with high crystallinity. Microwave-assisted synthesis uses microwave irradiation to rapidly heat and decompose iron precursors in the presence of dextran, providing a fast and energy-efficient method for synthesizing IONP-dextran nanocomposites.

IONP-dextran nanocomposites exhibit excellent stability and biocompatibility in physiological environments. The dextran functionalization prevents the aggregation of IONPs and provides a hydrophilic layer that improves their dispersibility in aqueous media. The superparamagnetic properties of IONPs are retained in the nanocomposites, allowing for their use in various biomedical applications, such as MRI contrast enhancement, magnetic hyperthermia, and targeted drug delivery. In MRI applications, IONP-dextran nanocomposites can be used as T2 contrast agents to enhance the negative contrast of tissues, improving the detection and diagnosis of various diseases, such as cancer and cardiovascular diseases (Avasthi et al. 2020). The dextran functionalization improves the circulation time and reduces the non-specific uptake of IONPs by the reticuloendothelial system, allowing for longer imaging windows and better contrast enhancement. In magnetic hyperthermia applications, IONP-dextran nanocomposites can generate localized heat upon exposure to an alternating magnetic field, selectively killing cancer cells. The dextran functionalization can be further modified with targeting ligands to improve the specific delivery of nanocomposites to tumor sites, minimizing the damage to healthy tissues.

2.7 Carbon Nanotube-Based Nanocomposites

Carbon nanotubes (CNTs) refer to hollow graphene cylindrical nanostructures with diameter dimensions typically under 100 nm and length scales ranging from hundreds of nanometers to several millimeters. Based on several concentric walls, single-walled (SWCNTs) and multi-walled (MWCNTs) variants exist, displaying exceptional tensile strength (~100 GPa) and electrical conductivity (104 to 106 S/cm) alongside high thermal stability and hydrophobicity (Aqel et al. 2012). Aggregation risks from van der Waals forces often complicate processing and applications for pure CNTs. This motivated significant efforts toward fabricating CNT nanocomposites by incorporating them as reinforcing phases within varied matrices. Polymers, ceramics, noble metals, and bioactive constituents are shared hosts for building multifunctional composites for diverse biomedical contexts spanning sensing, drug delivery, tissue scaffolds, and medical devices (Mohd Nurazzi et al. 2021).

SWCNTs synthesized via chemical vapor deposition or electric arc discharge methods allow the selection of diameters and properties based on intended roles. Aggregates complicate liquid phase dispersions. So, functionalization covalently using carboxyl, hydroxyl, and amine groups or non-covalently with surfactants like sodium dodecyl sulfate facilitates handling and allows further bioconjugation. Wrapping stabilizers, moreover, aid in de-bundling nanotubes for subsequent composite buildup. Well-dispersed CNTs readily get incorporated as reinforcing fillers within diverse matrices via techniques encompassing solution casting, melt mixing, and in-situ polymerization (Kharlamova et al. 2022). Growing CNT forests on substrate surfaces using chemical vapor deposition marks another route granting firm interfacial attachment. Desired dispersion homogeneity, density, orientations, and networking dictate resultant properties. However, characterizing distribution patterns across bulk composites remains challenging. Microscopy methods help but remain restricted to surface distributions alone. So electrical, mechanical, and thermal assessments act as bulk proxies for network formations through percolation theory models, for example.

Biomedical Applications

Carbon nanotubes are now actively developed across contexts like antibacterial wound dressings, neural growth substrates, biosensors, bone scaffolds, and drug delivery vehicles. Selected cases highlight multifaceted possibilities and ongoing promise.

Antimicrobial Wound Healing: CNTs directly damage microbial cell walls by disrupting membranes. Composites with silver nanoparticles, chitosan biopolymers, and other matrices boost antibacterial action for accelerated wound healing (Nandhini et al. 2024). Photothermal excitation capabilities from near-infrared irradiation further aid in controlling sterile inflammation, which is beneficial for tissue repair cascades. Such CNT-impregnated pads successfully healed wounds in diabetic murine models.

Neural Interfaces and Scaffolds: The biocompatibility, conductivity, and nanoscale mimic complex neural network morphologies provide cell signaling capabilities to guide nerve tissue regeneration. Conducting CNT lattices with growth factor decoration assists neuronal differentiation and synapse formation for applications like retinal implants or cortical neuroprosthetics (Filippi et al. 2024).

Biosensing: Extremely high surface area-to-volume ratios yields unprecedented, immobilized probe packing densities for ultrasensitive fluorescence, chemiresistor, or electrochemical mode analyte detection. Glucose sensors rely on changes in electrical properties from redox reactions or binding events (Speranza 2021).

Drug Delivery: High payload capacities due to surface sorption enable encapsulation of diverse therapeutic compounds for targeted, sustained release to tumors and pathologic sites. Stimuli-triggered liberation responding to light, pH, or external electromagnetic field changes provides spatiotemporal dosage control (Tian et al. 2022).

2.7.1 Carbon Nanotube-Polyethylene Glycol (CNT-PEG) Nanocomposites

CNT-PEG nanocomposites combine the properties of CNTs with the hydrophilicity, biocompatibility, and non-toxicity of PEG. These nanocomposites can be synthesized using covalent functionalization or non-covalent wrapping methods. Covalent functionalization involves chemically attaching PEG chains to the surface of CNTs. At the same time, non-covalent wrapping involves the physical adsorption of PEG onto CNTs through hydrophobic interactions or π-π stacking (Shar et al. 2023). CNT-PEG nanocomposites exhibit improved dispersibility in aqueous media and enhanced biocompatibility compared to pristine CNTs. They have been explored as drug delivery vehicles, cancer therapy agents, and biosensors for detecting various biomolecules (Tang et al. 2021).

2.7.2 Carbon Nanotube-Chitosan (CNT-Chitosan) Nanocomposites

CNT-chitosan nanocomposites combine the properties of CNTs with the biocompatibility, biodegradability, and antimicrobial properties of chitosan. These nanocomposites can be synthesized using solution mixing, in-situ polymerization, or electrostatic interaction methods (Rao et al. 2022). CNT-chitosan nanocomposites exhibit improved mechanical properties, electrical conductivity, and thermal stability compared to pure chitosan. They have been explored for tissue engineering applications, wound healing, drug delivery, and biosensing.

2.7.3 Carbon Nanotube-Polylactic Acid (CNT-PLA) Nanocomposites

CNT-PLA nanocomposites combine the properties of CNTs with the biodegradability and biocompatibility of PLA. These nanocomposites can be synthesized using solution mixing, melt blending, or in-situ polymerization methods. CNT-PLA nanocomposites exhibit improved mechanical properties, electrical conductivity, and thermal stability compared to pure PLA (Shar et al. 2023). They have been explored for tissue engineering applications, drug delivery, and biosensing.

2.8 Chitosan-Based Nanocomposites

Chitosan refers to a linear polysaccharide composed of randomly distributed β-(1–4)-linked D-glucosamine and N-acetyl-D-glucosamine derived via partial deacetylation of chitin, the main structural element in crustacean shells, and fungal cell walls. Unlike parent chitin, protonating abundant-free amine groups at slightly acidic to neutral pH allows chitosan salts to dissolve readily, yielding cationic polyelectrolyte solutions (Aranaz et al. 2021). Subsequent properties like biocompatibility, biodegradability to non-toxic metabolites in mammals by human enzymes like lysozyme, hypoallergenic, bioadhesive, and intrinsic antimicrobial properties have made chitosan popular across diverse biomedical contexts, including wound healing, drug delivery, and tissue engineering. Chitosan scaffolds are readily generated from solutions via versatile techniques, including freeze-drying, electrospinning, gas foaming, and 3D printing, providing control over geometry, porosity, pore architecture, and crystallinity. Higher deacetylation degrees yield enhanced mechanical properties but impede solubility above pH ~ 6.5 as deprotonation reduces cationic charge density (Levengood and Zhang 2014). MW reaches 50–1000 kDa, which dictates polymer chain entanglements and viscosities at concentrations. Thus, selecting appropriate grades aids optimal composite manufacture.

Wound Dressing Materials: Chitosan biopolymers are inherently non-toxic, hemostat, and bacteriostatic, promoting fibroblast proliferation to accelerate dermal healing and tissue regeneration with moderate exudate absorption capacity. Nanocomposite fabrics impregnated with metallic silver or zinc oxide further combat bacterial colonization, including antibiotic-resistant strains (Singh et al. 2017). Photothermal therapy relevance arises through infrared-absorbing carbon nanotube incorporation, enabling wireless external stimulation of beneficial heat and reactive oxygen production. Such multifunctional and stimuli-responsive capacities portend smart bandages that reactively modulate local wound microenvironments, though costs remain high currently.

Bone Regeneration Scaffolds: The cationic polysaccharide structure in chitosan nanofibers or membranes mimics sulfated glycosaminoglycans, an essential bone extracellular matrix (ECM) component. This intrinsically promotes osteogenic progenitor cell adhesion and differentiation phenomenologically similar to Heparin. Coating chitosan scaffolds with hydroxyapatite nanocrystals further emulates the composite architecture of natural bone-driving osteogenesis and calcified matrix deposition in vitro and preclinical models (Soriente et al. 2022). Additional doping elements like silica or silver enhance antimicrobial qualities, reducing graft rejection risks from infection. Degradation synchronized to progressive regeneration also remains realizable by tuning nanofiller solubility rates. Thus, chitosan nano-scaffolds continue showing tremendous promise, augmenting bone defect healing substantially faster than untreated controls.

Anti-cancer Drug Delivery: Chitosan nanoparticle stealth coatings over lipid vesicles cutaneously administering chemotherapeutics suppress tumor growth up to 94% relative to free drugs in melanoma models. Efficient interstitial colloidal transport likely assists in enhancing bioavailability. Folic acid functionalization additionally induces active targeting to enter folate-hungry malignant cells preferentially. Sustained and localized delivery thus limits systemic toxicity (Sachdeva et al. 2023). External ultrasound-mediated liberation focusing dosage timing using integrated phase change perfluorocarbon nano-droplets introduces another spatiotemporal trigger strategy currently undergoing fine-tuning. Chitosan nanocarriers display versatility, enhancing tumor-localized bioavailability for numerous potent cytotoxics otherwise challenged by dose-limiting adverse effects and minimal selectivity.

Gene Therapy Vectors: Gene vectors facilitate nucleic acid therapeutic delivery by stabilizing vulnerable payloads, facilitating cell transport and endosomal escape into cytosol/nucleus. Given protonated amines at physiological pH, chitosan nanoparticles condense and shield anionic DNA/RNA strands into complexes via electrostatic interactions for transfection while displaying negligible innate toxicity (Santos-Carballal et al. 2018). Mannose functionalization promotes selective uptake by antigen-presenting dendritic cells and macrophages, boosting vaccine potency. Photothermally triggered endo/lysosomal disruption using integrated graphene or gold nanomaterials also prevents degradation, enhancing cytosolic mRNA deliveries. Chitosan nano complexes continue garnering attention as versatile gene delivery vectors balancing efficacy and safety requisite for clinical translations.

2.9 DNA-Based Nanocomposites

DNA-based nanocomposites have emerged as a promising class of materials for various biomedical applications owing to their unique properties, such as programmable self-assembly, molecular recognition, and biocompatibility. These nanocomposites consist of DNA molecules conjugated with inorganic nanomaterials, such as gold nanoparticles, carbon nanotubes, or graphene oxide, where the DNA component provides specific binding and targeting capabilities (Jiang et al. 2016). In contrast, the inorganic component offers additional functionalities, such as optical, electrical, or mechanical properties. By designing the DNA sequences rationally and selecting the appropriate inorganic nanomaterials, researchers can create

DNA-based nanocomposites with tailored properties for specific applications, such as biosensing, drug delivery, and tissue engineering (Hu and Fan 2022).

DNA-based nanocomposites, including DNA-gold nanoparticle conjugates, DNA-carbon nanotube conjugates, and DNA-graphene oxide conjugates, have emerged as promising materials for various biomedical applications, such as biosensing, drug delivery, gene therapy, and tissue engineering (Abu-Salah et al. 2010). By combining the molecular recognition properties of DNA with the unique optical, electrical, and mechanical properties of inorganic nanomaterials, these nanocomposites offer new opportunities for developing intelligent and responsive nanomaterials with tailored properties and functions. The synthesis methods for these nanocomposites, such as direct adsorption, thiol-gold chemistry, linker-mediated conjugation, non-covalent wrapping, covalent functionalization, and non-covalent adsorption, allow for the controlled assembly and functionalization of the nanomaterials with DNA molecules. The resulting nanocomposites exhibit unique properties, such as programmable self-assembly, molecular recognition, optical and electrical modulation, and enhanced stability and biocompatibility, which can be exploited for various biomedical applications (Amina and Guo 2020).

DNA-gold nanoparticle conjugates have been explored for biosensing, drug delivery, gene regulation, and immunomodulation, leveraging the plasmonic properties and ease of functionalization of gold nanoparticles. DNA-carbon nanotube conjugates have been investigated for biosensing, drug delivery, cancer therapy, and tissue engineering, exploiting carbon nanotubes' high aspect ratio, electrical conductivity, and near-infrared absorption. DNA-graphene oxide conjugates have been studied for biosensing, drug delivery, gene therapy, and tissue engineering, leveraging the high surface area, fluorescence quenching ability, and electrical conductivity of graphene oxide (Cheung et al. 2010). Despite the significant progress in developing DNA-based nanocomposites, several challenges still need to be solved for their successful translation into clinical applications. These include the potential toxicity and immunogenicity of the nanomaterials, the stability and biodegradability of the conjugates in physiological environments, the specificity and efficacy of the targeting and delivery, and the scalability and reproducibility of the synthesis methods.

2.10 DNA-Gold Nanoparticle Conjugates

DNA-gold nanoparticle conjugates are a class of nanomaterials that combine the molecular recognition properties of DNA with the unique optical and electronic properties of gold nanoparticles (AuNPs). AuNPs have been widely used in biomedical applications due to their biocompatibility, surface plasmon resonance, and ease of functionalization. DNA-gold nanoparticle conjugates exhibit unique optical and electronic properties that arise from combining DNA and AuNPs. The surface plasmon resonance of AuNPs can be modulated by binding target molecules to the DNA aptamers or probes on the nanoparticle surface, resulting in a colorimetric or spectroscopic signal. This property has been exploited to develop sensitive and specific biosensors for detecting various analytes, such as proteins, nucleic acids, and small molecules. The programmable self-assembly of DNA-AuNP conjugates has been used to create complex nanostructures, such as DNA origami and superlattices, with potential applications in drug delivery and tissue engineering (Ferrari 2023).

DNA-gold nanoparticle conjugates have been explored for various biomedical applications, such as cancer therapy, gene regulation, and immunomodulation. In cancer therapy, DNA-AuNP conjugates can deliver anticancer drugs or small interfering RNAs (siRNAs) specifically to tumor cells by targeting cancer-specific receptors or antigens (Sun et al. 2023). The photothermal properties of AuNPs can also be selectively exploited to kill cancer cells upon irradiation with near-infrared light. In gene regulation, DNA-AuNP conjugates can modulate gene expression by delivering antisense oligonucleotides or siRNAs that bind to complementary mRNA sequences and inhibit their translation. In immunomodulation, DNA-AuNP conjugates can deliver immunostimulatory or immunosuppressive oligonucleotides to immune cells, such as dendritic cells or T cells, to enhance or suppress immune responses.

2.11 DNA-Carbon Nanotube Conjugates

DNA-carbon nanotube conjugates are a class of nanomaterials that combine the molecular recognition properties of DNA with the unique mechanical, electrical, and optical properties of carbon nanotubes (CNTs). CNTs are cylindrical nanostructures composed of rolled-up sheets of graphene, with diameters ranging from a few nanometers to several hundred nanometers and lengths up to several centimeters. They exhibit exceptional strength, high electrical and thermal conductivity, and unique optical properties, making them attractive for various biomedical applications. The conjugation of DNA molecules to the surface of CNTs allows for the dispersion and functionalization of CNTs and the specific binding and sensing of target molecules (Gupta et al. 2019).

DNA-carbon nanotube conjugates exhibit unique mechanical, electrical, and optical properties from combining DNA and CNTs. The hybridization of complementary DNA strands on the surface of CNTs can result in the assembly of CNTs into complex nanostructures, such as bundles, networks, and arrays, with potential applications in nanoelectronics and nanomechanics. The electrical conductivity of CNTs can be modulated by the binding of target molecules to the DNA aptamers or probes on the nanotube surface, resulting in a change in the resistance or capacitance of the nanotube (Hughes et al. 2007). This property has been exploited to develop label-free and real-time biosensors for detecting various analytes, such as proteins, nucleic acids, and small molecules. DNA-carbon nanotube conjugates have been explored for various biomedical applications, such as drug delivery, cancer therapy, and tissue engineering. In drug delivery, DNA-CNT conjugates can encapsulate and deliver hydrophobic drugs, such as anticancer agents or antibiotics, to specific cells or tissues by targeting cell surface receptors or antigens. The high aspect ratio and needle-like shape of CNTs can facilitate their cellular uptake and intracellular trafficking. In cancer therapy, DNA-CNT conjugates can deliver siRNAs or anticancer drugs specifically to tumor cells, exploiting tumor vasculature's enhanced permeability and retention effect (Zare et al. 2021). The near-infrared absorption properties of CNTs can also be exploited for the photothermal ablation of cancer cells upon irradiation with near-infrared light. In tissue engineering, DNA-CNT conjugates can create conductive scaffolds that mimic the electrical properties of native tissues, such as cardiac or neural tissues, and promote cell growth and differentiation.

2.11.1 DNA-Graphene Oxide Conjugates

DNA-graphene oxide (GO) conjugates are a class of nanomaterials that combine the molecular recognition properties of DNA with the unique optical, electrical, and mechanical properties of GO. These conjugates can be synthesized using various methods, including non-covalent adsorption, covalent functionalization, and linker-mediated conjugation (Liu et al. 2015). Non-covalent adsorption involves the physical adsorption of DNA molecules onto the GO surface through π-π stacking interactions. In contrast, covalent functionalization involves the chemical attachment of DNA molecules to the GO surface through covalent bonds. Linker-mediated conjugation uses bifunctional linkers to connect DNA molecules to the GO surface, providing greater flexibility in the design of the conjugates (Whitfield et al. 2021).

DNA-GO conjugates exhibit unique properties arising from the combination of DNA and GO. The hybridization of complementary DNA strands on the GO surface can result in the assembly of GO sheets into complex nanostructures, such as multilayered films, hydrogels, and aerogels. The fluorescence quenching ability of GO can be exploited for the development of fluorescent biosensors (Liu et al. 2010). At the same time, the electrical conductivity of GO can be modulated by the binding of target molecules to the DNA aptamers or probes on the GO surface, enabling the development of label-free and real-time biosensors. These conjugates have been explored for various biomedical applications, including drug delivery, gene therapy, and tissue engineering. In drug delivery, DNA-GO conjugates can load and deliver hydrophobic drugs to specific cells or tissues by targeting cell surface receptors or antigens. In gene therapy, they can deliver therapeutic genes or siRNAs to specific cells or tissues, exploiting GO's ability to protect nucleic acids from enzymatic degradation and enhance their intracellular delivery. In tissue engineering, DNA-GO conjugates can create conductive and mechanically robust scaffolds that mimic the properties of native tissues and promote cell adhesion, proliferation, and differentiation (Jin and Ye 2007).

2.12 Graphene-Based Nanocomposites

Graphene is a single atomic layer of sp2 hybridized carbon atoms in a hexagonal arrangement. First isolated in 2004 by Geim and Novoselov, graphene's unique attributes stem from this pristine-layered sheet structure. High electrical and thermal conductivities, mechanical stiffness (~1 TPa), strength (~125 GPa), surface area (theoretical limit 2630 m^2/g), and biocompatibility enable diverse biomedical uses. However, the hydrophobic nature and tendency to agglomerate hamper solubility (Zhang et al. 2022b). Furthermore, unmodified graphene demonstrates dose-dependent cytotoxicity and risks triggering hypersensitivity reactions in vivo. This propelled the emergence of graphene derivatives via oxidation and functionalization. Graphene oxide (GO), reduced graphene oxide (rGO), functionalized graphene sheets (FG), and graphene quantum dots (GQDs) are explored for fabricating nanocomposites—multiphase materials with enhanced or tailored properties by combining interpenetrating networks of dissimilar constituents. Synthetic or natural polymers, hydroxyapatite, silver, chitosan, and carbon nanotubes serve as integrated matrices for biomedical composites with graphene derivatives in pharmaceutical applications, including drug delivery, bone tissue engineering, and biosensing (Saurav et al. 2024). Graphene nanocomposites demonstrate considerable promise for biomedical engineering by uniting the beneficial aspects of both classes of precursors—the mechanical resilience and electronic properties of graphene nanostructures with the processability and tunability arising from functionalized polymer, ceramic, or bio-organic matrices (Bei et al. 2019). Surface modification of graphene, controlled integration of detecting probes or bioactive, and multilayer fabrication allow tailoring toward drug delivery, tissue regeneration, and electrochemical biosensing applications.

Drug Delivery and Therapeutics: Graphene surface chemistry and submicrometer lateral dimensions assist large drug loading capacities, potentially surpassing conventional carriers by 10–100 times. π-π interactions between planar sheets and drug molecules with delocalized electrons favor physical sorption. Functional groups on edges and basal planes enable covalent linkages to diverse therapeutic agents. Reduced graphene readily complex with aromatic cancer drugs by non-covalent capture to transport into tumor cells. pH-sensitive bonds promote site-specific, triggered release profiles. SiRNA oligonucleotides bind covalently to GO's surfaces for intracellular gene therapies (Lai and Wong 2021). Unprecedented intracellular biosensing and targeting are feasible using graphene fluorescence quenching or Förster resonance energy transfer to surface-bound payloads. When dispersing nanosized payloads between biopolymer layers instead, extended-release behaviors occur attributable to diffusional resistance rather than surface interactions alone. The elevated surface areas accelerate dissolution rates based on the Noyes–Whitney equation. Any dissolved drugs subsequently encounter increased absorption driven by a very high adsorptive capacity. Temperature or pH-responsive graphene-polymer unions amplify stimuli-responsive profiles and tunability. Contrast agents like Gd(III), iron oxide nanoparticles, and fluorophores, when conjugated to graphene oxide, yield T1, T2, and fluorescent contrast for Theranostics (Jampilek and Kralova 2021). Despite lacking solid bonds, the long circulation half-lives in the bloodstream signify sustained in vivo activities.

Bone Tissue and Cartilage Regeneration: Bone features a natural 3D nHA (nano-hydroxyapatite) matrix that gets disrupted in diseases like osteoporosis, defects, or fractures affecting overall skeletal stability and structure—ideal implants aid regeneration by delivering cells and specific bioactive factors. Graphene nanoparticles mixed with nHA-based scaffolds enhance osteogenesis by signaling bone marrow stromal cells toward osteoblast lineages and upregulating collagen-I expression. Osteocyte attachment improves with rGO (Elkhenany et al. 2017). The use of graphene-based membranes for guided bone and nerve regeneration is also being explored. In chondrogenic applications, GO nanosheets are layered with sulfated glycosaminoglycans to mimic native cartilage extracellular matrix (ECM) and induce favorable genetic expression profiles in administered stem cells. Drug release capacities also imply that antimicrobials can accompany any cell vehicles. Metal ions like copper or magnesium introduced into such composites boost antibacterial action.

Biosensing and Diagnostics: The distinctive electronic and electrochemical properties support excellent performance as transducers in sensors for biomolecules, metabolites, or biomarkers of pathologies. The 2D morphology with every atom exposed maximizes surface area for reactions or binding events—capability grants tunable conductivity and charge density—a critical lever for electrochemical analysis. Edge-oxygen and basal plane defects generate redox centers that amplify electron transfer rates. All these contribute to ultralow detection limits reaching even single molecule scales in some demonstrations with specific aptamer recognition mediating homogeneous assays. The receptor-binding events at sensor interfaces distinctly alter localized charge densities and surface potentials to enable precise quantification relying on electrochemical impedance spectroscopy or I-V conductance measurements (Zhu 2017). Similar chemo-resistor responses also allow graphene oxide-based gas sensing devices. Some nanocomposite sensors, however, demand careful architecture engineering into multilayered stratified fashions, with each stratum tailored toward specific functions—passivation, selective transport, affinity capture, transduction, and so on. With further testing of specificity and sensing performance in natural clinical settings, graphene nanostructures can profoundly enhance point-of-care testing and diagnostics as a versatile next-generation platform technology.

2.12.1 Graphene-Polymer Nanocomposites

Graphene-polymer nanocomposites, such as graphene oxide-polyethylene glycol (GO-PEG), graphene oxide-chitosan (GO-chitosan), and graphene oxide-polylactic acid (GO-PLA), have emerged as promising materials for biomedical applications. These nanocomposites combine the unique properties of graphene, including high mechanical strength, excellent electrical and thermal conductivity, and large surface area, with the biocompatibility and functionality of polymers. Graphene oxide (GO), a derivative of graphene containing oxygen-containing functional groups, is widely used in synthesizing these nanocomposites due to its improved hydrophilicity and dispersibility compared to pristine graphene (Feng and Wang 2022). Synthesis methods for graphene-polymer nanocomposites, such as covalent functionalization, non-covalent wrapping, solution mixing, and in-situ polymerization, allow for tailoring their properties for specific biomedical applications, including drug delivery, tissue engineering, and biosensing. Despite the significant progress in developing these nanocomposites, challenges remain in their clinical translation, requiring further research to optimize synthesis methods, characterize long-term stability and degradation behavior in vivo, and evaluate safety and efficacy in large animal models and clinical trials. With continued research and development, graphene-polymer nanocomposites have the potential to revolutionize the treatment of various diseases and injuries, improving patient outcomes and quality of life.

- Graphene oxide-polyethylene glycol (GO-PEG)
- Graphene oxide-chitosan (GO-chitosan)
- Graphene oxide-polylactic acid (GO-PLA).

2.12.1.1 Graphene Oxide-Polyethylene Glycol (GO-PEG) Nanocomposites

Graphene oxide-polyethylene glycol (GO-PEG) nanocomposites have been explored for various biomedical applications due to their improved dispersibility and biocompatibility compared to pristine GO. PEG, a hydrophilic, biocompatible, and non-toxic polymer, can be covalently or non-covalently functionalized on the surface of GO. Covalent functionalization involves the chemical attachment of PEG chains to GO through the reaction of functional groups (Ghosh and Chatterjee 2020). In contrast, non-covalent wrapping involves the physical adsorption of PEG onto GO through hydrophobic interactions or hydrogen bonding. GO-PEG nanocomposites exhibit reduced aggregation and improved stability in physiological environments. Due to their high drug loading capacity and improved circulation time, they have been investigated as drug delivery vehicles for the controlled release of therapeutic agents, such as anticancer drugs, antibiotics, and growth factors. Additionally, GO-PEG nanocomposites have been explored for cancer therapy applications, such as photothermal therapy and targeted drug delivery, leveraging the strong optical absorption properties of GO in the near-infrared region. PEG functionalization also improves the tumor-targeting ability of GO by reducing its non-specific uptake by healthy cells. Furthermore, GO-PEG nanocomposites have been studied as biosensors for detecting various biomolecules, taking advantage of the excellent electrical properties of GO.

2.12.1.2 Graphene Oxide-Chitosan (GO-Chitosan) Nanocomposites

Graphene oxide-chitosan (GO-chitosan) nanocomposites have been investigated for various biomedical applications due to the unique properties of chitosan, a natural polysaccharide known for its biocompatibility, biodegradability, and antimicrobial properties. These nanocomposites can be synthesized using solution mixing, in-situ polymerization, or electrostatic interaction methods. Solution mixing involves dispersing GO in a chitosan solution, while in-situ polymerization involves the polymerization of chitosan monomers in the presence of GO. Electrostatic interaction involves the self-assembly of positively charged chitosan and negatively charged GO through electrostatic forces. GO-chitosan nanocomposites exhibit improved mechanical properties, thermal stability, and antimicrobial activity compared to pure chitosan. Incorporating GO into the chitosan matrix enhances the stiffness and strength of the nanocomposite, making it suitable for tissue engineering applications. The antimicrobial properties of chitosan, combined with the large surface area of GO, make GO-chitosan nanocomposites attractive for wound-healing applications, where they can prevent bacterial infections and promote tissue regeneration (Ghosh and Chatterjee 2020). Additionally, GO-chitosan nanocomposites have been explored as drug delivery vehicles for the controlled release of therapeutic agents. The porous structure of chitosan allows for drug encapsulation, while GO provides a high surface area for drug loading and functionalization with targeting ligands. Furthermore, these nanocomposites have been investigated as biosensors for detecting various biomolecules, leveraging the excellent electrical properties of GO and the biocompatibility of chitosan.

2.12.1.3 Graphene Oxide-Polylactic Acid (GO-PLA) Nanocomposites

Graphene oxide-polylactic acid (GO-PLA) nanocomposites have been investigated for various biomedical applications due to PLA's biodegradability, biocompatibility, and excellent mechanical properties. These nanocomposites can be synthesized using solution mixing, melt blending, or in-situ polymerization methods. Solution mixing involves dispersing GO in a PLA solution, while melt blending involves mixing GO with molten PLA under high-shear conditions. In-situ polymerization consists of the polymerization of PLA monomers in the presence of GO. GO-PLA nanocomposites exhibit improved mechanical properties, thermal stability, and barrier properties compared to pure PLA (Campos et al. 2015). The incorporation of GO enhances the stiffness and strength of the nanocomposite, making it suitable for tissue engineering applications. The barrier properties of GO can reduce the permeability of the nanocomposite to gases and liquids, which is beneficial for food packaging and drug delivery applications. These nanocomposites have been explored for various tissue engineering applications, such as bone, cartilage, and neural tissue regeneration. They can be fabricated into porous scaffolds that facilitate cell infiltration, nutrient transport, and vascularization. Additionally, GO-PLA nanocomposites have been investigated as drug delivery vehicles for controlled release and as biosensors for detecting various biomolecules, leveraging the high surface area and excellent electrical properties of GO (Andoh et al. 2024).

2.13 Lipid-Based Nanocomposites

Lipid-based nanocomposites consist of a lipid matrix or core combined with polymeric materials, where the lipid component provides a hydrophobic environment for drug encapsulation and helps to improve the bioavailability and stability of the encapsulated drugs (Lu et al. 2021). In contrast, the polymeric component offers additional functionalities, such as controlled release, targeted delivery, and enhanced mechanical properties. By carefully selecting the lipid and polymeric components and optimizing their ratios, researchers can design lipid-based nanocomposites with tailored properties for specific biomedical applications, such as drug delivery, gene therapy, and diagnostic imaging (Waheed et al. 2024).

2.13.1 Liposome-Polymer Hybrids

Liposome-polymer hybrids are a promising drug delivery system that combines the advantages of liposomes and polymeric materials to overcome the limitations of conventional liposomes, such as low stability, rapid clearance, and limited drug release control. These hybrids exhibit improved stability, longer circulation time, and enhanced drug delivery compared to traditional liposomes (Sriwidodo et al. 2022). Incorporating polymers, such as polyethylene glycol (PEG), provides a steric barrier that reduces protein adsorption and prevents immune system recognition, resulting in prolonged circulation time and improved drug accumulation at the target site. The polymeric component can also be functionalized with targeting ligands to target specific cells or tissues actively (Suk et al. 2016). Liposome-polymer hybrids have been explored for various biomedical applications, including cancer therapy, gene delivery, and vaccine development. In cancer therapy, these hybrids can deliver chemotherapeutic drugs specifically to tumor cells, reducing systemic toxicity and improving therapeutic efficacy. They can encapsulate and protect nucleic acids for gene delivery, enhancing cellular uptake and transfection efficiency. In vaccine development, liposome-polymer hybrids can be used as adjuvants to enhance the immunogenicity of antigens and promote robust immune responses (Sanati and Amin 2024). Overall, liposome-polymer hybrids offer a versatile and promising platform for drug delivery, with the potential to improve the efficacy and safety of various therapeutic and diagnostic applications.

2.13.2 Solid Lipid Nanoparticle-Polymer Hybrids

Solid lipid nanoparticle-polymer hybrids are a promising drug delivery system that combines the advantages of solid lipid nanoparticles (SLNs) and polymeric materials to overcome the limitations of conventional SLNs, such as limited drug loading capacity and drug expulsion during storage. These hybrids exhibit improved drug loading capacity, controlled drug release, and enhanced stability compared to traditional SLNs. Incorporating polymeric materials, such as PEG or chitosan, provides additional functional properties, including stealth behavior, mucoadhesion, and permeation enhancement (Ghasemiyeh and Mohammadi-Samani 2018). The polymeric component can also be used to modify the surface of the SLNs, enabling the attachment of targeting ligands or imaging agents. Solid lipid nanoparticle-polymer hybrids have been explored for various biomedical applications, such as oral drug delivery, topical drug delivery, and brain targeting. In oral drug delivery, these hybrids can enhance the bioavailability of poorly water-soluble drugs by improving their solubilization and intestinal permeation. They can provide a non-invasive approach to treating dermatological conditions for topical drug delivery. In brain targeting, these hybrids can deliver drugs across the blood–brain barrier by exploiting endogenous transport mechanisms and surface modification with targeting ligands. Overall, solid lipid nanoparticle-polymer hybrids offer a versatile platform for drug delivery, with the potential to improve the efficacy and safety of various therapeutic applications. In conclusion, lipid-based nanocomposites hold great promise for advancing the field of drug delivery and biomedical engineering (Sivadasan et al. 2021). With continued research and development, these materials have the potential to revolutionize the treatment of various diseases and improve patient outcomes and quality of life. The combination of the biocompatibility and drug encapsulation properties of lipids with the functional versatility of polymers opens up new opportunities for designing intelligent and responsive drug delivery systems that can address the complex challenges in modern medicine.

2.14 Protein-Based Nanocomposites

Protein-based nanocomposites consist of a protein matrix reinforced with inorganic nanoparticles, where the protein component provides a favorable environment for cell adhesion, proliferation, and differentiation. In contrast, inorganic nanoparticles offer mechanical strength, bioactivity, and additional functionalities (Wang et al. 2020). By carefully selecting the protein and inorganic components and optimizing their ratios, researchers can design protein-based nanocomposites with tailored properties for specific biomedical applications, such as bone tissue engineering, wound healing, and drug delivery systems.

2.14.1 Collagen-Hydroxyapatite Nanocomposites

Collagen-hydroxyapatite nanocomposites are promising materials for bone tissue engineering applications due to their excellent biocompatibility, biodegradability, and ability to mimic the composition and structure of natural bone. Collagen, the most abundant protein in the extracellular matrix, provides structural support and promotes cell adhesion and proliferation (Li et al. 2021). Hydroxyapatite, the primary inorganic component of bone and teeth, enhances the mechanical strength and osteoconductivity of the nanocomposite. Collagen-hydroxyapatite nanocomposites can be fabricated into porous scaffolds for repairing significant bone defects, hydrogels for delivering cells and growth factors, and films for coating metallic implants to improve their biocompatibility and osseointegration. The combination of collagen and hydroxyapatite results in a nanocomposite material that closely mimics the composition and structure of natural bone, providing a favorable environment for cell adhesion, proliferation, and differentiation. With their excellent properties and versatility, collagen-hydroxyapatite nanocomposites have the potential to revolutionize bone tissue engineering and improve patient outcomes (Teng et al. 2008).

2.14.2 Silk Fibroin-Hydroxyapatite Nanocomposites

Silk fibroin, a protein derived from silkworm cocoons, has a unique structure consisting of highly ordered β-sheet domains and less ordered amorphous regions, contributing to its high tensile strength and elasticity. Combining silk fibroin with hydroxyapatite results in a nanocomposite material with enhanced mechanical properties and bioactivity compared to pure silk fibroin (Nguyen et al. 2019). Silk fibroin-hydroxyapatite nanocomposites are promising materials for biomedical applications due to their excellent biocompatibility, biodegradability, and mechanical properties. These nanocomposites exhibit excellent biocompatibility, osteoconductivity, and mechanical properties, making them suitable for biomedical applications, such as bone tissue engineering, dental implants, and drug delivery systems. In bone tissue engineering applications, silk fibroin-hydroxyapatite nanocomposites can be fabricated into porous scaffolds that provide mechanical support and promote bone regeneration. Incorporating hydroxyapatite nanoparticles enhances the osteoconductivity of the scaffold, facilitating bone cell attachment and proliferation (Saleem et al. 2020). In dental implant applications, these nanocomposites can be used as coating materials for titanium implants, improving their biocompatibility and osseointegration. The controlled degradation of silk fibroin can also be utilized for the sustained release of therapeutic agents, such as antibiotics or growth factors, to promote tissue healing and prevent infections (Paun et al. 2023).

2.14.3 Elastin-Silver Nanoparticle Nanocomposites

Elastin-silver nanoparticle nanocomposites are a promising material for biomedical applications, combining elastin's biocompatibility and elasticity with silver nanoparticles' broad-spectrum antimicrobial activity. Elastin, a structural protein found in connective tissues, provides elasticity and promotes cell adhesion and proliferation. Silver nanoparticles have gained attention due to their antimicrobial properties, attributed to the release of silver ions and their interaction with bacterial cell membranes (Anh et al. 2013). These nanocomposites can be synthesized using various methods, including in-situ reduction, ex-situ blending, and electrospinning. In-situ reduction allows for the controlled synthesis of silver nanoparticles within the elastin matrix, while ex-situ blending involves mixing pre-synthesized silver nanoparticles with elastin solutions. Electrospinning fabricates nanofibers with high surface area and porosity (Freitas et al. 2024).

Elastin-silver nanoparticle nanocomposites exhibit excellent biocompatibility, antimicrobial activity, and mechanical properties. The elastin matrix promotes cell adhesion and proliferation, while the silver nanoparticles confer broad-spectrum antimicrobial activity. These nanocomposites have been explored for applications such as wound dressings, tissue engineering scaffolds, and antimicrobial coatings (Anh et al. 2013). Wound dressings provide a moist and antimicrobial environment for healing, preventing infections, and promoting tissue regeneration. In tissue engineering, they can be fabricated into porous scaffolds for regenerating soft tissues like skin, blood vessels, and cartilage, with antimicrobial activity preventing bacterial colonization.

2.15 Challenges and Future Perspectives

Biocompatible nanocomposites represent a revolutionary frontier in biomedical applications, encompassing a complex landscape of challenges and opportunities. The field stands at a critical juncture where technological capabilities are rapidly expanding, yet fundamental challenges persist. These materials offer unprecedented possibilities for personalized medicine, regenerative therapies, and advanced diagnostic approaches, but their translation from laboratory success to clinical reality requires careful navigation of various technical, biological, and regulatory considerations. The emergence of theranostic nanocomposites, which combine therapeutic and diagnostic capabilities, exemplifies the innovative potential in this field. These sophisticated materials can potentially transform healthcare delivery through targeted, multifunctional approaches to treatment and monitoring. The integration of nanocomposites with advanced manufacturing technologies, particularly 3D printing, is opening new avenues for creating patient-specific medical solutions. However, the path from promising research to practical medical applications requires addressing numerous interconnected challenges while capitalizing on emerging technological opportunities. The importance of interdisciplinary collaboration between materials scientists, biomedical engineers, clinicians, and regulatory experts cannot be overstated in this endeavor. Success in this field will require coordinated efforts across multiple disciplines to develop solutions that are not only technologically advanced but also practically implementable in clinical settings.

2.15.1 Critical Analysis of Biocompatible Nanocomposites

Despite the promising potential of biocompatible nanocomposites in medical applications, several significant challenges and limitations warrant careful consideration. Manufacturing challenges present a primary concern, particularly the difficulty of achieving uniform nanofiller dispersion throughout the matrix material. This challenge is compounded by batch-to-batch variability, which affects reproducibility and makes quality control exceptionally complex. The scale-up of production processes remains cost-intensive, requiring substantial capital investment in specialized equipment and expertise. From a biological perspective, there are serious concerns about the long-term effects of nanocomposite materials in biological systems. The limited understanding of how these materials accumulate in tissues over time and the potential for nanoparticle aggregation in biological environments raises important safety considerations (Velu et al. 2019). The uncertain impact on cellular metabolism and variable degradation rates can significantly affect therapeutic efficacy, making precise treatment planning challenging.

Clinical translation faces barriers, including complex regulatory pathways that must be navigated before new materials can be approved for medical use. The lack of standardized characterization methods makes it difficult to compare different nanocomposite systems effectively, while limited real-world performance data and complex cost–benefit considerations in healthcare settings further complicate implementation. Technical limitations also pose significant challenges, particularly the inherent trade-offs in material properties, for instance, enhanced strength often comes at the cost of reduced flexibility (Malik et al. 2023). Processing temperature constraints can affect drug incorporation in drug delivery applications, while storage stability concerns and the limited shelf life of some formulations create practical challenges for clinical use.

Economic considerations present another significant barrier to widespread adoption, as high production costs, complex manufacturing equipment requirements, and the need for specialized expertise drive up the overall cost of nanocomposite-based medical solutions. Market acceptance challenges persist, particularly given the relatively high costs of conventional materials and treatments. These challenges collectively highlight the need for continued research and development to optimize manufacturing processes, enhance our understanding of biological interactions, streamline regulatory pathways, and improve cost-effectiveness before biocompatible nanocomposites can achieve their full potential in medical applications.

2.15.2 Future Research Directions and Potential Breakthroughs

The future of biocompatible nanocomposites is poised for significant advancement through emerging technologies and innovative approaches. Artificial intelligence and machine learning are revolutionizing nanocomposite design by enabling optimal material composition prediction, modeling biological interactions, and automating synthesis protocols. Advanced manufacturing technologies, particularly 4D printing of responsive nanocomposites and continuous flow synthesis methods, open new possibilities for creating sophisticated materials while emphasizing environmentally conscious green manufacturing approaches.

The development of smart materials represents another frontier, with self-healing nanocomposites, programmable drug release systems, and bio-responsive materials showing tremendous promise for next-generation medical applications. In clinical applications, breakthrough developments are anticipated in personalized implant manufacturing, integration of real-time monitoring capabilities, and enhanced therapeutic efficacy through targeted delivery systems. Material science advancements are focused on developing novel hybrid compositions that offer enhanced stability and improved biocompatibility. At the same time, innovations in processing technologies aim to establish scalable, cost-effective production methods with robust quality control measures.

Research priorities increasingly focus on comprehensive safety studies, including long-term toxicity assessment, detailed biodistribution analysis, and thorough evaluation of immune responses to these materials. Standardization efforts are gaining momentum, emphasizing developing unified characterization protocols, establishing clear performance metrics, and creating quality benchmarks to facilitate regulatory approval and clinical adoption. The optimization of regulatory pathways, coupled with detailed cost–benefit analyses and real-world performance studies, is crucial for successful clinical translation. This multifaceted approach to future research and development suggests that biocompatible nanocomposites will continue to evolve, offering increasingly sophisticated solutions for medical challenges while addressing current safety, efficacy, and manufacturability limitations. The convergence of these various technological advances and research priorities promises to accelerate the development of more effective, safer, and accessible nanocomposite-based medical solutions.

Table1 Classification of nanocomposites

Natural Polymer-Based Nanocomposites			
	Collagen-based nanocomposites		
		Collagen/hydroxyapatite	Bone tissue engineering, wound healing
		Collagen/chitosan	Drug delivery, skin tissue engineering
		Collagen/gold nanoparticles	Biosensors, photothermal therapy
	Gelatin-based nanocomposites		
		Gelatin/hydroxyapatite	Bone regeneration, drug delivery
		Gelatin/magnetic nanoparticles	Targeted drug delivery, hyperthermia
		Gelatin/silica	Tissue engineering, wound healing
Synthetic Polymer-Based Nanocomposites			
	PLA-based nanocomposites		
		PLA/hydroxyapatite	Bone tissue engineering, drug delivery
		PLA/carbon nanotubes	Scaffolds with enhanced mechanical properties
		PLA/nanoclay	Improved barrier properties for drug delivery
	PGA-based nanocomposites		
		PGA/chitosan	Tissue engineering scaffolds, drug delivery
		PGA/nanosilver	Antimicrobial sutures, wound dressings
		PGA/graphene	Enhanced mechanical properties, electrical conductivity
	PEG-based nanocomposites		
		PEG/gold nanoparticles	Drug delivery, cancer therapy
		PEG/liposomes	Targeted drug delivery, gene therapy
		PEG/quantum dots	Imaging agents, biosensors
	Ceramic-Based Nanocomposites		
	Hydroxyapatite-based nanocomposites		
		Hydroxyapatite/collagen	Bone tissue engineering, dental implants
		Hydroxyapatite/polymer	Bone grafts, drug delivery
		Hydroxyapatite/carbon nanotubes	Enhanced mechanical properties, bone regeneration

(continued)

Table1 (continued)

	Calcium phosphate-based nanocomposites		
		Calcium phosphate/polymer	Bone cement, drug delivery
		Calcium phosphate/nanoclay	Improved bioactivity, bone regeneration
		Calcium phosphate/silica	Drug delivery, bone tissue engineering
Other Nanocomposites			
	Carbon nanotube-based nanocomposites		
		Carbon nanotubes/polymer	Tissue engineering scaffolds, biosensors
		Carbon nanotubes/ceramic	Bone tissue engineering, drug delivery
	Graphene-based nanocomposites		
		Graphene/polymer	Tissue engineering scaffolds, biosensors
		Graphene/ceramic	Bone tissue engineering, drug delivery
	Gold nanoparticle-based nanocomposites		
		Gold nanoparticles/polymer: Drug delivery, cancer therapy	
		Gold nanoparticles/silica: Biosensors, photothermal therapy	

References

Abazari S, Shamsipur A, Bakhsheshi-Rad HR, Ismail AF, Sharif S, Razzaghi M et al (2020) Carbon nanotubes (CNTs)-reinforced magnesium-based matrix composites: a comprehensive review. Materials (Basel) 13(19). https://doi.org/10.3390/ma13194421

Abbas M, Alqahtani MS, Alhifzi R (2023) Recent developments in polymer nanocomposites for bone regeneration. Int J Mol Sci 24(4). https://doi.org/10.3390/ijms24043312

Abu-Salah KM, Ansari AA, Alrokayan SA (2010) DNA-based applications in nanobiotechnology. J Biomed Biotechnol 2010:715295. https://doi.org/10.1155/2010/715295

Alle M, Sharma G, Lee SH, Kim JC (2022) Next-generation engineered nanogold for multimodal cancer therapy and imaging: a clinical perspectives. J Nanobiotechnology 20(1):222. https://doi.org/10.1186/s12951-022-01402-z

Alshangiti DM, El-Damhougy TK, Zaher A, Madani M, Mohamady GM (2023) Revolutionizing biomedicine: advancements, applications, and prospects of nanocomposite macromolecular carbohydrate-based hydrogel biomaterials: a review. RSC Adv 13(50):35251–35291. https://doi.org/10.1039/d3ra07391b

Alven S, Aderibigbe BA (2024) Chitosan-based scaffolds incorporated with silver nanoparticles for the treatment of infected wounds. Pharmaceutics 16(3). https://doi.org/10.3390/pharmaceutics16030327

Amina SJ, Guo B (2020) A review on the synthesis and functionalization of gold nanoparticles as a drug delivery vehicle. Int J Nanomedicine 15:9823–9857. https://doi.org/10.2147/IJN.S279094

Andoh V, Ocansey DKW, Naveed H, Wang N, Chen L, Chen K et al (2024) The advancing role of nanocomposites in cancer diagnosis and treatment. Int J Nanomedicine 19:6099–6126. https://doi.org/10.2147/IJN.S471360

Anh TT, Xing M, Le DH, Sugawara-Narutaki A, Fong E (2013) Elastin-based silver-binding proteins with antibacterial capabilities. Nanomedicine (Lond) 8(4):567–575. https://doi.org/10.2217/nnm.13.47

Anjum S, Rahman F, Pandey P, Arya DK, Alam M, Rajinikanth PS et al (2022) Electrospun biomimetic nanofibrous scaffolds: a promising prospect for bone tissue engineering and regenerative medicine. Int J Mol Sci 23(16). https://doi.org/10.3390/ijms23169206

Aqel A, El-Nour KMMA, Ammar RAA, Al-Warthan A (2012) Carbon nanotubes, science and technology part (I) structure, synthesis and characterisation. Arab J Chem 5(1):1–23. https://doi.org/10.1016/j.arabjc.2010.08.022

Aranaz I, Alcantara AR, Civera MC, Arias C, Elorza B, Heras Caballero A, et al (2021) Chitosan: an overview of its properties and applications. Polymers (Basel) 13(19). https://doi.org/10.3390/polym13193256

Avasthi A, Caro C, Pozo-Torres E, Leal MP, Garcia-Martin ML (2020) Magnetic nanoparticles as MRI contrast agents. Top Curr Chem (Cham) 378(3):40. https://doi.org/10.1007/s41061-020-00302-w

Azmy E, Al-Kholy MRZ, Fattouh M, Kenawi LMM, Helal MA (2022) Impact of nanoparticles additions on the strength of dental composite resin. Int J Biomater 2022:1165431. https://doi.org/10.1155/2022/1165431

Baruah A, Newar R, Das S, Kalita N, Nath M, Ghosh P et al (2024) Biomedical applications of graphene-based nanomaterials: recent progress, challenges, and prospects in highly sensitive biosensors. Discov Nano 19(1):103. https://doi.org/10.1186/s11671-024-04032-6

Bauso LV, La Fauci V, Longo C, Calabrese G (2024) Bone tissue engineering and nanotechnology: a promising combination for bone regeneration. Biology (Basel) 13(4). https://doi.org/10.3390/biology13040237

Bei HP, Yang Y, Zhang Q, Tian Y, Luo X, Yang M, et al (2019) Graphene-based nanocomposites for neural tissue engineering. Molecules 24(4). https://doi.org/10.3390/molecules24040658

Bhuckory S, Kays JC, Dennis AM (2019) In vivo biosensing using resonance energy transfer. Biosensors (Basel) 9(2). https://doi.org/10.3390/bios9020076

Bikiaris ND, Koumentakou I, Samiotaki C, Meimaroglou D, Varytimidou D, Karatza A et al (2023) Recent advances in the investigation of poly(lactic acid) (PLA) nanocomposites: incorporation of various nanofillers and their properties and applications. Polymers 15(5):1196

Bruna T, Maldonado-Bravo F, Jara P, Caro N (2021) Silver nanoparticles and their antibacterial applications. Int J Mol Sci 22(13). https://doi.org/10.3390/ijms22137202

Campos JM, Ferraria AM, Botelho do Rego AM, Ribeiro MR, Barros-Timmons A (2015) Studies on PLA grafting onto graphene oxide and its effect on the ensuing composite films. Mater Chem Phys 166:122–132. https://doi.org/10.1016/j.matchemphys.2015.09.036

Canales D, Moyano D, Alvarez F, Grande-Tovar CD, Valencia-Llano CH, Peponi L et al (2023) Preparation and characterization of novel poly (lactic acid)/calcium oxide nanocomposites by electrospinning as a potential bone tissue scaffold. Biomater Adv. 153:213578. https://doi.org/10.1016/j.bioadv.2023.213578

Chehelgerdi M, Chehelgerdi M, Allela OQB, Pecho RDC, Jayasankar N, Rao DP et al (2023) Progressing nanotechnology to improve targeted cancer treatment: overcoming hurdles in its clinical implementation. Mol Cancer 22(1):169. https://doi.org/10.1186/s12943-023-01865-0

Chen J, Ashames A, Buabeid MA, Fahelelbom KM, Ijaz M, Murtaza G (2020) Nanocomposites drug delivery systems for the healing of bone fractures. Int J Pharm 585:119477. https://doi.org/10.1016/j.ijpharm.2020.119477

Cheung W, Pontoriero F, Taratula O, Chen AM, He H (2010) DNA and carbon nanotubes as medicine. Adv Drug Deliv Rev 62(6):633–649. https://doi.org/10.1016/j.addr.2010.03.007

Cinteza LO, Scomoroscenco C, Voicu SN, Nistor CL, Nitu SG, Trica B, et al (2018) Chitosan-stabilized ag nanoparticles with superior biocompatibility and their synergistic antibacterial effect in mixtures with essential oils. Nanomaterials (Basel) 8(10). https://doi.org/10.3390/nano8100826

Darakai V, Punsawad C, Jitonnom J, Nisoa M, Rattanakit P (2024) Microwave-assisted ultrafine silver nanoparticle synthesis using Mitragyna speciosa for antimalarial applications. Green Process Synth 13(1). https://doi.org/10.1515/gps-2023-0257

Diez-Pascual AM, Rahdar A (2022) LbL nano-assemblies: a versatile tool for biomedical and healthcare applications. Nanomaterials (Basel) 12(6). https://doi.org/10.3390/nano12060949

Dykman LA, Khlebtsov NG (2019) Gold nanoparticles in chemo-, immuno-, and combined therapy: review [Invited]. Biomed Opt Express 10(7):3152–3182. https://doi.org/10.1364/BOE.10.003152

El-Hammadi MM, Arias JL (2022) Recent advances in the surface functionalization of PLGA-based nanomedicines. Nanomaterials (Basel) 12(3). https://doi.org/10.3390/nano12030354

Elkhenany H, Bourdo S, Hecht S, Donnell R, Gerard D, Abdelwahed R et al (2017) Graphene nanoparticles as osteoinductive and osteoconductive platform for stem cell and bone regeneration. Nanomedicine 13(7):2117–2126. https://doi.org/10.1016/j.nano.2017.05.009

Eshgh NA, Meftahi A, Khajavi R, Aljabali AAA, Barhoum A (2022) Nanocelluloses for tissue engineering and biomedical scaffolds. In: Barhoum A (ed) Handbook of nanocelluloses. Springer International Publishing, Cham, pp 709–743

Falsafi SR, Topuz F, Bajer D, Mohebi Z, Shafieiuon M, Heydari H et al (2023) Metal nanoparticles and carbohydrate polymers team up to improve biomedical outcomes. Biomed Pharmacother 168:115695. https://doi.org/10.1016/j.biopha.2023.115695

Fang J, Li P, Lu X, Fang L, Lu X, Ren F (2019) A strong, tough, and osteoconductive hydroxyapatite mineralized polyacrylamide/dextran hydrogel for bone tissue regeneration. Acta Biomater 88:503–513. https://doi.org/10.1016/j.actbio.2019.02.019

Farzin A, Etesami SA, Quint J, Memic A, Tamayol A (2020) Magnetic nanoparticles in cancer therapy and diagnosis. Adv Healthc Mater 9(9):e1901058. https://doi.org/10.1002/adhm.201901058

Feng W, Wang Z (2022) Biomedical applications of chitosan-graphene oxide nanocomposites. iScience 25(1):103629. https://doi.org/10.1016/j.isci.2021.103629

Ferrari E (2023) Gold nanoparticle-based plasmonic biosensors. Biosensors (Basel) 13(3). https://doi.org/10.3390/bios13030411

Filippi M, Balciunaite A, Katzschmann R (2024) Biohybrid nanointerfaces for neuromodulation. Nano Today 54:102094. https://doi.org/10.1016/j.nantod.2023.102094

de Freitas CF, Souza PR, Jacinto GS, Braga TL, Ricken YS, Souza GK, et al (2024) Silver nanoparticles in situ synthesized and incorporated in uniaxial and core-shell electrospun nanofibers to inhibit coronavirus. Pharmaceutics 16(2). https://doi.org/10.3390/pharmaceutics16020268

Gao W, Zhang Y, Zhang Q, Zhang L (2016) Nanoparticle-hydrogel: a hybrid biomaterial system for localized drug delivery. Ann Biomed Eng 44(6):2049–2061. https://doi.org/10.1007/s10439-016-1583-9

Ghasemiyeh P, Mohammadi-Samani S (2018) Solid lipid nanoparticles and nanostructured lipid carriers as novel drug delivery systems: applications, advantages and disadvantages. Res Pharm Sci 13(4):288–303. https://doi.org/10.4103/1735-5362.235156

Ghosh S, Chatterjee K (2020) Poly(Ethylene Glycol) functionalized graphene oxide in tissue engineering: a review on recent advances. Int J Nanomedicine 15:5991–6006. https://doi.org/10.2147/IJN.S249717

Goldberg M, Langer R, Jia X (2007) Nanostructured materials for applications in drug delivery and tissue engineering. J Biomater Sci Polym Ed 18(3):241–268. https://doi.org/10.1163/156856207779996931

Gupta N, Gupta SM, Sharma SK (2019) Carbon nanotubes: synthesis, properties and engineering applications. Carbon Lett 29(5):419–447. https://doi.org/10.1007/s42823-019-00068-2

Hauser AK, Mathias R, Anderson KW, Hilt JZ (2015) The effects of synthesis method on the physical and chemical properties of dextran coated iron oxide nanoparticles. Mater Chem Phys 160:177–186. https://doi.org/10.1016/j.matchemphys.2015.04.022

Hoang Thi TT, Pilkington EH, Nguyen DH, Lee JS, Park KD, Truong NP (2020) The importance of poly(ethylene glycol) alternatives for overcoming PEG immunogenicity in drug delivery and bioconjugation. Polymers (Basel) 12(2):298. https://doi.org/10.3390/polym12020298

Hu Y, Fan C (2022) Nanocomposite DNA hydrogels emerging as programmable and bioinstructive materials systems. Chem 8(6):1554–1566. https://doi.org/10.1016/j.chempr.2022.04.003

Hughes ME, Brandin E, Golovchenko JA (2007) Optical absorption of DNA−carbon nanotube structures. Nano Lett 7(5):1191–1194. https://doi.org/10.1021/nl062906u

Ielo I, Calabrese G, De Luca G, Conoci S (2022) Recent advances in hydroxyapatite-based biocomposites for bone tissue regeneration in orthopedics. Int J Mol Sci 23(17). https://doi.org/10.3390/ijms23179721

Jampilek J, Kralova K (2021) Advances in drug delivery nanosystems using graphene-based materials and carbon nanotubes. Materials 14(5):1059

Jiang D, England CG, Cai W (2016) DNA nanomaterials for preclinical imaging and drug delivery. J Control Release 239:27–38. https://doi.org/10.1016/j.jconrel.2016.08.013

Jin S, Ye K (2007) Nanoparticle-mediated drug delivery and gene therapy. Biotechnol Prog 23(1):32–41. https://doi.org/10.1021/bp060348j

Jokerst JV, Lobovkina T, Zare RN, Gambhir SS (2011) Nanoparticle PEGylation for imaging and therapy. Nanomedicine (Lond) 6(4):715–728. https://doi.org/10.2217/nnm.11.19

Kamaly N, Yameen B, Wu J, Farokhzad OC (2016) Degradable controlled-release polymers and polymeric nanoparticles: mechanisms of controlling drug release. Chem Rev 116(4):2602–2663. https://doi.org/10.1021/acs.chemrev.5b00346

Kayan GÖ, Kayan A (2023) Polycaprolactone composites/blends and their applications especially in water treatment. Chem Eng 7(6):104

Khane Y, Albukhaty S, Sulaiman GM, Fennich F, Bensalah B, Hafsi Z et al (2024) Fabrication, characterization and application of biocompatible nanocomposites: a review. Eur Polymer J 214:113187. https://doi.org/10.1016/j.eurpolymj.2024.113187

Kharlamova MV, Paukov M, Burdanova MG (2022) Nanotube functionalization: investigation, methods and demonstrated applications. Materials (Basel) 15(15). https://doi.org/10.3390/ma15155386

Kyriakides TR, Raj A, Tseng TH, Xiao H, Nguyen R, Mohammed FS, et al (2021) Biocompatibility of nanomaterials and their immunological properties. Biomed Mater 16(4). https://doi.org/10.1088/1748-605X/abe5fa

Lai WF, Wong WT (2021) Use of graphene-based materials as carriers of bioactive agents. Asian J Pharm Sci 16(5):577–588. https://doi.org/10.1016/j.ajps.2020.11.004

Levengood SL, Zhang M (2014) Chitosan-based scaffolds for bone tissue engineering. J Mater Chem B 2(21):3161–3184. https://doi.org/10.1039/C4TB00027G

Li Y, Lin R, Wang L, Huang J, Wu H, Cheng G et al (2015) PEG-b-AGE polymer coated magnetic nanoparticle probes with facile functionalization and anti-fouling properties for reducing non-specific uptake and improving biomarker targeting. J Mater Chem B 3(17):3591–3603. https://doi.org/10.1039/C4TB01828A

Li Y, Liu Y, Li R, Bai H, Zhu Z, Zhu L et al (2021) Collagen-based biomaterials for bone tissue engineering. Mater des 210:110049. https://doi.org/10.1016/j.matdes.2021.110049

Liu F, Choi JY, Seo TS (2010) Graphene oxide arrays for detecting specific DNA hybridization by fluorescence resonance energy transfer. Biosens Bioelectron 25(10):2361–2365. https://doi.org/10.1016/j.bios.2010.02.022

Liu M, Zhang W, Chang D, Zhang Q, Brennan JD, Li Y (2015) Integrating graphene oxide, functional DNA and nucleic-acid-manipulating strategies for amplified biosensing. TrAC, Trends Anal Chem 74:120–129. https://doi.org/10.1016/j.trac.2015.03.027

Liu X, Kouassi KGW, Vanbever R, Dumoulin M (2022) Impact of the PEG length and PEGylation site on the structural, thermodynamic, thermal, and proteolytic stability of mono-PEGylated alpha-1 antitrypsin. Protein Sci 31(9):e4392. https://doi.org/10.1002/pro.4392

Lu H, Zhang S, Wang J, Chen Q (2021) A review on polymer and lipid-based nanocarriers and its application to nano-pharmaceutical and food-based systems. Front Nutr 8:783831. https://doi.org/10.3389/fnut.2021.783831

Ma R, Su Y, Cao R, Wang K, Yang P (2023) Enhanced osteogenic activity and bone repair ability of PLGA/MBG scaffolds doped with ZIF-8 nanoparticles loaded with BMP-2. Int J Nanomedicine 18:5055–5072. https://doi.org/10.2147/IJN.S423985

Makadia HK, Siegel SJ (2011) Poly lactic-co-glycolic acid (PLGA) as biodegradable controlled drug delivery carrier. Polymers (Basel) 3(3):1377–1397. https://doi.org/10.3390/polym3031377

Malik S, Singh J, Goyat R, Saharan Y, Chaudhry V, Umar A et al (2023) Nanomaterials-based biosensor and their applications: a review. Heliyon 9(9):e19929. https://doi.org/10.1016/j.heliyon.2023.e19929

Malik S, Muhammad K, Waheed Y (2023) Emerging applications of nanotechnology in healthcare and medicine. Molecules 28(18). https://doi.org/10.3390/molecules28186624

Materón EM, Miyazaki CM, Carr O, Joshi N, Picciani PHS, Dalmaschio CJ et al (2021) Magnetic nanoparticles in biomedical applications: a review. Appl Surf Sci Adv 6:100163. https://doi.org/10.1016/j.apsadv.2021.100163

Milosevic M, Stojanovic DB, Simic V, Grkovic M, Bjelovic M, Uskokovic PS et al (2020) Preparation and modeling of three-layered PCL/PLGA/PCL fibrous scaffolds for prolonged drug release. Sci Rep 10(1):11126. https://doi.org/10.1038/s41598-020-68117-9

Mladenovic M, Jaric S, Mundzic M, Pavlovic A, Bobrinetskiy I, Knezevic NZ (2024) Biosensors for cancer biomarkers based on mesoporous silica nanoparticles. Biosensors (Basel) 14(7):326. https://doi.org/10.3390/bios14070326

Mo X, Zhang D, Liu K, Zhao X, Li X, Wang W (2023) Nano-hydroxyapatite composite scaffolds loaded with bioactive factors and drugs for bone tissue engineering. Int J Mol Sci 24(2). https://doi.org/10.3390/ijms24021291

Mody VV, Siwale R, Singh A, Mody HR (2010) Introduction to metallic nanoparticles. J Pharm Bioallied Sci 2(4):282–289. https://doi.org/10.4103/0975-7406.72127

Mohd Nurazzi N, Asyraf MRM, Khalina A, Abdullah N, Sabaruddin FA, Kamarudin SH et al (2021) Fabrication, functionalization, and application of carbon nanotube-reinforced polymer composite: an overview. Polymers (Basel) 13(7):1047. https://doi.org/10.3390/polym13071047

Mourdikoudis S, Kostopoulou A, LaGrow AP (2021) Magnetic nanoparticle composites: synergistic effects and applications. Adv Sci (Weinh) 8(12):2004951. https://doi.org/10.1002/advs.202004951

Nandhini J, Karthikeyan E, Rajeshkumar S (2024) Nanomaterials for wound healing: current status and futuristic frontier. Biomed Technol 6:26–45. https://doi.org/10.1016/j.bmt.2023.10.001

Nguyen TP, Nguyen QV, Nguyen VH, Le TH, Huynh VQN, Vo DN, et al (2019) Silk fibroin-based biomaterials for biomedical applications: a review. Polymers (Basel) 11(12). https://doi.org/10.3390/polym11121933

Nguyen MD, Tran HV, Xu S, Lee TR (2021) Fe(3)O(4) Nanoparticles: structures, synthesis, magnetic properties, surface functionalization, and emerging applications. Appl Sci (Basel) 11(23). https://doi.org/10.3390/app112311301

Njuguna J, Ansari F, Sachse S, Zhu H, Rodriguez VM (2014) Nanomaterials, nanofillers, and nanocomposites: types and properties. In: Njuguna J, Pielichowski K, Zhu H (eds) Health and environmental safety of nanomaterials. Woodhead Publishing, pp 3–27

Okkeh M, Bloise N, Restivo E, De Vita L, Pallavicini P, Visai L (2021) Gold nanoparticles: can they be the next magic bullet for multidrug-resistant bacteria? Nanomaterials (Basel) 11(2). https://doi.org/10.3390/nano11020312

Omidian H, Wilson RL, Babanejad N (2023) Bioinspired polymers: transformative applications in biomedicine and regenerative medicine. Life (Basel) 13(8). https://doi.org/10.3390/life13081673

Ornaghi HL, Jr., Monticeli FM, Agnol LD (2023) A review on polymers for biomedical applications on hard and soft tissues and prosthetic limbs. Polymers (Basel) 15(19). https://doi.org/10.3390/polym15194034

Ozkan E, Mondal A, Douglass M, Hopkins SP, Garren M, Devine R et al (2022) Bioinspired ultra-low fouling coatings on medical devices to prevent device-associated infections and thrombosis. J Colloid Interface Sci 608(Pt 1):1015–1024. https://doi.org/10.1016/j.jcis.2021.09.183

Padin-Gonzalez E, Lancaster P, Bottini M, Gasco P, Tran L, Fadeel B et al (2022) Understanding the role and impact of poly (Ethylene Glycol) (PEG) on nanoparticle formulation: implications for COVID-19 vaccines. Front Bioeng Biotechnol 10:882363. https://doi.org/10.3389/fbioe.2022.882363

Paun AG, Dumitriu C, Ungureanu C, Popescu S (2023) Silk fibroin/ZnO coated TiO(2) nanotubes for improved antimicrobial effect of ti dental implants. Materials (Basel) 16(17). https://doi.org/10.3390/ma16175855

Pielichowska K, Blazewicz S (2010) Bioactive polymer/hydroxyapatite (Nano)composites for bone tissue regeneration. In: Abe A, Dusek K, Kobayashi S (eds) Biopolymers. Advances in polymer science. Springer Berlin Heidelberg, Berlin, Heidelberg, pp 97–207

Pushpalatha C, Gayathri VS, Sowmya SV, Augustine D, Alamoudi A, Zidane B et al (2023) Nanohydroxyapatite in dentistry: a comprehensive review. Saudi Dent J. 35(6):741–752. https://doi.org/10.1016/j.sdentj.2023.05.018

Rao KSVK, Vani TJS, Lakshmi BA, Hemalatha D, Rao KM (2022) Surface modified carbon nanotubes for bone tissue engineering. Surface modified carbon nanotubes volume 2: industrial applications. ACS Symposium Series, vol 1425. American Chemical Society, pp 1–17

Rashid AB, Haque M, Islam SMM, Uddin Labib KMR (2024) Nanotechnology-enhanced fiber-reinforced polymer composites: recent advancements on processing techniques and applications. Heliyon 10(2):e24692. https://doi.org/10.1016/j.heliyon.2024.e24692

Ren Z, Chen H, Wang R, Zhu M (2024) Comparative assessments of dental resin composites: a focus on dense microhybrid materials. ACS Biomater Sci Eng 10(6):3718–3726. https://doi.org/10.1021/acsbiomaterials.4c00403

Rocha CV, Goncalves V, da Silva MC, Banobre-Lopez M, Gallo J (2022) PLGA-based composites for various biomedical applications. Int J Mol Sci 23(4). https://doi.org/10.3390/ijms23042034

Sachdeva B, Sachdeva P, Negi A, Ghosh S, Han S, Dewanjee S, et al (2023) Chitosan nanoparticles-based cancer drug delivery: application and challenges. Mar Drugs 21(4). https://doi.org/10.3390/md21040211

Saleem M, Rasheed S, Yougen C (2020) Silk fibroin/hydroxyapatite scaffold: a highly compatible material for bone regeneration. Sci Technol Adv Mater 21(1):242–266. https://doi.org/10.1080/14686996.2020.1748520

Saleemi MA, Kong YL, Yong PVC, Wong EH (2022) An overview of antimicrobial properties of carbon nanotubes-based nanocomposites. Adv Pharm Bull 12(3):449–465. https://doi.org/10.34172/apb.2022.049

Sanati M, Amin YS (2024) Liposome-integrated hydrogel hybrids: promising platforms for cancer therapy and tissue regeneration. J Control Release 368:703–727. https://doi.org/10.1016/j.jconrel.2024.03.008

Santos-Carballal B, Fernandez Fernandez E, Goycoolea FM (2018) Chitosan in non-viral gene delivery: role of structure, characterization methods, and insights in cancer and rare diseases therapies. Polymers (Basel) 10(4). https://doi.org/10.3390/polym10040444

Sarfraz N, Khan I (2021) Plasmonic gold nanoparticles (AuNPs): properties, synthesis and their advanced energy, environmental and biomedical applications. Chem Asian J 16(7):720–742. https://doi.org/10.1002/asia.202001202

Satchanska G, Davidova S, Petrov PD (2024) Natural and synthetic polymers for biomedical and environmental applications. Polymers (Basel) 16(8):1159. https://doi.org/10.3390/polym16081159

Saurav S, Sharma P, Kumar A, Tabassum Z, Girdhar M, Mamidi N et al (2024) Harnessing natural polymers for nano-scaffolds in bone tissue engineering: a comprehensive overview of bone disease treatment. Curr Issues Mol Biol 46(1):585–611. https://doi.org/10.3390/cimb46010038

Shar A, Shar A, Joung D (2023) Carbon nanotube nanocomposite scaffolds: advances in fabrication and applications for tissue regeneration and cancer therapy. Front Bioeng Biotechnol 11:1299166. https://doi.org/10.3389/fbioe.2023.1299166

Shen CC, Hsu SH, Chang KB, Yeh CA, Chang HC, Tang CM et al (2021) Physical gold nanoparticle-decorated polyethylene glycol-hydroxyapatite composites guide osteogenesis and angiogenesis of mesenchymal stem cells. Biomedicines 9(11). https://doi.org/10.3390/biomedicines9111632

Singh R, Shitiz K, Singh A (2017) Chitin and chitosan: biopolymers for wound management. Int Wound J 14(6):1276–1289. https://doi.org/10.1111/iwj.12797

Sivadasan D, Sultan MH, Madkhali O, Almoshari Y, Thangavel N (2021) Polymeric lipid hybrid nanoparticles (PLNs) as emerging drug delivery platform-a comprehensive review of their properties, preparation methods, and therapeutic applications. Pharmaceutics 13(8). https://doi.org/10.3390/pharmaceutics13081291

Soriente A, Fasolino I, Gomez-Sanchez A, Prokhorov E, Buonocore GG, Luna-Barcenas G et al (2022) Chitosan/hydroxyapatite nanocomposite scaffolds to modulate osteogenic and inflammatory response. J Biomed Mater Res A 110(2):266–272. https://doi.org/10.1002/jbm.a.37283

Speranza G (2021) Carbon nanomaterials: synthesis, functionalization and sensing applications. Nanomaterials (Basel) 11(4):967. https://doi.org/10.3390/nano11040967

Sriwidodo, Umar AK, Wathoni N, Zothantluanga JH, Das S, Luckanagul JA (2022) Liposome-polymer complex for drug delivery system and vaccine stabilization. Heliyon 8(2):e08934. https://doi.org/10.1016/j.heliyon.2022.e08934

Suk JS, Xu Q, Kim N, Hanes J, Ensign LM (2016) PEGylation as a strategy for improving nanoparticle-based drug and gene delivery. Adv Drug Deliv Rev 99(Pt A):28–51. https://doi.org/10.1016/j.addr.2015.09.012

Sultana A, Zare M, Thomas V, Kumar TSS, Ramakrishna S (2022) Nano-based drug delivery systems: conventional drug delivery routes, recent developments and future prospects. Med Drug Discov 15:100134. https://doi.org/10.1016/j.medidd.2022.100134

Sun L, Liu H, Ye Y, Lei Y, Islam R, Tan S et al (2023) Smart nanoparticles for cancer therapy. Signal Transduct Target Ther 8(1):418. https://doi.org/10.1038/s41392-023-01642-x

Sun J, Cai S, Yang W, Leng H, Ge Z, Liu T (2024) Microrobots based on smart materials with their manufacturing methods and applications. Inventions 9(3):67

Tang L, Xiao Q, Mei Y, He S, Zhang Z, Wang R et al (2021) Insights on functionalized carbon nanotubes for cancer theranostics. J Nanobiotechnology 19(1):423. https://doi.org/10.1186/s12951-021-01174-y

Teng S-H, Lee E-J, Wang P, Kim H-E (2008) Collagen/hydroxyapatite composite nanofibers by electrospinning. Mater Lett 62(17–18):3055–3058. https://doi.org/10.1016/j.matlet.2008.01.104

Tian H, Zhang T, Qin S, Huang Z, Zhou L, Shi J et al (2022) Enhancing the therapeutic efficacy of nanoparticles for cancer treatment using versatile targeted strategies. J Hematol Oncol 15(1):132. https://doi.org/10.1186/s13045-022-01320-5

Velu R, Calais T, Jayakumar A, Raspall F (2019) A comprehensive review on bio-nanomaterials for medical implants and feasibility studies on fabrication of such implants by additive manufacturing technique. Materials (Basel) 13(1). https://doi.org/10.3390/ma13010092

Vial S, Reis RL, Oliveira JM (2017) Recent advances using gold nanoparticles as a promising multimodal tool for tissue engineering and regenerative medicine. Curr Opin Solid State Mater Sci 21(2):92–112. https://doi.org/10.1016/j.cossms.2016.03.006

Wagner AM, Knipe JM, Orive G, Peppas NA (2019) Quantum dots in biomedical applications. Acta Biomater 94:44–63. https://doi.org/10.1016/j.actbio.2019.05.022

Waheed I, Ali A, Tabassum H, Khatoon N, Lai WF, Zhou X (2024) Lipid-based nanoparticles as drug delivery carriers for cancer therapy. Front Oncol 14:1296091. https://doi.org/10.3389/fonc.2024.1296091

Wang Z, Kang S, Cao S, Krecker M, Tsukruk VV, Singamaneni S (2020) Protein-based functional nanocomposites. MRS Bull 45(12):1017–1026. https://doi.org/10.1557/mrs.2020.302

Wang W, Zhou X, Wang H, Zhou G, Yu X (2024) Fabrication and evaluation of PCL/PLGA/beta-TCP spiral-structured scaffolds for bone tissue engineering. Bioengineering (Basel) 11(7):732. https://doi.org/10.3390/bioengineering11070732

Whitfield CJ, Zhang M, Winterwerber P, Wu Y, Ng DYW, Weil T (2021) Functional DNA-polymer conjugates. Chem Rev 121(18):11030–11084. https://doi.org/10.1021/acs.chemrev.0c01074

Wu Y, Gao X, Wu J, Zhou T, Nguyen TT, Wang Y (2023) Biodegradable polylactic acid and its composites: characteristics, processing, and sustainable applications in sports. Polymers (Basel) 15(14):3096. https://doi.org/10.3390/polym15143096

Xue J, Chen F, Bai M, Yu X, Wei J, Huang P et al (2017) DNA-mediated assembly of gold nanoparticles and applications in bioanalysis. ChemNanoMat 3(10):725–735. https://doi.org/10.1002/cnma.201700177

Yu X, Tang X, Gohil SV, Laurencin CT (2015) Biomaterials for bone regenerative engineering. Adv Healthc Mater 4(9):1268–1285. https://doi.org/10.1002/adhm.201400760

Zare H, Ahmadi S, Ghasemi A, Ghanbari M, Rabiee N, Bagherzadeh M et al (2021) Carbon nanotubes: smart drug/gene delivery carriers. Int J Nanomedicine 16:1681–1706. https://doi.org/10.2147/IJN.S299448

Zastulka A, Clichici S, Tomoaia-Cotisel M, Mocanu A, Roman C, Olteanu CD et al (2023) Recent trends in hydroxyapatite supplementation for osteoregenerative purposes. Materials (Basel) 16(3). https://doi.org/10.3390/ma16031303

Zhang D, Liu L, Wang J, Zhang H, Zhang Z, Xing G et al (2022a) Drug-loaded PEG-PLGA nanoparticles for cancer treatment. Front Pharmacol 13:990505. https://doi.org/10.3389/fphar.2022.990505

Zhang F, Yang K, Liu G, Chen Y, Wang M, Li S et al (2022b) Recent advances on graphene: synthesis, properties and applications. Compos A Appl Sci Manuf 160:107051. https://doi.org/10.1016/j.compositesa.2022.107051

Zhao X, Hu DA, Wu D, He F, Wang H, Huang L et al (2021) Applications of biocompatible scaffold materials in stem cell-based cartilage tissue engineering. Front Bioeng Biotechnol 9:603444. https://doi.org/10.3389/fbioe.2021.603444

Zhou W, Gao X, Liu D, Chen X (2015) Gold nanoparticles for in vitro diagnostics. Chem Rev 115(19):10575–10636. https://doi.org/10.1021/acs.chemrev.5b00100

Zhou R, Zhang M, Xi J, Li J, Ma R, Ren L et al (2022) Gold nanorods-based photothermal therapy: interactions between biostructure, nanomaterial, and near-infrared irradiation. Nanoscale Res Lett 17(1):68. https://doi.org/10.1186/s11671-022-03706-3

Zhu J (2010) Bioactive modification of poly(ethylene glycol) hydrogels for tissue engineering. Biomaterials 31(17):4639–4656. https://doi.org/10.1016/j.biomaterials.2010.02.044

Zhu Z (2017) An overview of carbon nanotubes and graphene for biosensing applications. Nanomicro Lett 9(3):25. https://doi.org/10.1007/s40820-017-0128-6

Chapter 3
Advances of Nano-Biocomposites in Medical Applications

Fathima Shajahan and Priya Thomas

Abstract Nano-biocomposite scaffolds have emerged as a promising avenue for revolutionizing healthcare practices in various medical specialties. This chapter provides an in-depth exploration of the recent advancements in nano-biocomposites and their diverse applications across orthopedics, cardiovascular, ophthalmology, dental treatment, and cancer therapy. The integration of nanotechnology with biocompatible materials has led to significant improvements in tissue engineering and regenerative medicine. These innovative composites offer enhanced mechanical properties, improved biocompatibility, controlled drug delivery capabilities, and the ability to mimic the natural extracellular matrix environment within the body. Additionally, they hold great potential for addressing specific clinical challenges such as bone regeneration after trauma or disease-related defects; promoting vascularization for ischemic tissue repair; supporting corneal tissue regeneration; enhancing oral implant stability; and delivering targeted therapeutics for cancer treatment. As research continues to expand in this field, nano-biocomposite scaffolds are expected to play an increasingly pivotal role in shaping the future of medical interventions.

Keywords Nano-biocomposites · Medical applications · Implantable devices · Orthopedics · Cardiovascular · Dental · Ophthalmic

3.1 Introduction

Nanocomposites are a class of materials comprising at least two phases with at least one dimension on the nanoscale, i.e., the matrix phase and the reinforcing phase, which might be fibers, sheets, or particles incorporated in the matrix phase, which could be metallic, ceramic, and polymer (Hafeez 2022). Bionanocomposites

F. Shajahan
Department of Civil Cross-Disciplinary, University of Alberta, Edmonton, Canada
e-mail: shajahan@ualberta.ca

P. Thomas (✉)
Department of Oral Pathology, Annoor Dental College, Muvattupuzha, Kerala, India
e-mail: priyathomask@gmail.com

R. Balakrishnan et al. (eds.), *Biocompatible Nanocomposites*, Advanced Structured Materials 239, https://doi.org/10.1007/978-981-96-7450-3_3

are hybrid materials in which one of the components is obtained from biological sources like protein, polymers, or other biomolecules. The combination of nanomaterials with its superior properties (such as higher surface area, reactivity) together with biocompatible matrices are considered ideal material for variety of medical applications in various branches including orthopedic, cardiovascular, ophthalmic, and dental. However, there are a number of notable drawbacks to nanomaterials, such as irreversible deformation, poor biological adherence, quick elimination, and burst release. The integration of both has demonstrated the possibility to enhance the mechanical, chemical, and physical properties of the resulting composites (Huang et al. 2022; Calais et al. 2018; Velu et al. 2019). The benefits of nanocomposites include small fill size, i.e., high surface area to volume ratio, enhanced mechanical properties with high ductility, and enhanced visual characteristics (light transmission is better).

Nanoparticles can be coupled with a wide range of materials, including metals, ceramics, polymers, and their composites. To combine biocompatibility with features like density, elasticity, fracture, and wear resistance, they must be carefully selected and designed (Velu et al. 2019). Strong corrosion resistance of stainless steel, Co-Cr alloys, and Ti alloys make it suitable for orthopedic implants. (Klee and Höcker 1999). Noble metals such as gold, silver are intriguing options for dental applications as well. Biodegradable natural and synthetic polymers, offer advantages such as biocompatibility, biodegradability, hydrophilicity, and solubility (Vlierberghe et al. 2011). Natural polymers such as chitosan, gelatin, heparin, and collagen and synthetic polymers such as PCL (poly-ε-caprolactone), PEG (Polyethylene glycol), PAA (Polyvinyl alcohol) has been extensively used and studied (Vlierberghe et al. 2011). Bio-ceramics are of particular importance for medical implants and for tissue engineering applications they could be bioinert ceramics, bioactive ceramics, and bioresorbable ceramics (Shanmugam and Sahadevan 2024).

3.1.1 Importance of Nano-Biocomposites in Addressing Current Medical Challenges

With their ability to precisely control drug release for targeted treatment, scaffold structures that mimic natural extracellular matrices to facilitate tissue regeneration, and stem cell delivery for tissue repair, nano-biocomposites have enormous promise in the fields of drug delivery, tissue engineering, and regenerative medicine. Nevertheless, there are still issues with their practical translation, including making sure they are biocompatible, matching the rates of degradation with the kinetics of tissue regeneration, and increasing production without sacrificing intended qualities. Overcoming these obstacles and realizing the full potential of nano-biocomposites in medical applications requires bridging the gap between materials science and medical disciplines (Nanocomposites 2024).

This chapter serves as a comprehensive overview of the recent advancements in nano-biocomposite scaffolds and their diverse applications across different medical specialties. We will start by discussing nano-biocomposites as implantable device, bone tissue engineering, and drug delivery for orthopedic applications followed by cardiovascular, dental, ophthalmic and cancer therapy applications.

3.2 Nano-Biocomposites for Implantable Devices

Devices or prostheses used to replace, enhance, or support biological structures are surgically placed into the human body as medical implants (Ahmed et al. 2012). Implants are now necessary instruments for treating long-term conditions, regaining function, and enhancing quality of life. However, despite all their advantages, medical implants have drawbacks and come with a host of difficulties in terms of their manufacture, design, and application. A major challenge is the possibility of failure of medical implants, which can result in problems and unfavorable health issues. Infection, mechanical wear, corrosion, and tissue rejection are just a few of the many variables that might cause this (Ramezani and Ripin 2023). Even with the most sophisticated implant designs and materials, these issues might arise, necessitating repeated procedures and implant replacements. Researchers have been investigating how nanocomposites might improve the functionality of medical implants to overcome these difficulties (Levana et al. 2023).

Orthopedic implants are used to treat degenerative joint disorders, fractures, and arthritis by replacing or repairing damaged bones and joints (Saeidi et al. 2020). These implants could include cartilage implants for osteoarthritis, screws, pins, and bone plates used to heal fractures, or joint replacements like hip and knee replacements (Ramezani and Ripin 2023).

Devices called cardiovascular implants are used to treat conditions related to the heart, such as arrhythmias, heart failure, and coronary artery disease (CAD). These could include heart valves, defibrillators, pacemakers, and stents. Cardiovascular implants are made to control heart rhythm, blood flow, and to mechanically support the valves and/or arteries. However, thrombosis, infection, or mechanical malfunction can result in the failure of cardiovascular implants. By improving the hemo-compatibility and biocompatibility, nanocomposite materials have shown great promise (Ramezani and Ripin 2023).

Dental implants are used to replace missing teeth. They are often made of titanium or other biocompatible materials and surgically inserted into the mandible (Pasang et al. 2016). Dental implants offer a solid and lasting base for dental restorations, including bridges, crowns, and dentures (Herrero-Climent et al. 2023). Implant failure can happen by fracture, infection, and loosening.

The most common kind of nanocomposites utilized in medical implant applications are polymer based. They are created by mixing nanoscale materials into a polymer matrix, such as graphene, carbon nanotubes, and clay nanoparticles. Polymer-based nanocomposites are highly advantageous because to their low density, superior biocompatibility, and ease of production and tunable biological and mechanical characteristics. (Mostakhdemin et al. 2021). The primary drawback of polymer-based nanocomposites is their comparatively poor mechanical strength and therefore polymer-based nanocomposites with enhanced mechanical properties have been developed.

Another kind of nanocomposite that has drawn interest for use in medical implants is metal-based nanocomposites. Usually, to create metal-based nanocomposites, nanoscale particles like carbon nanotubes or titanium dioxide are mixed with a metal matrix like titanium or stainless steel. These kinds of nanocomposites are advantageous because of their high stiffness, strength, and resistance to wear. Major drawback is its potential for toxicity by the release of metal ions leading to adverse biological reactions. (Li et al. 2023). However, novel metal-based nanocomposites with enhanced biocompatibility have shown promise in lowering the toxicity risk.

Another class of nanocomposites that has drawn interest for use in medical implants are those based on ceramic materials. Typically, to create ceramic-based nanocomposites, hydroxyapatite or alumina nanoparticles are mixed with zirconia or alumina to form a ceramic matrix. Ceramic-based nanocomposites are very strong, wear resistant, and biocompatible. These are only a few of their many benefits. The brittleness of ceramic-based nanocomposites is one of their main drawbacks. But new developments in ceramic-based nanocomposite technology have produced materials with improved strength and toughness (Ramezani and Ripin 2023).

3.3 Nano-Biocomposites in Orthopedics

3.3.1 Nano-Biocomposite Implants for Bone Tissue Engineering

The nanocomposite structure of natural bone tissue confers suitable physical and biological characteristics. For the biomaterial to encourage the regeneration of bone tissue, it must closely resemble real bone tissue. Since no single material can perfectly mimic the composition, structure, and properties of native bone, nanocomposites represent the greatest choice for bone tissue regeneration. For the several stages of bone tissue regeneration, they can offer the necessary matrix environment, incorporate desired biological characteristics, and supply many growth factors in a controlled, sequential manner. One of the main obstacles to treating bone abnormalities is the treatment of critical-sized lesions. (Sahoo et al. 2013).

He et al.'s project involved designing conductive PEEK/G (Polyetheretherketone—Graphene) nanocomposites with a bioactive antibacterial STAC-HA coating deposited through electrophoresis. The bioactive covering demonstrated antibacterial properties against S. aureus and E. coli while significantly improving bone tissue regeneration. Additionally, this composite effectively facilitated photothermal conversion effect whose temperature profile can be varied by adjusting NIR laser power density and irradiation period. Its photothermal characteristics have been successfully established for in vivo tumor inhibition. This versatile material combination is desirable for bone regeneration as well as potential applications in addressing significant bone defects caused by osteosarcoma and osteomyelitis (He et al. 2020). Perumal et al. coated the AZ31 magnesium mesh cage with a bilayered nanocomposite (PCL/nHA) via an electrospinning process. This biodegradable implant stimulated neovascularization, chondrocyte-to-osteoblastic gradient cells, and new bone formation. The implanted nanocomposite exhibited controlled deterioration without impeding bone formation while maintaining animals' limb stability. Compared to untreated implants, the coated ones stimulated neovascularization, chondrocyte-to-osteoblastic gradient cells, and new bone formation confirmed by micro-CT and X-ray (Perumal et al. 2020).

Using the SH technique, Bagherifard et al. produced nanostructured TCP-WS-Zr(tricalcium phosphate-zircon), which improved mechanical characteristics by adding more Zr. When compared to TCP single phase, in vivo examination showed that composite coatings had reduced calcium ion release, which enhanced the osseointegration ratio and promoted bone regeneration (Bagherifard et al. 2020). Chen et al. developed multiwalled carbon nanotubes/carrageenan-chitosan@ Ce and Sr substituted hydroxyapatite biocomposite (FMWCNT/KCG@MHAp) nanocomposite coating on Ti implants, improving mechanical strength and cell response compared to MHAp alone. Animal studies showed presence of osteoblasts and fibrous tissue formation without inflammatory reactions (Chen et al. 2020). Murugesan et al. functionalized graphene oxide (rGO) with Pd NPs and PPy, effectively preventing bacterial biofilms from K.pneumoniae E.Coli, B.subtilis, and P.aeruginosa colonization or adhesion. The NC shows potential as a material for bone implantation due to its antimicrobial properties (Murugesan et al. 2020). Zhou et al. prepared dexamethasone-loaded mesoporous silica nanoparticles (DMSNs/M-CAG composite scaffold with highly porous structure) using electrochemical deposition and freeze-drying for bone tissue engineering. Moreover, they created porous biocomposite scaffolds based on chitosan, alginate, and gelatin that had a calcium phosphate coating on the surface. The hybrid mineralized scaffolds showed a significant increase in mechanical strength and a reduction in swelling properties when compared to pristine CAG scaffolds. In vitro research also revealed that mesenchymal stem cells from rat bone marrow grew and differentiated osteogenically on these biocomposite scaffolds, and in vivo tests showed that the biocomposite scaffolds with mineral deposition and DEX loading had better new bone forming ability in rat calvarial deficiency models than either pure CAG scaffold or single mineralized scaffold (Zhou et al. 2020). Hydroxyapatite/silver (HA/Ag) nanocomposites were fabricated as implantation biomaterials by Martinez Sanmiguel et al., showing bioactivity verified through apatite formations on simulated bodily fluid, along with unaffected MG-63 cell viability. Additionally, these nanocomposites showed non-cytotoxic concentrations of silver nanoparticles, exhibiting antibacterial activity against Candida albicans, Staphylococcus aureus, and Escherichia coli; additionally, they suppressed nitric oxide release and IL-1 and TNF-α secretion in lipopolysaccharide-treated cells. The results suggest potential applications for HA/Ag nanocomposites as implantation biomaterials for various clinical procedures due to their antibacterial properties, low cytotoxicity, and anti-inflammatory effects (Martínez-Sanmiguel et al. 2019). Yu Ling Li et al.'s study involved developing nanohydroxyapatite/polyamide 66 (nHA/PA66) bone screws reinforced with yttria-stabilized tetragonal zirconia. Their findings revealed enhanced adhesion, proliferation, and osteogenic differentiation capabilities of MC3T3-E1 cells when implanted into rabbits' femoral condyles, furthermore, the screw implants resulted in improved trabecular bone development at the interfaces between implants and bones, suggesting great potential for nHA/PA66 composites reinforced with YTZ utilize within bone tissue engineering (Li et al. 2020).

Different amounts of layered double hydroxide within poly(ε-caprolactone) scaffolds were successfully synthesized using electrospinning in another investigation by Dehkordi et al. When exposed to simulated bodily fluid, the inclusion of LDH enhanced mechanical tensile strength while encouraging biomineralization, demonstrating the bioactive characteristics of these nanocomposite scaffolds. Prakash et al. manufactured graphene oxide/hydroapatite/gold nanocomposite(GO/HAP/Au) using a straightforward hydrothermal process combined with gel casting method to create polymeric films incorporating this composite which showed great potential use for bone regeneration due its extreme biocompatibility (Prakash et al. 2020). To study the angiogenesis potential of polymethyl methacrylate-mineralized collagen (nanohydroxyapatite/collagen; nHAC)/Mg-Ca composite materials on human umbilical vein endothelial cells and the osteogenesis effect in vivo, Jiaxin et al. The results demonstrated a considerable increase in HUVEC adhesion, proliferation, migratory motility, and VEGF expression in the magnesium matrix when PMMA-nHAC reinforcing materials were used. In vivo micro-CT analysis revealed the greatest bone growth with coated samples, without inducing localized or systemic cytotoxic effects. Thus, the PMMA-nHAC/Mg-Ca composite implant shows potential as a Mg-based metal composite material for medical devices or biomaterials (Bao et al. 2022). In another study, the osteoconductive bulk metallic glass Mg66Zn29Ca5 and the osteogenic growth peptide OGP-containing MSNs (mesoporous silica nanoparticles) known as MSN@PEG/PEI-OGP were coupled by Yun Shin Chu et al. Superior compressive strength makes Mg66Zn29Ca5 BMG an excellent choice for bone implants. In MC3T3-E1 cells, metal ions generated from the breakdown of Mg66Zn29Ca5 BMG enhanced ALP activity and calcium deposition. Furthermore, MSN@PEG/PEI-OGP demonstrated effective cellular absorption, encouraging osteogenesis, and cell migration. The histological and in vivo micro-CT findings demonstrated that MSN@PEG/PEI-OGP might promote osteogenesis and the growth of new bone surrounding the implant site. Therefore, it can be concluded that coating MSN@PEG/PEI-OGP onto the Mg66Zn29Ca5 BMG interference screw could result in a useful application to satisfy the needs of clinical research (Chu et al. 2022). Hu et al. recently developed a novel oxygenated nanocomposite thin film, titanium nanocomposite-coated TaON-Ag. This coated titanium exhibited broad-spectrum antibacterial activity without impeding fracture healing both in vivo or in vitro on a rat femoral fracture model (Hu et al. 2020).

In a very recent studies, Logesh et al. used ultrasonication for f-MWCNT/BCP composites coated on 316L stainless steel substrates with demonstrated corrosion resistance qualities, suggesting potential for bone tissue restoration (Logesh et al. 2024). Zhou et al. developed an injectable nanocomposite hydrogel with strong bone adhesion and ultrasound responsiveness. The hydrogel combines amine-modified barium titanate piezoelectric nanoparticles with bio-adhesive gelatin-chondroitin sulfate biopolymer. This combination provides injectability, adaptability to uneven defect morphologies, and threefold increased bone adhesion. Furthermore, the nanocomposite generates electrical stimulation in response to ultrasonic cavitation which promotes bone regeneration. Studies show that electromagnetic stimulation from the hydrogel accelerated bone repair and stimulated osteogenic differentiation of BMSCs in both cell tests and animal research. Transcriptome analysis revealed activation of calcium ion channels by electrical stimulation as well as downstream signaling pathways such as PI3K/AKT and MEK/ERK leading to upregulation of the Runx2 gene—ultimately regulating osteogenic differentiation fate of BMSCs. This novel approach holds promise for clinical development of ultrasound-powered hydrogels for repairing broken bones fragments and accelerating healing processes (Zhou et al. 2024) (Table 3.1).

3.3.2 Nano-Biocomposite as Drug Delivery System for Orthopedic Applications

Drug delivery systems are designed to safeguard, move, and release a medication in a predetermined and regulated way. Nanoscale systems offer multiple benefits over conventional medications. They boast a substantial ratio of surface area to volume which facilitates effective encapsulation of the drug and aids in navigating biological barriers for targeted delivery, and they enhance both the solubility and the stability of drugs within biological environments. Moreover, their surfaces can be easily altered to attach biological markers that assist in detecting, imaging, diagnosing, and administering drugs. These systems allow for precise control over drug release rates and bioavailability, minimize unfavorable side effects, and are of a size that permits interactions with biological entities to influence their function (Chen et al. 2019; Yang et al. 2020).

Drug delivery systems can be broadly categorized as targeted and sustained drug delivery systems, each with specific applications. Sustained drug delivery systems are the focus of research for the treatment of bone fractures and include ceramic implants as well as biocompatible and biodegradable composites made of bioactive materials (Chen et al. 2020). Ji et al. created novel porous graphene oxide/chitosan-HAp nanocomposite films (GO/CTS-HA) for aspirin distribution by combining a layer-by-layer approach with a biomimetic mineralization methodology. The in vitro release of loaded aspirin from these nanocomposite films showed a sustained-releasing pattern. Additionally, the (GO/CTS-HA)n nanocomposite films

Table 3.1 Nano-biocomposites for bone tissue engineering

Study	Material/Coating effects	Effects
He et al. (2020)	PEEK/G nanocomposites with antibacterial STAC-HA coating deposited through electrophoresis	Antibacterial properties against S. aureus and E. coli Improved bone tissue regeneration Effective photothermal conversion effect for in vivo tumor inhibition
Perumal et al. (2020)	A bilayered nanocomposite (PCL/nHA) was electrospun onto an AZ31 magnesium mesh cage	Stimulated neovascularization, chondrocyte-to-osteoblastic gradient cells, and new bone formation Controlled deterioration without impeding bone formation or limb stability
Bagherifard et al. (2020)	Nanostructured TCP-WS-Zr coatings	Enhanced mechanical properties -Lower calcium ion release leading to improved osseointegration ratio and bone regeneration
Chen et al. (2020)	FMWCNT/KCG@MHAp nanocomposite coating on Ti implants	Improved mechanical strength -Promoted cell response without inflammatory reactions
Murugesan et al. (2020)	Graphene oxide (rGO) functionalized with Pd NPs and PPy	-Effective prevention of bacterial biofilms colonization/adhesion -Potential as material for antimicrobial implantation applications configured/coated to prevent bacteria from entering the body
Zhou et al. (2020)	DMSNs/M-CAG highly porous composite scaffold and chitosan-alginate-gelatin-based porous biocomposite scaffolds with calcium phosphate coating on the surface	Increased mechanical strength, reduced swelling properties, improved growth and osteogenic differentiation of MSCs from rat bone marrow, superior new bone forming ability in rat calvarial deficiency models
Martinez Sanmiguel et al. (2019)	Hydroxyapatite/silver (HA/Ag) nanocomposites	Bioactivity verified through apatite formations on simulated bodily fluid, non-cytotoxic levels of silver nanoparticles, antibacterial efficacy against Candida albicans, Staphylococcus aureus, and Escherichia coli, Inhibited release of nitric oxide and secretion of IL-1and TNF-α
Li et al. (2020)	Nanohydroxyapatite/polyamide 66(NHA/PA66)bone screws reinforced with yttria-stabilized tetragonal zirconia	Bone density enhancement at interfaces between implants and bones
Dehkordi et al. (2022)	Layered double hydroxide within poly(ε-caprolactone) scaffolds fabricated through electrospinning	I Increased mechanical tensile strength and promotion of biomineralization in simulated body fluid, indicating bioactive features

(continued)

Table 3.1 (continued)

Study	Material/Coating effects	Effects
Prakash et al. (2020)	Graphene oxide/hydroapatite/gold nanocomposite(GO/HAP/Au) using a straightforward hydrothermal process combined with gel casting method to create chitosan polymeric films incorporating this composite	Great potential use for bone regeneration due to its extreme biocompatibility
Bao et al. (2022)	PMMA-nHAC/Mg-Ca composite materials	Increased HUVEC adhesion, proliferation, migratory motility, VEGF expression in Mg matrix; greatest bone growth with coated samples
Yun Shin Chu et al. (2022)	Mg66Zn29Ca5 bulk metallic glass and MSN@PEG/PEI-OGP	Increased ALP activity and calcium deposition in MC3T3-E1 cells; promoted osteogenesis and new bone growth surrounding the implant site
Hu et al. (2020)	Titanium nanocomposite-coated TaON-Ag thin film	Broad-spectrum antibacterial activity without impeding fracture healing

demonstrated promising in vitro compatibility and provided a suitable environment for cultivating mMSCs (mouse mesenchymal stem cells), thus indicating potential application in bone tissue engineering (Ji et al. 2016). Agarose-encapsulated mesoporous carbonated HAp nanocomposite powder was created by Kolanthai et al. with the purpose of being used as an orthopedic drug delivery carrier matrix and as a filler for bone voids. They produced mesoporous, nanosized carbonated hydroxyapatite/agarose(CHAp/agarose) powder composites using a solvothermal approach. These composites showed improved release profiles for amoxicillin and 5-fluorouracil in comparison with calcined samples, as well as a high specific surface area. Additionally, these composites showed significant antibacterial activity against S. aureus, S. epidermidis, and E. Coli (Kolanthai et al. 2017).

A three-dimensional nanocomposite of cellulose-hydroxyapatite integrated with a zeolitic imidazolate framework ZIF-8 (HA/DMOF) loaded with dexamethasone was developed as a local drug delivery system for bone tissue engineering by Sarkar et al. The study involved the creation and description in situ of metal–organic frameworks enclosed in dexamethasone, which were then successfully combined with hydroxyapatite-carboxymethyl cellulose nanocomposite to form the HA/DMOF localized drug delivery system. The HA/DMOF nanocomposite exhibited compressive strengths and moduli within the range of cancellous bone at 16.3 ± 1.57 MPa and 0.54 ± 0.073 GPa, respectively. The in vitro release behavior of dexamethasone from both DMOFs and HA/DMOFs was studied, revealing an extremely slow release from HA/DMOF lasting four weeks compared to DMOF nanoparticles. Additionally, cytocompatibility assessment showed that the synthesized HA/DMOF nanocomposite is compatible with pre-osteoblast MC3T3 cells, resulting in increased extracellular mineralization capacity and ALP activity when cells were grown with it (Sarkar et al. 2019). Furthermore, Zhang et al. created a self-healing hydrogel using LAP nanoplatelet-mediated BMP-2 administration for stem cell-enhanced bone repair. The hydrogels exhibited improved mechanical and rheological properties due to LAP nanoplatelets. Experiments showed that electrostatic immobilization on LAP nanoplatelets maintained bioactivity while reducing burst release. HD(LAP@BMP2's prolonged release promoted proliferation, injection into defect locations encouraged development new bone. This suggests HD/LAP@BMP2 offers non-invasive stem cell treatment and potential drug delivery platform for bone regeneration (Zhang et al. 2020).

Sukhodub et al. synthesized hydroxyapatite–alginate beads loaded into single walled carbon nanotubes. The composites were created in the shape of beads and are a three-dimensional matrix made of calcium-ion-crosslinked alginate macromolecules, 0.05% and 0.5% SWCNT particles, and HA immobilized in the matrix. The mechanical characteristics of SWCNT-loaded HA-alginate composites (maximal relative deformation, compressive strength, and Young's modulus) that were examined confirm that the SWCNT content is a determining factor. The HA–alginate–SWCNT system's structural modeling revealed that SWCNT physically adsorbed into HA–alginate through the formation of triple complexes stabilized by H-bonds and solvophobic/van der Waals interactions. The HPLC analysis shows that SWCNTs have an impact on the extended release of model drug (456 h). The assay for cell culture verifies biocompatibility and stimulation and thereby the potential as a drug delivery platform for bone regeneration. (Sukhodub et al. 2020).

The synthesis of mesoporous 70S bioactive glass/silk fibroin nanocomposites was studied by Moses and Mandal recently. By adding a bioactive biopolymer—silk fibroin—derived from B. mori and A.assama, the researchers were able to create biomimetic mesoporous nanocomposites that are similar to the organo-apatite found in bones. These nanocomposites allowed the loading of antibiotics (doxycycline and gentamicin) and promoted their quick release while maintaining bioactivity. Because the mesoporous nanocomposites were easy to modify, 3-(aminopropyl)-triethoxysilane could be tailored to the silanol network of the bioactive glass with more ease, increasing the drug's ability to load hydrophobic (dexamethasone). The alteration promoted dexamethasone's gradual, continuous release, which is ideal for regulating immunomodulation and mediating osteogenesis. Conformal coatings of these drug-loaded nanocomposites were made on stainless steel implants (EPD) using a basic electrophoretic deposition technique. While the antibiotic-loaded coatings showed antibacterial efficacy with bioactivity maintained for up to 28 days, it was shown that the coatings favored mesenchymal stem cell adherence and osteoinduction. Additionally, the immunomodulatory activities were determined, demonstrating that the dexamethasone-loaded coatings favored M2 macrophage biasness. These mesoporous biomimetic nanocomposites' adaptability ensures that scenario-specific medications can be loaded to help with local delivery using conformal EPD coatings created over metal implants to improve implant patency (Moses and Mandal 2022) (Table 3.2).

Table 3.2 Nano-biocomposites for drug delivery in orthopedics

Study	Material/Coating	Effects/Findings
Ji et al. (2016)	Layer-by-layer construction and biomimetic mineralization techniques combined to create porous graphene oxide/chitosan-HAp nanocomposite films (GO/CTS-HA) for the administration of aspirin	In vitro release showed sustained-releasing pattern, promising in vitro compatibility
Kolanthai et al. (2017)	Developed agarose-encapsulated mesoporous carbonated HAp nanocomposite powder as a bone void filler and orthopedic drug delivery carrier matrix using solvothermal technique	Composites exhibited high specific surface area, enhanced release profiles for amoxicillin and 5-fluorouracil compared to calcined samples, strong antibacterial efficacy against E.Coli, S.aureus, and S.epidermidis
Sarkar et al. it (2019)	a three-dimensional nanocomposite of cellulose-hydroxyapatite integrated with zeolitic imidazolate framework ZIF-8 (HA/DMOF) loaded with dexamethasone as local drug delivery system for bone tissue engineering	HA/DMOF exhibited compressive strengths within the range of cancellous bone at 16.3 ± 1·57 MPa, and moduli within the same range. Compatible with Increased extracellular mineralization capacity and ALP activity when cells(pre-osteoblast MC3T3) were grown with it. Slowed release lasting four weeks compared to DMOF nanoparticles
Zhang et al. (2020)	self-healing hydrogel using LAP nanoplatelet-mediated BMP2 administration for stem cell-enhanced bone repair	Electrostatic immobilization on LAP maintained bioactivity while reducing burst release. LAP@BMP prevented acute rejection, promoted proliferation, injection into defect locations encouraged development of new bone
Sukhodub et al. (2020)	Hydroxyapatite–alginate beads in SWCNTs I Created in the shape of three-dimensional calcium-ion-crosslinked alginate macromolecules, 0.05% and 0.5% SWCNT particles, and HA immobilized in the matrix	Confirmed that SWCNT content affects compressive strength, maximal relative deformation, Young's modulus; impact on extended drug release (456 h)
Moses and Mandal (2022)	Mesoporous 70S bioactive glass/silk fibroin nanocomposites. silk fibroin derived from B. mori and A.assama, antibiotics (doxycycline, gentamicin), 3-(aminopropyl)-triethoxysilane, dexamethasone to create biomimetic mesoporous nanocomposites similar to organo-apatite in bones for local drug delivery	The nanocomposites allow loading of antibiotics promoting quick release while maintaining bioactivity and allowing easy modification of silanol network with more ease. Conformal coatings on metal implants showed favorability toward stem cell adhesion/ osteoinduction along with antibacterial efficiency for up to 28 days. Additionally demonstrated M2 macrophage biasness

3.3.3 For Cartilage Tissue Engineering

In the work by Ghandforoushan et al., PLGA-collagen/PLGA-PEG-PLGA is used to create a biocompatible nanocomposite hydrogel containing transforming growth factor-β1. The results of using human dental pulp stem cells on the scaffolds showed great adhesion, proliferation, and differentiation. SEM images and RT-PCR assay data showing the expression of chondrogenic genes indicated that the TGF-β1 loaded PLGA-PEG-PLGA nanoparticles boosted chondroblast development in hDPSCs (Ghandforoushan et al. 2022)

3.3.4 Advantages, Challenges, and Future Perspectives for Nano-Biocomposites in Orthopedics

Nano-biocomposites show great potential for enhancing bone tissue engineering:

- Unique properties such as mechanical strength, biocompatibility, and bioactivity make them promising materials for developing advanced scaffolds and implants.
- Can promote bone regeneration and healing.
- Further research and development could play a significant role in improving bone tissue engineering outcomes. (Sahoo et al. 2013)

Nanomaterials perform better and are more biocompatible than conventional or microstructured materials in medical applications. However, there are still drawbacks with nanocomposites for tissue engineering, including:

- Questionable cytotoxicity
- Mechanical and corrosion issues
- Structural integrity concerns
- Long-term stability and service challenges
- Component stability problems (Sahoo et al. 2013)

Each nanocomposite combination should be evaluated independently using in vivo and cell culture methods to forecast the biological result. It is imperative and feasible to design test protocols that guarantee the secure production and application of nanomaterials. Creating universal safety standards for all nanomaterials is inappropriate due to their differences in composition, size, shape, surface chemistry, and crystal structure (Tautzenberger et al. 2012).

Numerous biomedical nanocomposites have demonstrated potential for better orthopedic applications, such as:

- Bone tissue engineering
- Orthopedic medication delivery

Differentiable nanocomposites are also being researched as orthopedic drug delivery vehicles to treat various illnesses, including:

- Arthrosis
- Osteomyelitis
- Osteoporosis
- Osteosarcoma
- Bone metastases

A variety of nanocomposites with enhanced bioactivity, cell attachment capacity, biocompatibility, and osteoconductivity are being developed as nano-biomaterials for use in bone tissue engineering applications to satisfy the dynamic reciprocity requirement (Hasnain and Nayak 2019).

3.4 Nano-Biocomposites in Cardiovascular Applications

3.4.1 Role of Nano-Biocomposites in Cardiovascular Implants, Such as Stents and Grafts

Stents are tubular devices utilized to support and preserve the patency of an anatomical lumen, such as a segment of a blood artery (Qi et al. 2013). Nanomaterials play a significant role in cardiovascular disease-related devices through two principal approaches. The first involves employing nanocoatings to enhance the biocompatibility of medical implants and facilitate their integration with surrounding tissues. The other approach focuses on utilizing nanostructured materials to replicate naturally occurring structures, thus improving the mechanical, biological, and physical characteristics of implants (Maleki Dizaj et al. 2019). One method for enhancing stent performance involves using nanocoatings that can be employed for sustained drug release by functioning as carriers for chemically labile pharmaceuticals. For instance, Nakano et al. developed metallic stents coated with chitosan have been shown to elute nanoparticles with cationic surface charges via electrodeposition brought about by the chitosan covering. This facilitates intracellular uptake by interacting with cellular membranes' negative charge. In vivo experiments comparing fluorescein isothiocyanate-encapsulated nanoparticle-coated stents versus those exclusively coated with FITC polymer demonstrated enhanced fluorescence in media layers and neo-intima (Nakano et al. 2004). Furthermore, developments such as Joo et al.'s "Ring Shaped Surface Tension Method" highlight innovative techniques used to coat PLGA nanoparticles onto stent surfaces effectively. Scanning electron microscopy images demonstrate even deposition and close adherence of nanoparticles on stent surfaces while providing strong layer structures due to inherent physical strength. (Joo May 2009).

Guan et al. evaluated the viability of tungsten carbide (WC) nanoparticles integrated into zinc-magnesium alloy for use in BRS (bioresorbable stents) applications.

The Zn-Mg alloy's fatigue resistance and thermal stability were enhanced by the addition of WC nanoparticles, which led to the creation of a Zn-0.5 Mg-WC nanocomposite with superior mechanical strength, a suitable corrosion rate, and deployment resilience. Its biocompatibility and efficacy as a BRS material need to be confirmed by more research in a large animal model (Guan et al. 2022).

3.4.2 Role of Nano-Biocomposites as Cardiovascular Patches

In their study, Li et al. described a green technique that uses decellularized bovine pericardia (dBPs) as the scaffolds for the patch to be functionalized and crosslinked. This procedure made use of the self-assembly of copper@tea polyphenol nanoparticles (Cu@TP NPs). In subcutaneous implantation models, Cu@TP-dBPs showed no signs of calcification and allowed host cell infiltration on Day 21. When comparing Cu@TP-dBP patches to Glut-dBPs (glutaraldehyde crosslinked decellularized bovine pericardias), it was proven that by Day 60, the former showed better extracellular matrix remodeling and ingrowth of functional blood vasculature in cardiac patch graft mice. These results suggested that, while keeping comparable biomechanics and biostability, Cu@TP-dBPs outperformed Glut-dBPs in a number of ways, including anticalcification, remodeling, and integration capabilities. Cu@TP-dBPs is a potentially helpful functional acellular patch for cardiovascular applications because of these qualities (Li et al. 2022). To test the biocompatibility of PU–CL nanoparticles, Zhu et al. co-sedimented them onto a polyurethane film that was saturated with ethanol solution, resulting in the creation of a polyurethane–cardiolipin/polyurethane composite film (PU–CL/PU). They went on to create a vascular patch called CLVP, which is made of PU–CL/PU/DCS. They investigated the vascular patch's functionality in vivo using a mouse model. The PU-CL/PU composite film demonstrated improved biocompatibility characteristics such as longer in vitro coagulation time, greater resistance to platelet adhesion, lower cytotoxicity, and increased affinity for endothelial progenitor cells. Additionally, the CLVP demonstrated enhanced antithrombogenic ability, maintaining the patency of the surgical artery without starting the coagulation cascade (Zhu et al. 2019). Sulthana et al.'s wet chemical method resulted in the creation of two types nanocomposite materials: Copper/Copper Oxide nanoparticle complex(Cu/CuO), graphene-copper. This study revealed an enzyme inhibitory effect on human angiotensin converting enzyme-2(hACE-2) due to complex stabilization leading to enzymatic inhibition. Hence, GO-Cu compound nanomaterials manifested remarkable high level inhibition against hACE-2. These discoveries have potential application as infection inhibitors or treatments for cardiovascular conditions (Sulthana et al. 2023) (Table 3.3).

Table 3.3 Nano-biocomposites for cardiovascular applications

Study	Material/Coating	Finding
Nakano et al. 2004)	Metallic stents coated with chitosan elute nanoparticles with cationic surface charges	Promoting intracellular uptake by interacting with cellular membranes' negative charge In vivo experiments showed enhanced fluorescence in media layers and neo-intima compared to stents exclusively coated with FITC polymer
Joo et al.'s. (2009)	Innovative technique used to effectively coat PLGA nanoparticles onto stent surfaces	SEM images demonstrate even deposition and close adherence of nanoparticles on stent surfaces, providing strong layer structures due to inherent physical strength
Guan et al. (2022)	Assessment of WC nanoparticles for BRS (bioresorbable stent)	When integrated into Zn–Mg alloy it improved fatigue resistance and thermal stability over an extended period, resulting in a Zn–0.Mg-WC nanocomposite material with excellent mechanical strength appropriate corrosion rate, resilience deployment
Li et al. (2022)	Functionalizing and crosslinking acellular patches using Cu@TP NPs on decellularized bovine pericardia (dBPs) as scaffolds	Absence of calcification in Cu@TP-dBPs - Enabled host cell infiltration at Day 21 Improved ECM remodeling and ingrowth of functional blood vessels at Day 60 compared to Glut-dBPs
Zhu et al. (2019)	Creation of PU–CL/PU composite film by co-sedimenting PU–CL nanoparticles onto a PU film saturated with ethanol solution,	Longer in vitro coagulation time Better nonhemolytic qualities - Increased resistance to platelet adhesion Decreased cytotoxicity - Greater affinity for endothelial progenitor cells Enhanced antithrombogenic capacity preserving surgical artery's patency without initiating the coagulation cascade
Sulthana et al. (2023)	Wet chemical method resulting in creation of Copper/Copper Oxide nanoparticle complex (Cu/CuO), graphene-copper materials	High level inhibition against hACE-2(human angiotensin converting enzyme-2)

3.4.3 Advanced Developments of Nano-Biomaterials in Stent Coating

The resolution of issues with in-stent restenosis through the use of a drug-polymer matrix to coat stents has brought about a new concern—late stent thrombosis. Another significant development involves the utilization of nanomaterials to mimic natural structures, leading to improved mechanical, biological, and physical characteristics of implants. Upon exiting the stent surface, associated nanoparticles enter injured epithelium and internalize into artery tissues. For drug delivery applications, researchers have utilized biodegradable polymers such as poly (lactide), poly (lactide-co-glycolide) copolymers, poly(ε-caprolactone), and poly (amino acids) (Shevach et al. 2013). Wang et al. successfully developed a novel configuration involving super-paramagnetic nano-silica @poly (lactic-co-glycolic acid)-coated with quercetin. This innovation aims to prevent atherosclerosis by carefully regulating drug delivery mechanisms (Wang et al. 2019). Similarly, Giannouli et al. demonstrated the effectiveness of quercetin-loaded PLGA for preventing atherosclerosis while maintaining good encapsulation efficiencies and appropriate drug release performance, hinting at their potential for reducing cardiovascular diseases (Giannouli et al. 2018). In another study, researchers described cyclic arginyl-glycyl-aspartic acid (cRGD)-surface functionalized peptide containing urokinase within liposome carriers which effectively targeted GPIIb/IIIa receptors on active platelets (Zhang et al. 2018). Furthermore, an innovative approach using lumbrokinase loaded targeted micelles was created by Pan et al. utilizing specific block copolymers that showed promising results in lowering bleeding risk according to investigations conducted on mice models (Pan et al. 2019).

3.4.4 Challenges and Future Considerations in Cardiovascular Applications

Despite the promising benefits of nanoparticles in biomedical applications, future studies must consider several factors:

- Nanoparticles may pose greater toxicity risks due to their enhanced mobility in the body compared to larger materials.
- Even with using known safe, biocompatible materials as the matrix, the nanoparticle nature can lead to harmful effects.
- Multiple studies have shown the cytotoxic potential of various nanoparticles, such as ultra-small superparamagnetic iron oxide nanoparticles causing platelet aggregation, thrombosis, DNA damage, and cardiac reactive oxygen species.
- Biocompatible nanomaterials like gold nanoparticles can also produce toxicity, including DNA damage.
- Nanomaterial toxicity profiles are crucial for tissue engineering and regenerative medicine, as both extracellular and intracellular toxicity can disrupt signaling pathways (Pan et al. 2019; Nemmar et al. 2016; Kang et al. 2010).

To address these concerns:

- Innovative methods for delivering nanoparticles and reducing their harmful effects should be developed.
- Advanced coating technologies should be implemented on an industrial scale.
- Thorough investigation of nanoparticle interactions and absorption by cells is necessary for therapeutic use.

- Stabilizing nanoparticles using techniques like utilizing stabilizers or regulating solvent parameters is essential due to their high surface area to volume ratio and tendency to aggregate. (Maleki Dizaj et al. 2019)

3.5 Nano-Biocomposites for Dental Applications

Nano-biocomposites is an emerging research in the field of tissue engineering be it as a dental material for restorations, crowns, impression material or for regeneration of lost dental tissues with tissue engineering. When a tooth or periodontium is infected, the main goals of treatment are to get rid of the pathogens, extract the decayed tissue, and replace it so that the oral cavity can function properly for a long time. For each of them, several approaches, including top-down, bottom-up, biomimetic, and functional approaches, were looked into. (Sreenivasalu et al. 2022). A biocompatible nanoproduct's creation requires an analysis of the architecture and physiology of the target site, which is the mouth cavity and teeth. The anatomical, structural, and chemical composition of human tooth tissues varies widely. Understanding each element and the distinct quantity of each material in each layer helps to shed light on how the nanomaterial should be shaped to exactly suit the target layer's requirements. From the outside to the interior, the tooth can be mainly divided into the enamel, dentin and pulp, cementum, and periodontal ligament (Wheeler's 2024). The most crystalline, calcified, and stiff type of hydroxyapatite (HA) with extraordinarily strong intermolecular forces is found in dental enamel (water, 4 weight percent organic content, and 96% inorganic matter) Conversely, dentin exhibits roughly 70% of the inorganic material. Because it is composed of HA biocomposites surrounding the collagen fibers, HA is the most widely utilized ingredient in dental goods, especially in nanoparticle (Palumbo 2011).

Researchers have been studying dental biomaterials for a long time. Metals, ceramics (which includes carbons, ceramic lenses, and glasses), metals, and polymers are the four primary categories for these (Bapat et al. 2019). It is possible to further categorize polymers into synthetic and natural compounds (including plant and animal products) (Yudaev et al. 2022). Almost any product can be produced using nanodentistry. The goal now is to develop a very effective bio-mimicking biomaterial, rather than just a supporting nanocomposite. There is an infinite number of dental items that are manufactured with nanotechnology, such as nanocomposites, nanoimpressions, anti-infective mouthwash, and the often used nanofillers (Bapat et al. 2019).

3.5.1 Preventive Dentistry

Preventive dentistry aims to provide care before dental caries develop into a full-fledged infection, serving as the primary goal of effective oral hygiene management. Dental preventative products prevent additional damage to the pulp by either repairing damaged teeth using remineralization procedures or by preventing the creation of biofilm. Therefore, nano-biocomposites that mimic native structures is promising. Dai et al. created a functional nanocomposite called epigallocatechin gallate-amorphous calcium phosphate, utilizing EGCG's antibacterial ability and ACP's remineralization impact to halt and reverse demineralization. The synthesized EGCG-ACP showed good biocompatibility with human gingival fibroblasts and L-929 cells, while effectively restoring the mechanical characteristics and microstructure of demineralized enamel under neutral conditions through prolonged release of ACP. Furthermore, in combination with ACP, EGCG successfully inhibited S. mutans proliferation and biofilm formation, demonstrating its capability to prevent demineralization development. Results from an in vivo experiment on rats in an acidic cariogenic environment supported the efficacy of EGCG-ACP in remineralizing demineralized enamel by leveraging the compound's pH-responsive properties for protons releasing ACP (Dai et al. 2023). Similarly, Huang et al. developed a new nanocomposite eggshell/Ag through green biosynthesis which was combined it pit and fissure sealants without significantly altering its mechanical characteristics eggshell/AG displayed potent broad-spectrum anti-dice activity. The resulting sealant exhibited strong antibacterial effects against common dental caries bacterial biofilms. Acceptable cytotoxicity demonstrated using gradient dilution extract furthermore no abnormalities were observed histology blood indices or local mucosal tissues Golden hamster oral contact model. These results point toward eggshell / Ag as offering promising potential for clinical applications due robust antibacterial activity outstanding biosafety both vitro when paired both in vitro and in vivo when paired with pit and fissure sealants (Huang et al. 2023).

3.5.1.1 Role of Silver and Gold Nanoparticles

The survival of microbial biofilms was greatly reduced when nanoparticulate silver was added to quaternary ammonium dimethacrylate, increasing the antibacterial efficacy. (Zhang et al. 2012). Additionally, Chang et al. demonstrated the application of Nano Ag in conjunction with quaternary ammonium dimethacrylate, resulting in reduced lactic acid production and increased antibacterial effect. This relationship yielded promising results (Cheng et al. 2015). Furthermore, Dias et al.'s investigation into titanium dioxide modification of Ag NPs revealed a decrease in biofilm adhesion and production (Dias et al. 2019). Magalhães et al. evaluated the effectiveness of three distinct marketing cements: Vitrebond, RelyX ARC, and Sealapex after adding Ag NPs. The comparison showed that the silver NPs groups had superior antibacterial capabilities compared to the group without integrated AgN Ps (Magalhães et al.

2012). Regarding gold nanoparticles coated aligners treated with antibiofilm properties against P.gingivalis; gold NP-associated aligners were proposed by Zhang et al. as a substitute for the current conventional periodontal therapy in treating infections (Zhang et al. 2020). While Kamath created a gold nanoparticle varnish to promote tooth remineralization, showing promising results (Kamath et al. 2020). Chlorhexidine was used by Kovtan's team to functionalize calcium phosphate nanoparticles as a paste or rinse solution endorsing its benefits such as aiding in enamel remineralization lost due to microbial activity along with having an antimicrobial effect on oral microbiota (Kovtun et al. 2012). Researchers have demonstrated that by closing dentin tubules that are open at the root surface, these calcium salt nanoparticles loaded with chlorhexidine have an antibiofilm effect in addition to a remineralization effect. In a cyclic artificial saliva/lactic acid environment, Liang et al. evaluated the effects of poly (amidoamine) dendrimers with bonding agents mixed with NACP, suggesting a significant rise after 20 days, validating their theory regarding enhanced dentin layer support (Liang et al. 2017).

3.5.2 Tooth Restoration

The next most important part of dentistry is tooth restoration, which comes after preventive care. There are several ways to achieve restoration, including crowns, bridges, implants, and fillings. While amalgam has been commonly used for filling cavities, it has drawbacks like mercury leaching and aesthetic concerns. More aesthetically pleasing options like glass ionomer cements and composites are available but they tend to be brittle with associated technique sensitivity. Sharifi et al. developed HA HP-gelatin/curcumin nanocomposites and assessed their antibacterial efficacy against Escherichia coli, Staphylococcus aureus, and Streptococcus mutans (Sharifi et al. 2021). Recently, Balu et al. tackled the difficult issue of adding graphene oxide, multi-walled carbon nanotubes, graphene oxide nano-ribbons, and Ag NPs to HA nanocomposites. They evaluated the activity against the bacteria S. aureus and E. coli using biostrain testing and looked into how the amount of carbon affected the final hardness of these nanocomposites. Moreover, the created nanocomposite offered regulated release of lidocaine, perhaps improving patient outcomes (Balu et al. 2021).

3.5.3 Endodontics

Endodontics is the discipline in dentistry that focuses on treating the dental pulp and its surrounding tissues. This treatment becomes necessary when an infection enters the tooth's inner layer, known as the pulp. In order to give teeth long-term stability, a mix of antimicrobial treatments and supporting procedures is performed. The painful and irreversible condition known as pulpitis necessitates a root canal operation to completely remove the infected pulp and replace it with an inner polymeric material. Hanafy et al. investigated the potential of two widely used dental biomaterials as odontogenic differentiation promoters: nano-HA and mineral trioxide aggregate. According to their findings, the treatment group's expression of genes associated with odontotomy differentiation was noticeably higher than that of the control groups (Hanafy et al. 2018). Sinjari et al. proposed using drug-encapsulated liposomes as a sophisticated nanotechnological method to restore homeostasis of dental pulp stem cells. They found that this approach promoted cell proliferation and reduced inflammatory indicators during 2-hydroxyethyl methacrylate tooth restoration (Sinjari et al. 2019). Tondnevis et al., utilizing polymers and freeze-drying methods, recommended creating a dental tissue scaffold incorporating nano-HA or nano fluoro HA/Chitosan which led to significant increases in cell proliferation due to inclusion of chitosan according to their results. These investigations highlighted the potential of chitosan nanoparticles in advancing endodontic treatments (Tondnevis et al. 2019).Finally, Bhaskar et al.'s recent laboratory study demonstrated profound effects on dental bioactivity and cell proliferation from CMC composite combined with eggshell-derived porous nano-HA by showing considerable elevation levels VEGF and dentine sialophosphoprotein (Baskar et al. 2022).

3.5.4 Prosthodontics

Prosthodontics involves the use of artificial materials to restore teeth and surrounding tissues. The qualities of currently available materials in prosthodontics have been enhanced by nanotechnology. For denture and maxillofacial prostheses, silicone elastomers and polymethylmethacrylate polymers are frequently utilized, and different nanoparticles are added to improve their characteristics. Multi-walled carbon nanotube fillers have been introduced to PMMA in removable dentures to enhance fit, tensile strength, and dimensional stability. Additionally, PMMA polymers are strengthened by metal-based NPs such as Ag and TiO2 for increased antibacterial activity and tensile strength (Ali et al. 2020). Similarly, silicone elastomers are supported by metal nanoparticles, a nanoscale silica cage, and polyhedral oligomeric silsesquioxanes to enhance their physical properties (Zafar 2020). Fixed prosthesis includes fixed prostheses made of nanomaterials like calcium phosphate, silica-based NPs, and carbon nanotubes that were utilized in creating improved restorative materials. Glass–ceramics using the sol–gel method has also shown promise in enhancing hardness and corrosion resistance (Yılmaz and Gul 2022). Furthermore, nanohybrid composite products exhibit great surface wear resistance using nano-sized alumina and glass components resulting excellent amalgam substitute (Saber et al. 2010). For dental implants, Ti phosphate/n-HA (TiP–Ca)composite coatings created through two-step hydrothermal treatment exhibits ultra-high bonding with Ti substrate (Liao et al. 2023). Anodic oxidation coated HOC-Ti implant surface with detailed structures that promote biocompatibility, osseointegration and new bone creation results while providing insight into mechanism underlying clinical applicability and establishing groundwork for future research on impact of oxygen atoms (Luo et al. 2023).

3.5.5 Orthodontics

Orthodontic brackets and archwires are crucial for achieving the appropriate tooth movement. However, the frictional forces between the wires and brackets can lead to delayed improvement in gripping and stability (Stefani et al. 2020). Addressing these issues through nanotechnology coatings has been explored. For instance, adding gold nanoparticles to orthodontic adhesives has shown enhanced antibacterial activity according to Akarajarasrod et al. (2021). Additionally, Alam et al.'s study compared standard orthodontic adhesives with those containing silver nanoparticles, revealing that more recent adhesives can effectively prevent enamel demineralization when compared to older ones (Alam et al. 2022).

The application of nanotechnology in prosthetics and orthodontics provides opportunities to prevent or minimize demineralization and demonstrates that increased anchoring results in better patient compliance.

3.5.6 Periodontology

Improving bone regeneration and promoting dental health is the main objective of periodontal therapy. Patients with periodontal diseases frequently experience dental hypersensitivity as a primary issue. Because of their demonstrated ability to plug dentin tubules and reduce nerve exposure, HA NPs are now used clinically in mouthwash and toothpaste (nano-XIM of FLUIDINOVA, S.A. Portugal). Additionally, these NPs have demonstrated improved adhesion to the tooth surface, encouraging the remineralization of enamel (Barone and Malpassi 1991; Gopinath et al. 2015).

In conclusion, the integration of nanotechnology into various aspects of dentistry, including prosthodontics, dental implants, and orthodontics, has shown significant promise in enhancing the performance and properties of dental materials. The use of nano-biocomposites has led to advancements in antibacterial activity, mechanical strength, and biocompatibility of dental materials. These developments hold great potential in improving clinical outcomes, patient comfort, and the overall effectiveness of dental treatments. As researchers continue to explore and refine the applications of nano-biocomposites in dentistry, the future looks promising for further innovations and enhancements in dental care (Table 3.4).

Table 3.4 Nano-biocomposites in dentistry

Study	Nanocomposite	Finding
Preventive and restorative Dentistry		
Dai et al. (2023)	EGCG-ACP	Halted and reversed demineralization by utilizing antibacterial ability of EGCG and remineralization impact of ACP Showed good biocompatibility with human gingival fibroblasts and L-929 cells Inhibited S. mutans proliferation and biofilm formation Demonstrated efficacy in remineralizing demineralized enamel in acidic cariogenic environment
Huang et al. (2023)	Eggshell/Ag	Displayed potent broad-spectrum anti-bacterial activity when combined with pit and fissure sealants Exhibited strong antibacterial effects against common dental caries bacterial biofilms Demonstrated acceptable cytotoxicity using gradient dilution extract Showed outstanding biosafety both in vitro and in vivo when paired with pit and fissure sealants
Chang et al. (2015)	Nanoparticulate silver to quaternary ammonium dimethacrylate	Decreased viability of microbial biofilms, improved antibacterial efficacy Additionally, reduced lactic acid production and increased antibacterial effect when combined with Nano Ag in conjunction with quaternary ammonium dimethacrylate
Dias et al.'s (2019)	Titanium dioxide modification of Ag NPs	Decrease in biofilm adhesion and production.l Investigated the effectiveness after adding Ag NPs to Vitrebond, RelyX ARC, and Sealapex cements. The comparison showed that the silver NPs groups had superior antibacterial capabilities compared to the group without integrated AgNPs
Zhang et al.'s (2020)	Gold nanoparticles coated aligners treated with antibiofilm properties against P.gingivalis	Gold NP-associated aligners suggested as an alternative treatment approach for infections from current standard periodontal therapy
Kovtan et al. (2012)	Chlorhexidine functionalized calcium phosphate nanoparticles as a paste or rinse solution	aiding in enamel remineralization lost due to microbial activity along with having an antimicrobial effect on oral microbiota

(continued)

Table 3.4 (continued)

Study	Nanocomposite	Finding
Liang et al. (2017)	Poly(amidoamine) dendrimer plus bonding agent effects combined with NACP	Under cyclic artificial saliva / lactic acid environment supporting significant increase after 20 days reinforcing their raised hypothesis regarding enhanced dentin layer support
Sharifi et al. (2021)	HA HP-gelatin/curcumin nanocomposites	Antibacterial efficacy against Escherichia coli, Staphylococcus aureus, and Streptococcus mutans
Balu et al. (2021)	HA nanocomposites with graphene oxide, multi-walled carbon nanotubes, graphene oxide nano-ribbons, and Ag NPs	Activity against S. aureus and E. coli using biostrain testing; Effect of carbon amount on the final hardness; Regulated release of lidocaine for improved patient outcomes
Endodontics		
Hanafy et al. (2018)	Nano-HA and mineral trioxide aggregate	Expression of genes associated with odontotomy differentiation was noticeably higher in the treatment group than in the control groups
Sinjari et al. (2019)	Curcumin-encapsulated liposomes	Promoted cell proliferation and reduced inflammatory indicators during 2-hydroxyethyl methacrylate tooth restoration
Tondnevis et al. (2019)	Dental tissue scaffold incorporating nano-HA or nano fluoro HA/ Chitosan	The addition of chitosan led to significant increases in cell proliferation due to their results
Bhaskar et al. (2022)	CMC composite combined with eggshell-derived porous nano-HA	Caused considerable elevation levels VEGF & dentine sialophosphoprotein resulting from profound effects on dental bioactivity and cell proliferation
Prosthodontics		
Ali et al. (2020)	Removable dentures with PMMA/ multi-walled carbon nanotube fillers	Enhance fit, tensile strength, and dimensional stability metal-based NPs such as Ag and TiO_2 increased antibacterial activity
Yilmaz and Gul (2022)	Fixed prostheses with silica-based NPs nanoscale silica cage Polyhedral oligomeric silsesquioxanes	Enhanced physical properties Glass–ceramics using sol–gel method hardness corrosion resistance

(continued)

Table 3.4 (continued)

Study	Nanocomposite	Finding
Liao et al. (2023)	Ti phosphate/ n-HATiP–Ca Ultra-high bonding with Ti substrate Anodic oxidation coated HOC-Ti implant surface	Biocompatibility osseointegration new bone creation
Orthodontics		
Akarajarasrod et al. (2021)	Orthodontic adhesives with gold nanoparticles	Enhanced antibacterial activity

3.6 Nano-Biocomposites in Ophthalmic Applications

Drug delivery to the eye faces challenges due to its short residence time, hindering drugs from reaching therapeutic levels. Only about 3% of administered topical eye drops penetrate the precorneal region within minutes, posing difficulties for treating corneal issues. Scientists are focusing on creating poly (urethane-urea) nanofibrous scaffolds with Zn-Al layered double hydroxide NPs loaded with vitamin C as replacements for injured corneal stroma. These composite nanofibers had longer release durations and possible applications in the treatment of corneal-related conditions, but they also shown decreased tensile strength and elongation at breakdown (Moghanizadeh-Ashkezari et al. 2019).

Researchers developed an azithromycin-loaded poly(lactic-co-glycolic acid) copolymer/pluronic NPs ocular insert for treating bacterial keratitis. The composite nanofibrous insert showed sustained drug release, increased penetration potential, and good biocompatibility. In vitro and ex vivo studies demonstrated mucoadhesiveness and improved pharmacokinetic profile (Noninvasive 2024). Additionally, Ag NPs incorporated in PMMA NFs exhibited effective antibacterial properties against both gram-positive and gram-negative bacteria due to unique surface characteristics. Potential applications include providing antimicrobial protection without compromising safety features related to the transparency of the prepared nanoparticles (Green synthesised 2024).

A noninvasive topical formulation for the efficient delivery of ocular medications was created by electrospinning polyamidoamine dendrimers and the anti-glaucoma drug brimonidine tartrate with PEO NFs. The composite NFs showed less toxicity to NIH 3T3 fibroblasts in in vitro experiments. Additionally, when these NFs were applied to normotensive rats, the medication's efficiency increased with daily use when compared to standard saline eye drops, suggesting that this scaffold may be used to treat glaucoma (Lancina et al. 2017). By encasing Amphotericin B into PLGA NPs or establishing a polyelectrolyte complex with Eudragit L polymer, another scaffold was created having NP-in-NF architecture and in situ gelling characteristics. These scaffolds were then independently incorporated into gellan gum/pullulan NFs using an electrospinning process. Amphotericin B polyelectrolyte complex-incorporating

NFs shown superior suppression against the fungal strain Issatchenkia orientalis, according to in vitro tests. Furthermore, compared to traditional eye drop formulations, the created scaffold was less hazardous to corneal cells, according to cellular investigations on a multilayered-stratified human cornea cell model. Furthermore, the Amphotericin B-containing composite nanofibrous scaffold effectively retained its antifungal efficacy after being sterilized by electron irradiation, making it a promising treatment for keratomycosis (Göttel 2020).

Nanofibrous nanocomposites using PCL and CMC polymers with different concentrations of Cobalt-doped HAp were developed for wound dressings. The fabricated materials were thoroughly characterized and found to have favorable properties. Biological evaluations showed that the nanofibrous nanocomposites were hemocompatible and biocompatible under optimal conditions. Animal studies demonstrated that these composites promoted wound healing, with the best outcomes observed with PCL/CMC/CoHAp treatment. Overall, this research highlights the potential of combining Cobalt-doped HAp with PCL/CMC nanofibers for bioactive wound dressing in wound management applications (Wang and Shi 2022).

Bentonite and hydroxypropyl methylcellulose-poloxamer enhanced trimetazidine's anti-inflammatory effects on rabbit eyes in a study by Swain et al. The sol formulation exhibited pseudoplastic shear thinning behavior, sustained release (~79–97%), corneal penetration (~79–83%) over 6 h. It also showed no ocular inflammation after carrageenan injection, unlike the untreated eyes (Swain et al. 2023). A hierarchical FNL-TSG drug delivery system with consistent and long-lasting drug release was designed by Jiao et alfor AMD treatment using molecular docking technology to screen natural active compounds effective in treating AMD (Jiao et al. 2023).

In conclusion, the nano-biocomposite scaffolds have shown potential for ophthalmic applications, particularly in the treatment of glaucoma and keratomycosis. The combination of Polyamidoamine dendrimers with anti-glaucoma medication demonstrated improved drug effectiveness and minimal toxicity on fibroblasts and corneal cells. A prospective treatment approach for keratomycosis, the Amphotericin B polyelectrolyte complex-incorporating NFs also shown greater suppression against fungal strains while retaining their antifungal efficacy post-sterilization (Table 3.5).

3.7 Nano-Biocomposites for Cancer Treatment

Recent developments in nanocomposite technology have opened up a wide range of potential applications, including improved T1-WI MRI (T1 weighted image) contrast for disease diagnosis processes, cancer synergistic photothermal therapy and MR imaging, and tumor-targeting delivery therapies. These advancements also offer benefits such as enhanced blood flow facilitation and photoacoustic imaging capability for directed anticancer therapy. The versatile nature of these materials suggests an effective expansion of biological uses in various medical scenarios.

Table 3.5 Nano-biocomposites for ophthalmic applications

Study	Nanocomposite	Findings
Khalil et al. (2024)	Azithromycin-loaded poly(lactic-co-glycolic acid) copolymer/pluronic NPs ocular insert for treating bacterial keratitis	Sustained drug release, increased penetration potential, good biocompatibility, mucoadhesiveness, and improved pharmacokinetic profile
Philip et al. (2024)	Ag NPs incorporated in PMMA NFs Silver nanoparticles and polymethyl methacrylate nanofibers	Effective antibacterial properties against gram-positive and gram-negative bacteria due to unique surface characteristics Potential applications include providing antimicrobial protection without compromising safety features related to the transparency of the prepared nanoparticles
Lancina et al. (2017)	Electrospun polyamidoamine dendrimers and brimonidine tartrate with PEO NFs for Noninvasive topical formulation for efficient delivery of ocular medications using	Less toxicity to NIH 3T3 fibroblasts in vitro experiments Increased medication efficiency compared to standard saline eye drops when applied to normotensive rats suggests that this scaffold may be used to treat glaucoma
Gottel et al. (2020)	Amphotericin B-poly(lactic-co-glycolic acid) nanoparticles, polyelectrolyte complex with Eudragit L polymer	Superior suppression against Issatchenkia orientalis according to in vitro tests Less hazardous than traditional eye drop formulations on corneal cells based on cellular investigations on a multilayered-stratified human cornea cell model Effective retention of antifungal efficacy after electron irradiation sterilization
Wang and Shi (2022)	PCL, CMC polymers, Cobalt-doped HAp Nanofibrous nanocomposites for wound dressings	Hemo-compatible and biocompatible under optimal conditions; Promotion of wound healing
Swain et al. (2023)	Bentonite and hydroxypropyl methylcellulose-poloxamer for anti-inflammatory effects on rabbit eyes	Sustained release (~79–97%); corneal penetration (~79–83%) over 6 h; No ocular inflammation after carrageenan injection compared to untreated eyes

(continued)

Table 3.5 (continued)

Study	Nanocomposite	Findings
Jiao et al. (2023)	Hierarchical FNL-TSG drug delivery system for AMD treatment using molecular docking technology	Potential application in glaucoma and keratomycosis treatment Combination of Polyamidoamine dendrimers with anti-glaucoma medication Improved drug effectiveness and minimal toxicity on fibroblasts and corneal cells Amphotericin B polyelectrolyte complex-incorporating NFs for keratomycosis treatment Greater suppression against fungal strains while retaining antifungal efficacy post-sterilization

Ji et al. successfully produced highly porous Mn3O4/PtOx NCs with improved T1-WI MRI contrast through ion-etching, laser-induced liquefaction, and laser ablation in liquid. The obtained porous Mn3O4/PtOx NCs had a relaxivity value of 20.48 mM^{-1} s^{-1}, significantly higher than that of pure Mn3O4 NPs in the literature and reported commercially available products like Teslascan and Magnevist. Furthermore, compared to Gd-DTPA, the Mn3O4/PtOX NPs demonstrated longer blood circulation time in vivo studies, advantageous for disease diagnosis processes (Ji et al. 2021). Ji et al. created a PDA-coated porous iron oxide nanocomposite for cancer synergistic photothermal therapy and MR imaging using a solvothermal technique followed by polydopamine coating sensitive to near-infrared light under alkaline conditions while being gently stirred (Jin et al. 2021). Wang et al. designed water-dispersible Covalent organic framework (COF@IR783) nanocomposites as potential tools for tumor-targeting delivery therapies due to their properties such as improved blood flow facilitation, enhanced photothermal treatment capabilities compared to cyanines alone in the near-infrared spectrum, and photoacoustic imaging capability directing anticancer therapy post-test on 4Tl cells. This research suggests an effective biological uses expansion for cleverly designed COFs (Wang et al 2019) (Table 3.6).

3.8 Challenges and Future Directions

Despite the significant advancements in nano-biocomposite-based medical devices, several challenges and limitations still exist. One primary challenge is the need for better understanding and control of the interactions between nanoparticles and biological systems, including addressing potential toxic effects on human health and the environment to ensure safety. Additionally, scalability and cost-effectiveness remain a hurdle for widespread medical applications. To overcome these challenges, continued research and innovation are crucial in designing nanoparticles

Table 3.6 Nano-biocomposite for cancer treatment and diagnosis

Study	Material	Findings
Ji et al. (2021)	Mn3O4/PtOx NCs using ion-etching, laser-induced liquefaction, and laser ablation in liquid	Highly porous nanocomposites with improved T1-WI MR contrast and higher relaxivity value than pure Mn3O4 NPs in the literature Longer blood circulation time compared to Gd-DTPA in vivo studies
Jin et al. (2021)	PDA-coated porous iron oxide nanocomposite loaded with doxorubicin	Suitable for cancer synergistic photothermal therapy and MR imaging
Wang et al. (2019)	COF@IR783 nanocomposite	Water-dispersible COFs Enhanced properties such as improved blood flow facilitation Enhanced photothermal treatment capabilities compared to cyanines alone in the near-infrared spectrum Photoacoustic imaging capability directing anticancer therapy post-test on 4Tl cells This research suggests an effective biological uses expansion for cleverly designed COFs

with enhanced biocompatibility. This will require interdisciplinary collaboration among materials scientists, biologists, and medical professionals. Moreover, establishing standardized protocols for assessing nanoparticle toxicity will be essential to guarantee safety. Future directions involve exploring nanotechnology-based delivery systems for drug delivery as well as integrating nanomaterials into bioresorbable stents and tissue engineering scaffolds for regenerative medicine purposes. Addressing potential toxic effects also requires stringent regulatory measures along with comprehensive risk assessments focusing on sustainability. In conclusion, addressing current challenges through strategic research focus areas such as interdisciplinary collaboration to enhance nanoparticle design, establishment of standardized toxicity assessment protocols; exploring future applications like targeted drug delivery using nanotechnology-based systems; and integration into tissue engineering offer promising avenues forward in revolutionizing medical device technology while ensuring safety standards are imperative to realize their full clinical impact.

3.9 Conclusion

In conclusion, the development and application of nano-biocomposites have proven to be a game-changer in the field of medical technologies. The findings and insights from various studies emphasize the importance of nano-biocomposites in revolutionizing medical technologies and significantly improving patient outcomes. These composite materials exhibit enhanced biocompatibility, targeting capabilities, and therapeutic effects, making them valuable assets in medicine.

The continuous research and innovation in this field are of utmost significance to further unlock the full potential of nano-biocomposites. Advancements will not only address existing challenges and limitations but also pave the way for the exploration of new applications and breakthroughs. These efforts will be pivotal in ensuring that nano-biocomposites continue to contribute to the advancement of healthcare and medical treatments.

References

Ahmed W, Elhissi A, Jackson M, Ahmed E (2012) Precision machining of medical devices. Des Manuf Med Devices 59–113. https://doi.org/10.1533/9781908818188.59

Akarajarasrod P et al (2021) Antibacterial effect of experimental orthodontic adhesives containing gold nanoparticles against Streptococcus mutans and Streptococcus sobrinus. Key Eng Mater 904:301–308. https://doi.org/10.4028/www.scientific.net/KEM.904.301

Alam MK, Alsuwailem R, Alfawzan AA (2022) Antibacterial activity and bond strength of silver nanoparticles modified orthodontic bracket adhesive: a systematic review and meta-analysis of in-vitro and in-vivo studies. Int J Adhes Adhes 113:103040. https://doi.org/10.1016/j.ijadhadh.2021.103040

Ali NA, Hussein SI, Asafa TB, Abd-Elnaiem AM (2020) Mechanical properties and electrical conductivity of Poly(methyl methacrylate)/multi-walled carbon nanotubes composites. Iran J Sci Technol Trans Sci 44(5):1567–1576. https://doi.org/10.1007/s40995-020-00948-7

Bagherifard A et al (2020) Improvement in osseointegration of tricalcium phosphate-zircon for orthopedic applications: an in vitro and in vivo evaluation. Med Biol Eng Comput 58(8):1681–1693. https://doi.org/10.1007/s11517-020-02157-1

Balu SK, Sampath V, Andra S, Alagar S, Manisha S, Vidyavathy (2021) Fabrication of carbon and silver nanomaterials incorporated hydroxyapatite nanocomposites: enhanced biological and mechanical performances for biomedical applications. Mater Sci Eng C Mater Biol Appl 128:112296. https://doi.org/10.1016/j.msec.2021.112296

Bao J, Sun X, Chen Z, Yang J, Wang C (2022) Study on the angiogenesis ability of Polymethyl methacrylate-mineralized collagen/Mg-Ca composite material in vitro and the bone formation effect in vivo. J Biomater Appl 37(5):814–828. https://doi.org/10.1177/08853282221121851

Bapat RA et al (2019) The use of nanoparticles as biomaterials in dentistry. Drug Discov Today 24(1):85–98. https://doi.org/10.1016/j.drudis.2018.08.012

Barone M, Malpassi M (1991) Clinical trial of a 15% supermicronized hydroxyapatite gel for dentin hypersensitivity. G Ital Endod 5(2):43–47

Baskar K et al (2022) Eggshell derived nano-hydroxyapatite incorporated carboxymethyl chitosan scaffold for dentine regeneration: a laboratory investigation. Int Endod J 55(1):89–102. https://doi.org/10.1111/iej.13644

Calais T, Bancaud A, Estève A, Rossi C (2018) Correlation between DNA self-assembly kinetics, microstructure, and thermal properties of Tunable highly energetic Al–CuO nanocomposites for micropyrotechnic applications. ACS Appl. Nano Mater. 1(9):4716–4725. https://doi.org/10.1021/acsanm.8b00939

Chen X, Zhu Q, Xu X, Shen S, Zhang Y, Mo R (2019) Drug delivery: sequentially site-specific delivery of apoptotic protein and tumor-suppressor gene for combination cancer therapy (Small 40/2019). Small 15(40):1970218. https://doi.org/10.1002/smll.201902998

Chen M, Zhang H, Shan S, Li Y, Li X, Peng D (2020) Fabrication of multiwalled carbon nanotubes/carrageenan-chitosan@ Ce and Sr substituted hydroxyapatite biocomposite coating on titanium: in vivo bone formation evaluations. J King Saud Univ-Sci 32(1):1175–1181. https://doi.org/10.1016/j.jksus.2019.11.006

Chen J, Ashames A, Buabeid MA, Fahelelbom KM, Ijaz M, Murtaza G (2020) Nanocomposites drug delivery systems for the healing of bone fractures. Int J Pharm 585:119477. https://doi.org/10.1016/j.ijpharm.2020.119477

Cheng L, Zhang K, Weir MD, Melo MAS, Zhou X, Xu HH (2015) Nanotechnology strategies for antibacterial and remineralizing composites and adhesives to tackle dental caries. Nanomedicine 10(4):627–641. https://doi.org/10.2217/nnm.14.191

Chu YS, Wong PC, Jang JSC, Chen CH, Wu SH (2022) Combining Mg–Zn–Ca bulk metallic glass with a mesoporous silica nanocomposite for bone tissue engineering. Pharmaceutics 14(5), art no 5. https://doi.org/10.3390/pharmaceutics14051078

Dehkordi AN et al (2022) Highly effective electrospun polycaprolactone/ layered double hydroxide nanofibrous scaffold for bone tissue engineering. J Drug Deliv Sci Technol 76:103827. https://doi.org/10.1016/j.jddst.2022.103827

Dai D, Wang J, Xie H, Zhang C (2023) An epigallocatechin gallate–amorphous calcium phosphate nanocomposite for caries prevention and demineralized enamel restoration. Mater Today Bio 21:100715. https://doi.org/10.1016/j.mtbio.2023.100715

Dias HB, Bernardi MIB, Bauab TM, Hernandes AC, de Souza Rastelli AN (2019) Titanium dioxide and modified titanium dioxide by silver nanoparticles as an anti biofilm filler content for composite resins. Dent Mater 35(2):e36–e46. https://doi.org/10.1016/j.dental.2018.11.002

Ghandforoushan P et al (2022) Enhancing the function of PLGA-collagen scaffold by incorporating TGF-β1-loaded PLGA-PEG-PLGA nanoparticles for cartilage tissue engineering using human dental pulp stem cells. Drug Deliv. and Transl. Res. 12(12):2960–2978. https://doi.org/10.1007/s13346-022-01161-2

Giannouli M, Karagkiozaki V, Pappa F, Moutsios I, Gravalidis C, Logothetidis S (2018) Fabrication of quercetin-loaded PLGA nanoparticles via electrohydrodynamic atomization for cardiovascular disease. Materials Today: Proceedings 5(8): 15998–16005, Part 2. https://doi.org/10.1016/j.matpr.2018.05.044

Gopinath NM, John J, Nagappan N, Prabhu S, Kumar ES (2015) Evaluation of dentifrice containing Nano-hydroxyapatite for dentinal hypersensitivity: a randomized controlled trial. J Int Oral Health 7(8):118–122

Göttel B et al. (2020) In situ gelling amphotericin B nanofibers: a new option for the treatment of keratomycosis. Front Bioeng Biotechnol **8**. https://doi.org/10.3389/fbioe.2020.600384

Guan Z et al (2022) Zn–Mg–WC Nanocomposites for Bioresorbable cardiovascular stents: microstructure, mechanical properties, fatigue, shelf life, and corrosion. ACS Biomater Sci Eng 8(1):328–339. https://doi.org/10.1021/acsbiomaterials.1c01358

Hafeez M (2022) Recent progress and overview of nanocomposites, in nanocomposite materials for biomedical and energy storage applications. IntechOpen. https://doi.org/10.5772/intechopen.102469

Hanafy AK, Shinaishin SF, Eldeen GN, Aly RM (2018) Nano hydroxyapatite & mineral trioxide aggregate efficiently promote odontogenic differentiation of dental pulp stem cells. Open Access Maced J Med Sci 6(9):1727–1731. https://doi.org/10.3889/oamjms.2018.368

Hasnain MS, Nayak AK (2019) 7–Nanocomposites for improved orthopedic and bone tissue engineering applications. In: applications of nanocomposite materials in orthopedics, Inamuddin AM, Asiri A Mohammad, Eds., In woodhead publishing series in biomaterials, Woodhead Publishing, pp 145–177. https://doi.org/10.1016/B978-0-12-813740-6.00008-9.

He M et al (2020) Conducting Polyetheretherketone nanocomposites with an electrophoretically deposited bioactive coating for bone tissue regeneration and multimodal therapeutic applications. ACS Appl Mater Interfaces 12(51):56924–56934. https://doi.org/10.1021/acsami.0c20145

Herrero-Climent M, Falcao A, Tondela J, Brizuela A, Rios-Carrasco B, Gil J (2023) Relevant aspects of the dental implant design on the insertion torque, resonance frequency analysis (RFA) and micromobility: an in vitro study. J Clin Med 12(3), art no 3. https://doi.org/10.3390/jcm12030855

Hu C-C, Chang C-H, Chang Y, Hsieh J-H, Ueng SW-N (2020) <p>Beneficial Effect of TaON-Ag Nanocomposite Titanium on Antibacterial Capacity in Orthopedic Application</p>. IJN 15:7889–7900. https://doi.org/10.2147/IJN.S264303

Huang S et al (2022) Nanocomposite hydrogels for biomedical applications. Bioeng and Transl Med 7(3):e10315. https://doi.org/10.1002/btm2.10315

Huang X et al (2023) Application study of novel Eggshell/Ag combined with pit and fissure sealants. IJN 18:2911–2922. https://doi.org/10.2147/IJN.S403974

Ji M et al (2016) A novel porous aspirin-loaded (GO/CTS-HA)n nanocomposite films: synthesis and multifunction for bone tissue engineering. Carbohyd Polym 153:124–132. https://doi.org/10.1016/j.carbpol.2016.07.078

Ji S et al (2021) Surface morphology and payload synergistically caused an enhancement of the longitudinal relaxivity of a Mn3O4/PtOx nanocomposite for magnetic resonance tumor imaging. Biomater Sci 9(7):2732–2742. https://doi.org/10.1039/D0BM01993C

Jia H et al (2011) A novel polymer-free paclitaxel-eluting stent with a nanoporous surface for rapid endothelialization and inhibition of intimal hyperplasia: comparison with a polymer-based sirolimus-eluting stent and bare metal stent in a porcine model. J Biomed Mater Res A 98(4):629–637. https://doi.org/10.1002/jbm.a.33151

Jiao X et al (2023) Nano-composite system of traditional Chinese medicine for ocular applications: molecular docking and three-dimensional modeling insight for intelligent drug evaluation. Drug Deliv Transl Res 13(12):3132–3144. https://doi.org/10.1007/s13346-023-01376-x

Jin Z et al (2021) Preparation of doxorubicin-loaded porous iron Oxide@ polydopamine nanocomposites for MR imaging and synergistic photothermal–chemotherapy of cancer. Colloids Surf, B 208:112107. https://doi.org/10.1016/j.colsurfb.2021.112107

Joo J, Nam HY, Nam SH, Baek IS, 박종상 (2009) A novel deposition method of PLGA nanoparticles on coronary stents. Bull Korean Chem Soc 30(5):1085–1087. https://doi.org/10.5012/BKCS.2009.30.5.1085

Julius A, Malakondaiah S, Pothireddy RB (2025) Polymer and nanocomposite fillers as advanced materials in biomedical applications. Nano Trends 100087. https://doi.org/10.1016/j.nwnano.2025.100087

Kamath KA, Nasim I, Rajeshkumar S (2020) Evaluation of the re-mineralization capacity of a gold nanoparticle-based dental varnish: an in vitro study. J Conserv Dent 23(4):390–394. https://doi.org/10.4103/JCD.JCD_315_20

Kang B, Mackey MA, El-Sayed MA (2010) Nuclear targeting of gold nanoparticles in cancer cells induces DNA damage, causing cytokinesis arrest and apoptosis. J Am Chem Soc 132(5):1517–1519. https://doi.org/10.1021/ja9102698

Khalil IA, Ali IH, El-Sherbiny IM (2019) Noninvasive biodegradable nanoparticles-in-nanofibers single-dose ocular insert: *in vitro, ex vivo* and *in vivo* evaluation. Nanomedicine 14(1):33–55. https://doi.org/10.2217/nnm-2018-0297

Klee D, Höcker H (1999) Polymers for biomedical applications: improvement of the interface compatibility. In: biomedical applications polymer blends. Eastmond GC, Höcker H, Klee D, Eds. In: advances in polymer science, Berlin, Heidelberg: Springer, pp 1–57. https://doi.org/10.1007/3-540-48838-3_1

Kolanthai E, Abinaya Sindu P, Thanigai Arul K, Sarath Chandra V, Manikandan E, Narayana Kalkura S (2017) Agarose encapsulated mesoporous carbonated hydroxyapatite nanocomposites powder for drug delivery. J Photochem Photobiol B: Biol 166:220–231. https://doi.org/10.1016/j.jphotobiol.2016.12.005

Kovtun A et al (2012) Chlorhexidine-loaded calcium phosphate nanoparticles for dental maintenance treatment: combination of mineralising and antibacterial effects. RSC Adv 2(3):870–875. https://doi.org/10.1039/C1RA00955A

Lancina MGI, Singh S, Kompella UB, Husain S, Yang H (2017) Fast dissolving dendrimer nanofiber mats as alternative to eye drops for more efficient Antiglaucoma drug delivery. ACS Biomater Sci Eng 3(8):1861–1868. https://doi.org/10.1021/acsbiomaterials.7b00319

Levana O et al (2023) Development of nanoclay-based nanocomposite surfaces with antibacterial properties for potential biomedical applications. J Ind Eng Chem 120:448–459. https://doi.org/10.1016/j.jiec.2022.12.052

Li Y-L et al (2020) In vitro and in vivo evaluations of nano-hydroxyapatite/polyamide 66/yttria-stabilized zirconia as a novel bioactive material for bone screws: Biocompatibility and bioactivity. J Biomater Appl 35(1):108–122. https://doi.org/10.1177/0885328220916618

Li H, Zheng Y, Ji X (2023) Synthesis and in vitro evaluation of Ca-P coating on biodegradable Zn alloys. J Mater Sci Technol 141:124–134. https://doi.org/10.1016/j.jmst.2022.10.003

Li Q et al. (2022) Crosslinking and functionalization of acellular patches via the self-assembly of copper@tea polyphenol nanoparticles. Regen Biomater 9:rbac030. https://doi.org/10.1093/rb/rbac030

Liang K, Weir MD, Reynolds MA, Zhou X, Li J, Xu HHK (2017) Poly (amido amine) and nano-calcium phosphate bonding agent to remineralize tooth dentin in cyclic artificial saliva/lactic acid. Mater Sci Eng C Mater Biol Appl 72:7–17. https://doi.org/10.1016/j.msec.2016.11.020

Liao Z et al (2023) In situ titanium phosphate formation on a titanium implant as ultrahigh bonding with nano-hydroxyapatite coating for rapid osseointegration. Biomater Sci 11(6):2230–2242. https://doi.org/10.1039/D2BM01886A

Logesh M, Lavanya K, Mabrouk KE, Soundhararajan R, Srinivasan H, Ballamurugan AM (2024) Evaluation of time-dependent corrosion inhibition rate for f-MWCNT-BCP composite coatings on 316L stainless steel in simulated body fluid for orthopedic implantation. Appl Biochem Biotechnol 196(3):1544–1557. https://doi.org/10.1007/s12010-023-04602-x

Luo H et al (2023) Fabrication of a micro/nanocomposite structure on the surface of high oxygen concentration titanium to promote bone formation. Biomater Adv 154:213631. https://doi.org/10.1016/j.bioadv.2023.213631

Magalhães APR et al (2012) Nanosilver application in dental cements. Int Sch Res Not 2012:e365438. https://doi.org/10.5402/2012/365438

Maleki Dizaj S et al. (2019) The application of nanomaterials in cardiovascular diseases: a review on drugs and devices. J Pharm Pharm Sci 22:501–515. https://doi.org/10.18433/jpps30456

Martínez-Sanmiguel JJ et al (2019) Anti-inflammatory and antimicrobial activity of bioactive hydroxyapatite/silver nanocomposites. J Biomater Appl 33(10):1314–1326. https://doi.org/10.1177/0885328219835995

Moghanizadeh-Ashkezari M et al (2019) Vitamin C loaded Poly(urethane-urea)/ZnAl-LDH Aligned scaffolds increase proliferation of corneal Keratocytes and up-regulate vimentin secretion. ACS Appl Mater Interfaces 11(39):35525–35539. https://doi.org/10.1021/acsami.9b07556

Moses JC, Mandal BB (2022) Mesoporous silk-bioactive glass nanocomposites as drug eluting multifunctional conformal coatings for improving Osseointegration and bactericidal properties of metal implants. ACS Appl Mater Interfaces 14(13):14961–14980. https://doi.org/10.1021/acsami.2c00093

Mostakhdemin M, Nand A, Ramezani M (2021) A novel assessment of microstructural and mechanical behaviour of bilayer silica-reinforced nanocomposite hydrogels as a candidate for artificial cartilage. J Mech Behav Biomed Mater 116:104333. https://doi.org/10.1016/j.jmbbm.2021.104333

Murugesan B et al (2020) Fabrication of palladium nanoparticles anchored polypyrrole functionalized reduced graphene oxide nanocomposite for antibiofilm associated orthopedic tissue engineering. Appl Surf Sci 510:145403. https://doi.org/10.1016/j.apsusc.2020.145403

Nakano K, Egashira K, Masuda S, Funakoshi K, Zhao G, Kimura S, ... Sunagawa K (2009) Formulation of nanoparticle-eluting stents by a cationic electrodeposition coating technology: efficient nano-drug delivery via bioabsorbable polymeric nanoparticle-eluting stents in porcine coronary arteries. JACC Cardiovasc Interv 2(4):277–283. https://doi.org/10.1016/j.jcin.2008.08.023

Nemmar A et al (2016) Ultrasmall superparamagnetic iron oxide nanoparticles acutely promote thrombosis and cardiac oxidative stress and DNA damage in mice. Part Fibre Toxicol 13(1):22. https://doi.org/10.1186/s12989-016-0132-x

Palumbo A (2011) The Anatomy and physiology of the healthy periodontium, in Gingival Diseases–Their Aetiology, prevention and treatment. IntechOpen. https://doi.org/10.5772/25275

Pan Y, Wang X, Yin Z (2019) Synthesis and evaluation of cationic polymeric micelles as carriers of lumbrokinase for targeted thrombolysis. Asian J Pharm Sci 14(2):144–153. https://doi.org/10.1016/j.ajps.2018.03.004

Pasang T et al (2016) Fatigue of commercially pure titanium dental implants. Accessed 9 Feb 2024. https://hdl.handle.net/10292/14152

Perumal G, Ramasamy B, Nandkumar M, Dhanasekaran AS, Ramasamy S, Doble M (2020) Bilayer nanostructure coated AZ31 magnesium alloy implants: in vivo reconstruction of critical-sized rabbit femoral segmental bone defect. Nanomedicine: Nanotechnol, Biol Med 29:102232. https://doi.org/10.1016/j.nano.2020.102232

Philip P, Jose T, KS S, Kuriakose S (2021) Green synthesised silver nanoparticles incorporated electrospun poly (methyl methacrylate) nanofibers with different architectures for ophthalmologic alternatives. J Bioact Compat Polym 36(2):93–110. https://doi.org/10.1177/0883911521997856

Prakash J, Prema D, Venkataprasanna KS, Balagangadharan K, Selvamurugan N, Venkatasubbu GD (2020) Nanocomposite chitosan film containing graphene oxide/hydroxyapatite/gold for bone tissue engineering. Int J Biol Macromol 154:62–71. https://doi.org/10.1016/j.ijbiomac.2020.03.095

Qi P, Maitz MF, Huang N (2013) Surface modification of cardiovascular materials and implants. Surf Coat Technol 233:80–90. https://doi.org/10.1016/j.surfcoat.2013.02.008

Ramezani M, Ripin ZM (2023) An overview of enhancing the performance of medical implants with nanocomposites. J Compos Sci 7(5), art no. 5. https://doi.org/10.3390/jcs7050199

Saber MH, Loomans BAC, El Zohairy A, Dörfer CE, El-Badrawy W (2010) Evaluation of proximal contact tightness of Class II resin composite restorations. Oper Dent 35(1):37–43. https://doi.org/10.2341/09-037L

Saeidi M et al (2020) The influence of an extra-articular implant on bone remodelling of the knee joint. Biomech Model Mechanobiol 19(1):37–46. https://doi.org/10.1007/s10237-019-01193-7

Sahoo NG, Pan YZ, Li L, He CB (2013) Nanocomposites for bone tissue regeneration. Nanomedicine 8(4):639–653. https://doi.org/10.2217/nnm.13.44

Sarkar C, Chowdhuri AR, Garai S, Chakraborty J, Sahu SK (2019) Three-dimensional cellulose-hydroxyapatite nanocomposite enriched with dexamethasone loaded metal–organic framework: a local drug delivery system for bone tissue engineering. Cellulose 26(12):7253–7269. https://doi.org/10.1007/s10570-019-02618-3

Shanmugam K, Sahadevan R (2018) Bioceramics—an introductory overview. Fundam Biomater: Ceram 1–46. https://doi.org/10.1016/B978-0-08-102203-0.00001-9

Sharifi S, Zaheri Khosroshahi A, Maleki Dizaj S, Rezaei Y (2021) Preparation, physicochemical assessment and the antimicrobial action of Hydroxyapatite-Gelatin/Curcumin Nanofibrous composites as a dental biomaterial. Biomimetics (Basel) 7(1):4. https://doi.org/10.3390/biomimetics7010004

Shevach M, Maoz BM, Feiner R, Shapira A, Dvir T (2013) Nanoengineering gold particle composite fibers for cardiac tissue engineering. J Mater Chem B 1(39):5210–5217. https://doi.org/10.1039/C3TB20584C

Sinjari B et al (2019) Curcumin/Liposome Nanotechnology as Delivery Platform for Anti-inflammatory Activities via NFkB/ERK/pERK Pathway in human dental Pulp Treated With 2-HydroxyEthyl MethAcrylate (HEMA). Front Physiol 10:633. https://doi.org/10.3389/fphys.2019.00633

Sreenivasalu PKP et al (2022) Nanomaterials in dentistry: current applications and future scope. Nanomaterials (Basel) 12(10):1676. https://doi.org/10.3390/nano12101676

Swain R et al (2023) Bentonite-in hypromellose-poloxamer sol-gel for corneal application of trimetazidine: Study of rheology and ocular anti inflammatory potential. Int J Biol Macromol 242:124628. https://doi.org/10.1016/j.ijbiomac.2023.124628

Sukhodub LB et al (2020) Single-walled carbon nanotubes loaded hydroxyapatite–alginate beads with enhanced mechanical properties and sustained drug release ability. Prog Biomater 9(1):1–14. https://doi.org/10.1007/s40204-020-00127-2

Sulthana S et al (2023) Synthetic graphene–copper nanocomposites interact with the hACE-2 enzyme and inhibit its biochemical activity. Nanoscale Adv. 6(1):188–196. https://doi.org/10.1039/D3NA00468F

De Stefani A, Bruno G, Preo G, Gracco A (2020) Application of nanotechnology in orthodontic materials: a state-of-the-art review. Dent J (Basel) 8(4):126. https://doi.org/10.3390/dj8040126

Tautzenberger A, Kovtun A, Ignatius A (2012) Nanoparticles and their potential for application in bone. Int J Nanomed 7:4545–4557. https://doi.org/10.2147/IJN.S34127

Tondnevis F, Ketabi MA, Fekrazad R, Sadeghi A, Abolhasani MM (2019) Using Chitosan Besides Nano Hydroxyapatite and Fluorohydroxyapatite boost dental pulp stem cell proliferation. J BiomimS, Biomater Biomed Eng 42:39–50. https://doi.org/10.4028/www.scientific.net/JBBBE.42.39

Van Vlierberghe S, Dubruel P, Schacht E (2011) Biopolymer-based hydrogels as scaffolds for tissue engineering applications: a review. Biomacromol 12(5):1387–1408. https://doi.org/10.1021/bm200083n

Velu R, Calais T, Jayakumar A, Raspall F (2019) A comprehensive review on bio-nanomaterials for medical implants and feasibility studies on fabrication of such implants by additive manufacturing technique. Materials (Basel) 13(1):92. https://doi.org/10.3390/ma13010092

Wang Y, Shi D (2022) In vitro and in vivo evaluations of nanofibrous nanocomposite based on carboxymethyl cellulose/polycaprolactone/cobalt-doped hydroxyapatite as the wound dressing materials. Arab J Chem 15(11):104270. https://doi.org/10.1016/j.arabjc.2022.104270

Wang Y, Zhang W, Zhang J, Sun W, Zhang R, Gu H (2013) Fabrication of a novel polymer-free nanostructured drug-eluting coating for cardiovascular stents. ACS Appl Mater Interfaces 5(20):10337–10345. https://doi.org/10.1021/am403365j

Wang K et al (2019) Cyanine-assisted exfoliation of covalent organic frameworks in nanocomposites for highly efficient Chemo-Photothermal Tumor therapy. ACS Appl Mater Interfaces 11(43):39503–39512. https://doi.org/10.1021/acsami.9b13544

Wang L et al (2019) Fabrication of superparamagnetic nano-silica@ quercetin-encapsulated PLGA nanocomposite: Potential application for cardiovascular diseases. J Photochem Photobiol, B 196:111508. https://doi.org/10.1016/j.jphotobiol.2019.05.005

Wheeler's Dental Anatomy, Physiology and Occlusion–11th Edition | Elsevier Shop. Accessed 12 Mar 2024. https://shop.elsevier.com/books/wheelers-dental-anatomy-physiology-and-occlusion/nelson/978-0-323-63878-4

Wu S, Weng Z, Liu X, Yeung KWK, Chu PK (2014) Functionalized TiO2 based nanomaterials for biomedical applications. Adv Funct Mater 24(35):5464–5481. https://doi.org/10.1002/adfm.201400706

Yang Y et al (2020) Trisulfide bond–mediated doxorubicin dimeric prodrug nanoassemblies with high drug loading, high self-assembly stability, and high tumor selectivity. Sci Adv 6(45):eabc1725. https://doi.org/10.1126/sciadv.abc1725

Yılmaz MN, Gul P (2022) Monomer release from dental restorative materials containing dimethacrylate resin after bleaching. Clin Oral Investig 26(6):4647–4662. https://doi.org/10.1007/s00784-022-04446-5

Yudaev P, Chuev V, Klyukin B, Kuskov A, Mezhuev Y, Chistyakov E (2022) Polymeric dental nanomaterials: antimicrobial action. Polymers 14(5), art no 5. https://doi.org/10.3390/polym14050864

Zafar MS (2020) Prosthodontic Applications of Polymethyl Methacrylate (PMMA): an update. Polymers (Basel) 12(10):2299. https://doi.org/10.3390/polym12102299

Zhang K, Melo MAS, Cheng L, Weir MD, Bai Y, Xu HHK (2012) Effect of quaternary ammonium and silver nanoparticle-containing adhesives on dentin bond strength and dental plaque microcosm biofilms. Dent Mater 28(8):842–852. https://doi.org/10.1016/j.dental.2012.04.027

Zhang N et al (2018) Cyclic RGD functionalized liposomes encapsulating urokinase for thrombolysis. Acta Biomater 70:227–236. https://doi.org/10.1016/j.actbio.2018.01.038

Zhang Y, Chen M, Dai Z, Cao H, Li J, Zhang W (2020) Sustained protein therapeutics enabled by self-healing nanocomposite hydrogels for non-invasive bone regeneration. Biomater Sci 8(2):682–693. https://doi.org/10.1039/C9BM01455A

Zhang M et al (2020) Biological safe gold nanoparticle-modified dental aligner prevents the Porphyromonas gingivalis Biofilm Formation. ACS Omega 5(30):18685–18692. https://doi.org/10.1021/acsomega.0c01532

Zhou X et al (2020) Incorporation of dexamethasone-loaded mesoporous silica nanoparticles into mineralized porous biocomposite scaffolds for improving osteogenic activity. Int J Biol Macromol 149:116–126. https://doi.org/10.1016/j.ijbiomac.2020.01.237

Zhou S et al (2024) Injectable ultrasound-powered bone-adhesive nanocomposite hydrogel for electrically accelerated irregular bone defect healing. J Nanobiotechnol 22(1):54. https://doi.org/10.1186/s12951-024-02320-y

Zhu H et al (2019) Polyurethane–Cardiolipin nanoparticle-modified Decellularized scaffold-based vascular patches for tissue engineering applications. ACS Appl Bio Mater 2(4):1696–1702. https://doi.org/10.1021/acsabm.9b00079

Chapter 4
Regenerative Medicine and Controlled Drug Delivery in Periodontics

R. M. Baiju, P. Athul Krishna, and S. Santhosh Kumar

Abstract Periodontal disease is the sixth prevalent disease affecting mankind and is multifactorial in its etiology. Treatment of periodontal disease is challenging as the ideal outcome is to regenerate lost supporting structures of teeth. Nanomedicine is gaining much interest in periodontal therapeutic applications due to its demonstrated potential in contributing to tissue regeneration. Controlling periodontal inflammation and arresting disease progression has been the mainstay in the clinical management of periodontal disease. Nanotechnology has shown promising progress in controlled drug delivery systems. This paper tries to narrate the current information in the field of nanotechnological applications in controlled drug delivery and regenerative medicine approaches particularly with respect to periodontal regeneration.

Keywords Regenerative medicine · Nanotechnology · Periodontics · Drug delivery

4.1 Introduction

In response to injury or insult, the regeneration capacity of human tissues appears to be limited. Owing to the recent regenerative engineering principles, the reconstruction of damaged or lost anatomic parts can be accomplished with better reproduction of functional demand and anatomic precision.

Organ and tissue loss due to injury or disease has prompted the development of treatment strategies aimed at regeneration, hence less reliance on transplants. The hurdles of the current transplant system include limited availability of donor tissues and immune rejection (Vacanti et al. 2014).

Regenerative medicine thus shares a pivotal role in healing or replacing tissues damaged by disease, trauma, age or congenital defects by bypassing the following obstacles through usage of tissue engineered substitutes (Mao and Mooney 2015).

R. M. Baiju (✉) · P. A. Krishna · S. S. Kumar
Government Dental College, Kottayam, Kerala, India
e-mail: baijurm2@gmail.com

R. Balakrishnan et al. (eds.), *Biocompatible Nanocomposites*, Advanced Structured Materials 239, https://doi.org/10.1007/978-981-96-7450-3_4

Table 4.1 Products for regenerative medicine approved by FDA

Category	Name	Biologic agent	Approved use
Biologics	LaViv Carticel Apligraf, GINTUIT Cord blood	Autologous fibroblasts Autologous chondrocytes Allogenic cultured fibroblasts and keratinocytes in bovine collagen Hematopoietic stem and progenitor cells	Improving appearance of nasolabial fold Cartilage defects from acute or repetitive trauma Mucogingival conditions and diabetic foot ulcers Hematopoietic and immunological reconstitution following myelo ablative treatment
Cell based medical devices	Dermagraft Celution	Allogenic fibroblasts Cell extraction	Diabetic foot ulcers Transfer of autologous adipose stem cells
Biopharmaceuticals	GEM 21S Regranex Infuse, Infuse bone graft, Inductos Osteogenic protein–1	PDGF-BB,tricalcium phosphate PDGF-BB BMP-2 BMP-7	Periodontal defects Lower extremity diabetic ulcers Tibia fracture and nonunion and lower spine fusion Tibia nonunion

Adapted from. Mao, A.S et.al. Regenerative medicine: Current therapies and future directions (Mao and Mooney 2015)

The field of regenerative medicine includes de novo generation of cells and tissues or enhancement of the innate healing response of body to promote regeneration (Kami and Gojo 2014).

Such therapies employ either autologous or allogeneic cells that are typically differentiated cells maintaining proliferative capacity. Carticel, the first FDA-approved biologic product in orthopedics, employs autologous chondrocytes for the management of focal articular cartilage defects (Mao and Mooney 2015; Dewan et al. 2014).

Other available therapies approved by Federation Dentaire Internationale (FDA) are listed in Table 4.1.

4.1.1 Drug Delivery Systems in Regenerative Medicine

Evolution of Commercial Drug Delivery Technology

A drug delivery system (DDS) is a device employed to enable the introduction of a therapeutic substance at the intended site of action within the body. Its efficacy and safety can be optimized by controlling the time, rate, and place of drug release in the body (Jain 2008).

Way back, small molecules drugs, such as antibiotics and steroids, were the primary therapeutics available. Their manipulation was primarily dependent on their physicochemical properties which largely influenced the drug bioavailability. With evolution of newer generation of therapeutics including antibodies, peptides, nucleic acids, and proteins, new drug delivery systems were also introduced. For any drug delivery, the objective is to maximize therapeutic efficacy by transporting and releasing the drug (passively or actively) to the target site in the body and by minimizing off-target accumulation of the drug (Vargason et al. 2021) (Fig. 4.1).

Controlled release formulations comprise those with sustained release, extended release, timed release, and others. Nowadays these terminologies are interchangeably used. Following the advances in drug delivery technologies over the years, the minor differences in the functions that different names demand have become irrelevant.

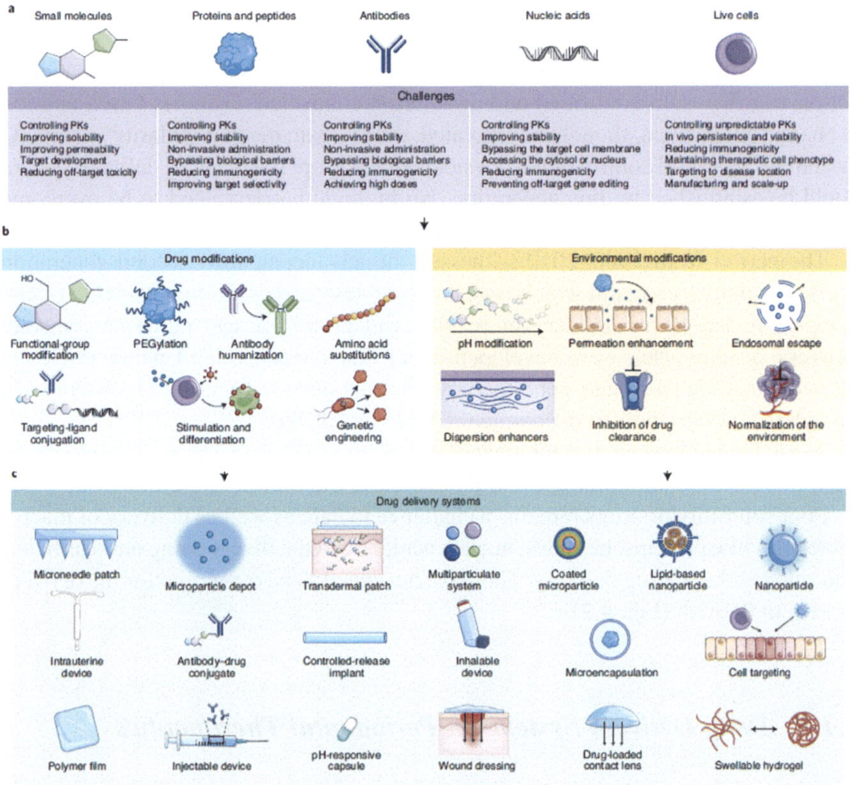

Fig. 4.1 **a** *Five generations of therapeutic approaches (*small molecules, proteins and peptides, antibodies, nucleic acids, and cell therapies) and their therapeutic challenges. **b, c**, The modification of the drug or its environment and **c**, the design of drug delivery systems. (*Adapted from The evolution of commercial drug delivery Technologies Vargason* et al. (2021))

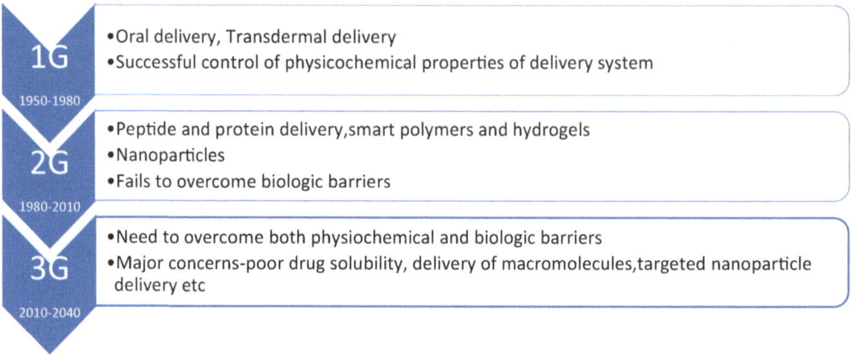

Fig. 4.2 Evolution of drug delivery systems: Modified and adapted from Yeon Hee Yun et al. (2015)

The drug delivery systems can be broadly categorized into three generations. The first-generation(1G) drug delivery system, encompassing oral and transdermal controlled release drugs, was formulated in the year 1950–1980.Their mechanism of action was diffusion controlled and dissolution controlled. Osmosis-based and ion exchange mechanism, though incorporated, did not gain much popularity among 1G systems. Successful control of the physiochemical properties of drug delivery system could be established by this generation. No biologic barriers need to be overcome by this system except for the limited gastrointestinal transit time (Yun et al. 2015).

The period from 1980–2010 witnessed the advancement of second-generation (2G) drug delivery system which was, however, less successful than the 1G system. Long-term depots for delivery of peptide and protein drugs, pulmonary insulin delivery systems, etc., were developed under this system. Use of nanoparticles in targeted drug delivery and gene therapy are also included under 2G system even though this system fails to overcome the biologic barriers.

The 3G era have seen development of many drug delivery systems. These systems need to overcome both biologic and physicochemical barriers for clinical success.

Poor solubility of drugs remains a challenge to date as well as delivery of macromolecules like proteins, peptides, nucleic acid, targeted delivery using nanoparticles, and self-regulated drug delivery. These barriers need to be overcome for the delivery system to flourish (Fig. 4.2).

4.1.2 Drug Delivery Systems in Periodontal Therapeutics

Periodontitis is an microbial infection of the teeth supporting tissues primarily treated traditionally using mechanical debridement such as scaling and root planning to remove dental plaque and calculus. On the other hand, systemic antibiotic administration has limitations and side effects associated with it that have contributed to an increased interest in localized drug delivery methods (Liang et al. 2020). In this regard, periodontal pockets containing gingival crevicular fluid serve as the perfect natural reservoirs for localized antimicrobial agents to be delivered. Adding the adjunct polymeric delivery approach to nanoformulations helps in increased persistence of therapeutic drug concentration at infection sites for longer periods and therefore reduces its requirement daily with minimum systemic side effects (Nair and Anoop 2012).

An example, local treatment with minocycline (Periocline), a 2% minocycline-HCl ointment in an adjunct to scaling and root planning (SRP), has been examined for the clinical morphologic results on recurrent periodontal pockets. This group showed significant improvement in the clinical parameters (probing pocket depth, and CAL) and also eradicated gram-negative periodontal pathogens (Nakagawa et al. 1991).

Furthermore, the controlled release of antimicrobials in periodontal therapy has been taken a look at using controlled release delivery systems unveiled favorable aspects such as biodegradability and maintained high antibiotic concentrations present within the gingival crevicular fluid (GCF) to enhance patient compliance (Southard et al. 1998).

4.1.3 Strategies for Drug Delivery

Several approaches are used to enhance drug delivery in periodontal therapeutics which include:

Sustained and Controlled Release: engineered to prolong pharmacologic effects, minimize systemic exposure, and reduce side effects (Schwach-Abdellaoui et al. 2000).

Site-Specific Delivery: This directs delivery of drug into the periodontal pocket and reduces systemic side effects (Schwach-Abdellaoui et al. 2000).

4.1.4 Drug Delivery Systems

Treatments using drug delivery systems have been developed to overcome a sustained pharmacodynamic effect in the local therapies of periodontal disease.

Solvent-controlled Systems: These systems depend on the permeability of the macromolecular matrix releasing the drug and allowing it to diffuse in a controlled manner (Schwach-Abdellaoui et al. 2000).

Reservoir Systems: Controlled drug release by diffusion through a polymeric membrane to yield zero-order or steady-state delivery profiles (Schwach-Abdellaoui et al. 2000).

Chemically Controlled Systems: This type works by degrading chemical bonds or polymer erosion that release drugs over time (Schwach-Abdellaoui et al. 2000).

4.1.5 Polymers Used

The choice of polymers in drug delivery systems is the most important parameter that determines how effective and biocompatible it can become.

Natural Polymers: Natural polymers are derived from renewable resources (proteins and collagen) and will biodegrade. However, delivery of such molecules may be hindered by their short half-lives and immunogenicity that can preclude in vivo use (Schwach-Abdellaoui et al. 2000).

Semi-Synthetic Polymers: These polymers are polymeric derivatives of natural cellulose and present a hybrid system between synthetic and naturally polymerized materials providing more control over the release profile (Schwach-Abdellaoui et al. 2000).

Synthetic Polymers: These have been synthetically designed to have pharmacological and biological requirements, offering the possibility of tailoring drug delivery systems according to treatment needs (Schwach-Abdellaoui et al. 2000).

4.1.6 Local Delivery Systems

Local drug delivery systems have proved their efficacy for periodontal therapy in many prevalence studies. For example, Pavia et al. reported that local tetracycline delivery following SRP has added benefits to treatment and gives superior clinical results (Pavia et al. 2003). It has given consistent results in the case of local delivery systems such as tetracycline, minocycline, and chlorhexidine, with an average clinical gain of 0.3 mm to 0.6 mm. However, the infrequent achievement of these outcomes renders their clinical significance debatable (Bonito et al. 2004).

Comparative studies of locally delivered antimicrobials have revealed that all the modalities are better than scaling and root planning alone, and tetracycline fibers work better in the treatment of persisting periodontal lesions compared to other systems that include in its category those of metronidazole gel and minocycline ointment demonstrate variable clinical gain (Position Paper 1999).

Other antimicrobial agents have been investigated for their effectiveness in periodontal therapy. For example, chlorhexidine is effective in plaque and gingivitis reduction, both in mouth rinses and post-operatively; however, it possesses limitations such as staining, disturbed taste, and less efficacy at deeper pockets. Sanguinarine has antiseptic action with poor solubility at physiological pH, while histatins possess antifungal and bactericidal activity but are characterized by low stability.

Tetracyclines are efficacious against both gram-negative and gram-positive bacteria. Their use can develop resistance, but they do carry the potential to elicit bacterial resistance. Metronidazole is another antibiotic effectively administered against anaerobes and spirochetes; such administration is rather common. Over time, its effectiveness will most probably decline.

Other agents like clindamycin and ofloxacin have been reported to be effective against few periodontal pathogens but the long-term effectiveness is yet to be established. Minocycline gels produce good reductions in probing depth and microbial load but these are not maintained at significant numbers when compared to results with scaling and root planning by itself, meaning that it has to improve further in combination therapy (Schwach-Abdellaoui et al. 2000).

Lastly, other degradable delivery systems available in the body are the Atrigel system with Doxycycline Hyclate. The biodegradable feature thus permits the controlled release over a period of seven days and imparts improved clinical performance, compared with previous systems, including those based upon Sanguinarine (Schwach-Abdellaoui et al. 2000).

Possibly for the first time, drug delivery so far had been possible, increasing in recent years. In doing so, the recent years have increased the use of the versatility of nanotechnology to provide targeted, controlled, and sustained release of drugs. Nanotechnology ensures more precision in drug delivery, bettering patient safety with minimal side effects. Some of the major advances are the use of colloidal systems such as liposomes, micelle encapsulation, dendrimers, and nanocrystals (Harini and Kaarthikeyan 2014). Another innovative approach that could be used in the targeting of multiple drug-resistant bacteria involved in periodontal infections is phage-based therapy together with nanotechnology (Bennett et al. 2023).

Nanoparticles have improved drug delivery systems considerably, hence making them indispensable in the field of periodontal therapeutics. These nanoparticles are designed in such a way that they are capable of carrying drugs to the site with more accuracy and effectiveness—especially in localized infections and inflammations (Mitchell et al. 2021). Personalized defect-specific antibiotic-laden polymeric scaffolds are examples of the kind that has been presented for the first time in the field of periodontal therapeutics. These can be used to load such antibiotics as metronidazole or Tetracycline and are targeted directly to such periodontal defects (Ferreira et al. 2021).

Novel nanofibers obtained through the electrospinning technique, for example, composites of polylactic acid with nano-hydroxyapatite and doxycycline, have shown great potential for enhancing the antimicrobial effect against pathogens (Andrei et al. 2022). It is, in fact, the development of these novel materials that makes the next significant step in the fight against periodontal diseases.

Nano-based Carriers: These nano-based carriers, including nanospheres, nanotubes, quantum dots, and gold nanoparticles, offer several advantages, including the ability to exercise better control over drug release rates, better stability, controlled biocompatibility, and improving a lot of biodegradability. These systems are designed to achieve better retention and permeability of the drug in the periodontal pocket to make sure for the transports in the epithelial lining of the pocket, which is a pivotal site for the action of any drug, for maximal effect and sustained antimicrobial activity (Nair and Anoop 2012; Harini and Kaarthikeyan 2014). Combining phage therapy with nanoparticles could offer a route to more stable and effective antimicrobial treatments (Bennett et al. 2023).

Liposomes: Small, spherical vesicles consisting of cholesterol and phospholipids are used as controlled release and targeting drug carriers. One of the most researched and versatile nano-based drug delivery systems in periodontal therapy, the biocompatibility of liposomes minimizes systemic side effects (Harini and Kaarthikeyan 2014; Sercombe et al. 2015). This includes enhanced antimicrobial activity against resistant strains like *Pseudomonas aeruginosa* using liposomes containing polymyxin B by reducing its toxicity (Alipour et al. 2008).

Liquid Crystals: Equally useful in sustained drug release and topical delivery in periodontal therapy, these systems express the properties of both liquids and crystalline states. By increasing the bioavailability of poorly soluble drugs, these systems provide long-term hydration (Harini and Kaarthikeyan 2014).

Nanorobots and Quantum Dots: Nanorobots and quantum dots represent new advances in the development of technology for periodontal therapy. Nanorobots will allow the performance of less invasive procedures with more precision, while quantum dots, or those already being researched to promote healing, are non-toxic indium phosphide particles, which help target inflamed tissues (Harini and Kaarthikeyan 2014).

4.2 Advanced Delivery Systems

The advanced delivery systems prepared using nanoparticles have proved to enhance treatment effectiveness in the domain of periodontal therapy. For instance, doxycycline-loaded PLGA microspheres show sustained drug release in periodontal pockets, significantly enhancing the therapeutic effect on periodontal pathogens, such as *P. gingivalis and S. aureus* (Liang et al. 2020).

Fibers and Films: This drug delivery system established sustained drug release directly into periodontal pockets. Sustained drug release for up to 75 days was realized using electrospun fibers loaded with tetracycline hydrochloride (Liang et al. 2020).

The electrospun nanofibers exhibit high surface area, porosity, supporting the adhesion of cells and adsorption of proteins. For example, the electrospun PLA/nano-HAP nanofibers loaded with doxycycline showed increased antimicrobial activity against the periodontal pathogens compared to the traditional antibiotics (Andrei et al. 2022; Thomas and Baiju 2022).

Gels and Hydrogels: Another popular dosage form in periodontal therapy is the injectable hydrogels or in situ forming gels; these are popular because of their easy application along with the sustained release of therapeutic moieties. Hydrogels laden with antibiotics such as tinidazole have demonstrated effective and sustained release into the periodontal pocket, thereby improving the treatment of periodontitis (Liang et al. 2020). The loading of doxycycline into electrospun fibers is also a controlled drug delivery system that could be transferred into gels or hydrogels for periodontal localized therapy (Andrei et al. 2022).

4.2.1 Ideal Properties of a Nanosystem for Periodontal Therapy

For the targeted periodontal therapy to be effective, the nanosystems loaded with the drug need to have some basic properties:

1. **High Antimicrobial Efficacy**: Since, doxycycline incorporated in the system should not nullify the antimicrobial potential of HAP. In this context, achievability in the case of doxycycline-loaded PLA/nano-HAP fibers has been reported (Andrei et al. 2022). This allows for an effective fight against periodontal pathogens.
2. **Targeted Pathogen Specificity**: The system, treating the specific pathogens in the targeted cause, should be sensitive, against the major important pathogens; those include *Aggregatibacter actinomycetemcomitans and Porphyromonas gingivalis* (Andrei et al. 2022).
3. **Controlled and Sustained Release**: Nanosystems must control the released therapeutic agents and assure sustained release over a long span of time to optimize their function within the periodontal environment (Harini and Kaarthikeyan 2014). For instance, other than the ability to stabilize the encapsulated drugs and offer controlled release that is liposomes can enhance efficacy in target delivery (Sercombe et al. 2015).
4. **Biocompatibility**: The nanosystem should be biocompatible to eliminate any adverse reaction, to integrate well into the body tissue. Therefore, ensure minimal side effects and an insignificant inflammatory response to guarantee patients' safety (Harini and Kaarthikeyan 2014; Zhao et al. 2022).
5. **Targeted Delivery**: This is an area where precision in targeting specific tissues or even cells is being the key advantage. An efficient nanosystem would have lessened systemic exposure and thus enhanced the therapeutic outcomes of the medicament being utilized, by specifically and selectively delivering drugs to the parts that need to be acted upon (Harini and Kaarthikeyan 2014).

New generation delivery systems, including but not limited to liposomes and nanoparticles, can be designed to enhance such targeting (Mitchell et al. 2021; Sercombe et al. 2015).

6. **Microparticle Properties**: In order for it to be appropriate for periodontal therapy, microparticles should be designed to offer sustained release and be biodegradable. The kinetics of release are influenced by the different types of polymeric materials that require strict control toward exerting the desired therapeutic effects (Kumar and Vijayalakshmi 2006).
7. **Sustained Release and Biodegradability**: Ideal drug delivery systems must provide for the maintenance of therapeutic effective drug concentration for prolonged periods. Besides, it should be biodegradable and accepted by the patients, thus ensuring both effectiveness and acceptability by the patients (Zhang et al. 2022).

8. **Efficacy and Clinical Relevance**: Nanosystems are intended to show significant clinical improvements in comparison with the marginal effectiveness presented by the current traditional adjuncts for antimicrobial application. It is effective and thus enhances efficacy and is clinically relevant in the quest for periodontal therapy improvement (Bonito et al. 2004).
9. **Controlled Release Mechanisms**: The mechanized controlled release mechanisms are effective aspects that would bring about sustainable therapeutic levels of drugs in the periodontal pocket. It avails benefits similar to a number of local delivery systems seen so far (Bonito et al. 2004).
10. **Safety and Biocompatibility**: Nanodevices must be non-pyrogenic and biocompatible to address safety concerns and to avoid inflammatory response for acceptance (Zhao et al. 2022).
11. **Precision and Efficiency**: Nanosystems are expected to render high precision in drug delivery and effective targeting for improvement in clinical outcomes and treatment efficiency (Zhao et al. 2022).
12. **Adaptability and Precision Medicine**: The ideal nanosystem would be one that can cross biological barriers and can be tailored for precision medicine (Mitchell et al. 2021). Such a system can easily biodegrade within the periodontal pocket and be a system that is able to exhibit constantly controlled and sustained drug release properties, sharing effectiveness of treatment for a longer time while reducing number of administrations (Nair and Anoop 2012).

4.3 Nano-Based Therapeutic Strategies for Periodontitis

The conventional intra pocket drug delivery has got inherent limitations due to poor solubility, retension of the agent and inferior penetration to the biofilm. Nanomedicine approach tries to overcome both physical and biochemical challenges offered by the oral environment and attempts to control the delivery of antimicrobial, anti-inflammatory or biological agent into the biofilm, pathogen, or host cells. Published evidence suggests promising potential of nanomedicine approaches for the intra pocket drug delivery in the management of periodontitis.

4.3.1 Antimicrobial Approaches

This new perspective into the delivery and use of antimicrobial agents could be offered by nanotechnology in the treatment of periodontitis. Such actions would be utilized not only against periodontal pathogens but also in the facilitation of tissue regeneration within the periodontal tissues (Zhang et al. 2022, 2023).

In that respect, the use of recent nano-based strategies, like a 1:3 MET/TCH combination packed defect-specific scaffolds, provides remarkable activity against periodontopathogens and, thus, has caused a substantial effect in reducing infection and, in parallel, in the promotion of new bone formation in periodontally compromised areas by the dual benefits of nanotechnology (Ferreira et al. 2021). In the same way, doxycycline-loaded PLA/nano-HAP nanofibers have been documented for the inhibition of periodontal pathogens, thus providing a targeted antimicrobial strategy without much systemic exposure to the drugs (Andrei et al. 2022).

In reality, it was demonstrated that triclosan-loaded nanoparticles, when applied, do help reduce the level of gingival inflammation, thereby confirming in general the benefits of using nanotechnology strategies to improve antimicrobial periodontal treatments (Harini and Kaarthikeyan 2014). As evidenced in the treatment involving localized delivery of minocycline-HCl ointment, where the eradication of the pathological bacteria was achieved concurrently with the reduction of the probing pocket depth correlate with a concomitant decrease in clinical attachment loss, the study demonstrates the utility for periodontal programs in the targeted therapy options where antimicrobial treatments are concerned (Nakagawa et al. 1991).

It is, therefore, safe to argue that the efficiency of the antimicrobial therapies is really ameliorated by the controlled release systems. For instance, tetracycline-loaded microparticles give sustained antibiotic delivery over a long time, which is crucial for managing chronic periodontal infections and supporting regenerative therapy (Kumar and Vijayalakshmi 2006). Southard (1998) also states that the benefit of such systems will also be in sustaining the effect because high levels of antimicrobials have been proven to be effective in GCF, thereby increasing the overall efficacy of treatment (Esposito et al. 1997).

The use of nanotechnology in conjunction with controlled release drug delivery systems in periodontal therapy opens new doors to enhance the treatment effectiveness, reduce the side effects, and hasten the tissue regeneration. However, further research is needed to develop the optimal conditions for these systems and their testing under clinical conditions.

4.3.2 Emerging Nano-Based Therapies

The new emerging nano-based therapies gives a significant upsurge to the treatment of periodontal diseases, and a novel approach to enhance the treatment efficiencies is by the means of the following attributed:

Nano-based systems have shown that the antimicrobial agents' effectiveness can be enhanced to a great extent, going beyond even what has at this point in time been realized with the conventional definite methods. These systems can help to efficiently overcome the limited improvements experienced with the conventional treatments by more targeted and sustained delivery.

4.3.3 *Enhanced Antimicrobial Efficacy*

Nano-based systems have shown that the potential to significantly improve the effectiveness of antimicrobial agents, besides the capabilities of traditional methods by providing more targeted and sustained delivery of antimicrobials (Bonito et al. 2004; Zhao et al. 2022).

4.3.4 *Nanoparticles for Antimicrobial Activity*

Nanoparticles like hydroxyapatite are being evaluated for their antimicrobial use, which would be extremely helpful for the periodontal therapy. These nanoparticles are known to target periodontal pathogens and reducing infections, thus promoting healthy oral environments (Zhao et al. 2022).

4.3.5 *Photodynamic Therapy*

Nanotechnology also offers potential to enhance photodynamic therapy, a technology in use of delivering light-activated compounds into the destruction of a pathogen. Further enhancement of therapeutic effectiveness, along with the treatment strategy, will be through the optimization of delivery with the help of photosensitizers, thereby offering more successful, precise, and effective approach in combating periodontal infections (Zhao et al. 2022).

4.3.6 *Electrospun Nanofibers*

Electrospun nanofibers are another advanced treatment modality for infections in periodontal problems. In fact, it can be designed even to incorporate antimicrobial agents, which will make it efficient for fighting infections. In addition, incorporation of such agents into the nanofibers will also enhance the result of treatment, as their release is better controlled and sustained (Thomas and Baiju 2022; Zhao et al. 2022) (Fig. 4.3).

4.3.7 *Immune Modulation*

Nanotechnology provides a ray of hope for tuning the immune response in periodontal diseases, likely better for therapeutic efficacy. This can be through the inclusion of

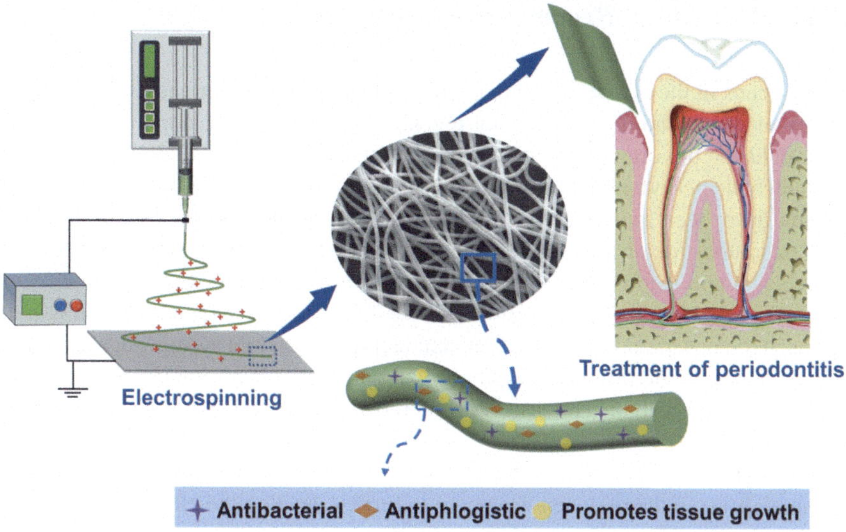

Fig. 4.3 Electrospun nanofibers (Adapted from Electrospun nanofibers for periodontal treatment: a recent progress, Zhao et al. (2022)

immune-interacting components in the nanofibers created mainly by electrospinning to help decrease inflammation concomitant with tissue regeneration (Thomas and Baiju 2022). Moreover, fungicidal agents incorporated into nanofiber materials have the capacity to modulate the immune response indirectly through a reduction of the pathogen load and related inflammation (Andrei et al. 2022).

4.3.8 Gene Delivery Using Bubble Liposomes

One such innovative methodology in the domain of nano-based periodontal therapies is gene delivery through bubble liposome application. This involves the use of ultrasound in combination with bubble liposomes during plasmid DNA delivery to the gingival tissues; this can modulate the immune response and increase the overall effect of periodontal therapy (Harini and Kaarthikeyan 2014). This offers a new way to treat periodontal diseases at the molecular level, opening new possibilities not only in treating periodontal disease strategies but also in the realization of regenerative treatment strategies at molecular level.

4.3.9 Periodontal Regenerative Strategies

Recent trends in the periodontal therapy have placed a greater emphasis in the applicability of nanotechnology in driving regenerative potential. Nanoparticles significantly enhance the drug penetration, drug kinetics, and controlled delivery optimal for predictable periodontal regeneration. A plethora of nano-based bioactive delivery systems have been explored including liposomes, lipid nanoparticles, nanopolymers, nanotubes, dendrimers, and micelles. Nanoparticles are particularly alluring for precision applications that involve a direct strategy, allowing the combination of two or more therapeutic agents within a single delivery system (Mitchell et al. 2021) such as in the new paradigm of various nanotechnology-based systems that integrate both nanoparticles and liposomes in optimizing the regenerative outcome and allow for a targeted and sustained release of the therapeutic agent (Harini and Kaarthikeyan 2014).

The use of liposome drug delivery that would hold the localized and sustained release of drugs in the tissue environment to further develop the efficiency of regenerative treatments focused on the reconstitution of the periodontal tissues while at the same time, minimizing the infection and inflammation (Sercombe et al. 2015). In addition, property modifications, such as the incorporation of other therapeutic agents into electrospun nanofibers, result in multi-drug systems or chemical post-processing for enhanced regenerative outcomes. These combinations help to address limitations existing in single-drug treatments and make a necessary contribution to contemporary periodontal therapy (Zhao et al. 2022).

In clinical practice, it has turned out that combined therapy including scaling and root planning with local minocycline application, which has demonstrated more favorable clinical results to scaling and root planning alone, is evidence to the importance of targeted delivery systems in enhanced periodontal regeneration along with microbial control (Nakagawa et al. 1991).

Despite the advancements, integration of advanced drug delivery systems has, nevertheless, continued with only modest advances over adjuncts using traditional therapies for regenerative therapy (Bonito et al. 2004). However, there have been continued advances, though, with emerging technologies showing promise. For example, HA nanoparticles are used in products such as Ostim® and VITOSSO in the treatment of bone defects and hold promise for regenerative potential in the treatment of periodontal bone defects as well (Kumar and Vijayalakshmi 2006).

Nanotechnology also has a good potential role in the differentiation of stem cells, which is the very crux of periodontal tissue regeneration. High surface area and biocompatibility of electrospun nanofibers will further facilitate adhesion and differentiation of stem cells for effective tissue repair. Moreover, novel nanotechnology may be used to target and inhibit the action of osteoclasts, which are a key factor in bone resorption in relation to periodontitis. Drug delivery systems containing statins such as simvastatin and rosuvastatin have applications in reducing bone resorption and promoting new bone formation at periodontal defect sites (Liang et al. 2020).

Several nanomedicine approaches have been analyzed in the realm of osseous regeneration. Chitosan, a derivative of chitin, has been widely employed for regeneration of periodontal osseous defects due to its beneficial antimicrobial non-toxic, biocompatible and osteoinductive properties. Chitosan can be retained for 4–6 weeks when employed as a barrier membrane in periodontal sites. Mota et al. reported superior outcomes when chitosan was combined with bioactive glass nanoparticles (Mota et al. 2012). Chitosan nanoparticles encapsulating platelet derived growth factor (PDGF) incorporated into a porous chitosan-collagen composite achieved sustained growth factor delivery over 6 weeks and demonstrated increased periodontal ligament cell proliferation (Peng et al. 2009).

Polytetrafluoroethylene membrane (PTFE) has been widely utilized for guided tissue regeneration (GTR). Exposure of PTFE membrane due to flap dehiscence is a usual complication in GTR surgery leading to inferior clinical outcomes. In an attempt to provide antimicrobial coatings for these membranes to protect it when exposed, silver nanoparticles have been incorporated to PTFE membranes by applying a mussel inspired approach (Nardo et al. 2021). Magnesium oxide nanoparticles when incorporated into polylactic acid membrane (PLA) demonstrated osteoinductive properties favoring periodontal tissue regeneration in rabbits (Bilal et al. 2022).

Aside from the established benefits of stimulating osteoblast activity, nanotechnology is also anticipated to have a positive impact on the differentiation of cementoblasts, a key step in the reparation of periodontal tissue, thus enhancing regeneration outcomes (Kumar and Vijayalakshmi 2006). This invasion of nanotechnology into older, regenerative approaches could make the latter more therapeutically effective in general; most significantly, this could translate into more potent enhancement of periodontal therapies.

An instance of an important related advancement was the development of defect-specific, antibiotic-carrying, tailor-made scaffolds by Ferreira et al. (2021) These scaffolds were designed as MET and TCH-delivery systems—a combination used to address the relative effectiveness of antibiotic agents. For the periodontal regeneration of anatomically complex defects, these scaffolds showed better tissue and bone regenerative abilities in experimental models, which also contributed to new bone formation with less loss and lower inflammatory response. This could mean that the integration of antibiotic delivery with regenerative scaffolds can effectively combine infection control with tissue regeneration (Ferreira et al. 2021).

Another interesting idea is in the field of bioactive agent delivery is the use of electrospun nanofibers in combination with the preparation of the scaffold with doxycycline. This combination does more than just control the spread of infection; it promotes a very friendly environment for tissue repair, hence supporting the regenerative aspects of periodontal therapy (Andrei et al. 2022).

4.4 Conclusion

This foray into nanotechnology for periodontal therapy represents an exciting new chapter in the management of periodontitis, one of the challenging yet relentless diseases of the oral cavity. With the integration of nanosystems in the treatment strategies, one will not only upgrade the methods for fighting infections but change the paradigm of periodontal care.

Nanosystems offer tremendous promise, especially in the field of doxycycline-loaded nanofibers. These go even further than just an improvement in the delivery of drugs; selective treatment of the specific pathogen is offered, along with the promotion of tissue regeneration. This indicates that the healing and restoration of the affected area are being dealt with, not just the symptoms.

Further innovation is derived from new therapies, such as photodynamic treatment and gene delivery using bubble liposomes. These offer new solutions to old problems, delivering better methods in treatment delivery, managing immunity, and making possible targeted, well-defined genes to make our approach to periodontal care more precise and more successful.

The realization of huge potential in combining both dimensions nanotechnologies and regenerative medicine is certainly an added boon. It is this combination that can really have a strong merging of the infection control approach into the tissue repair mechanisms of the body. This combined/ integrated approach could improve outcomes, providing more of a holistic solution to periodontal disease.

Although a lot of progress have been made, untapped ground remains. Much ongoing research for the optimization of these technologies and the true impact on patients remains to be seen. We are moving one step closer each day to a future where periodontal therapy is not just the management of disease, but truly wound healing and restoration of health.

In other words, in periodontal therapy, nanotechnology is not merely a technical advance. It is a leap toward more personalized, effective, and hopeful treatments. Through these, it is that we strive forward to a new periodontal care era that guarantees superior outcomes and a brighter future for patients.

References

Alipour M, Halwani M, Omri A, Suntres ZE (2008) Antimicrobial effectiveness of liposomal polymyxin B against resistant Gram-negative bacterial strains. Int J Pharm 355(1–2):293–298. https://doi.org/10.1016/j.ijpharm.2007.11.035

Andrei V, Fiţ NI, Matei I, Barabás R, Bizo LA, Cadar O, Boşca BA, Farkas N-I, Marincaş L, Muntean D-M et al (2022) In Vitro antimicrobial effect of novel Electrospun Polylactic Acid/ Hydroxyapatite Nanofibres loaded with doxycycline. Materials 15(18):6225. https://doi.org/10.3390/ma15186225

Bennett EM, Murray JW, Isalan M (2023) Phages and nanotechnology: new insights against multidrug-resistant bacteria. BioDesign Res. https://doi.org/10.34133/BDR.0004

Bilal B, Niazi R, Nadeem S, Farid MA, Nazir MS, Akhter T, Javed M, Mohyuddin A, Rauf A, Ali Z et al (2022) Fabrication of guided tissue regeneration membrane using Lignin-Mediated ZnO nanoparticles in biopolymer matrix for antimicrobial activity. Front Chem 10:837858. https://doi.org/10.3389/fchem.2022.837858

Bonito AJ, Lohr KN, Lux L, et al (2004) Effectiveness of antimicrobial adjuncts to scaling and root-planing therapy for periodontitis: summary. In: AHRQ Evidence Report Summaries. Rockville, MD: agency for healthcare research and quality (US); 1998–2005. Report no 88

Dewan AK, Gibson MA, Elisseeff JH, Trice ME (2014) Evolution of autologous chondrocyte repair and comparison to other cartilage repair techniques. Biomed Res Int 2014:272481. https://doi.org/10.1155/2014/272481

Esposito E, Cortesi R, Cervellati F, Menegatti E, Nastruzzi C (1997) Biodegradable microparticles for sustained delivery of tetracycline to the periodontal pocket: formulatory and drug release studies. J Microencapsul 14(2):175–187. https://doi.org/10.3109/02652049709015331

Ferreira JA, Kantorski KZ, Dubey N, Daghrery A, Fenno JC, Mishina Y, Chan H-L, Mendonça G, Bottino MC (2021) Personalized and defect-specific antibiotic-laden scaffolds for periodontal infection ablation. ACS Appl Mater Interfaces 13:49642–49657. https://doi.org/10.1021/acsami.1c11787

Harini G, Kaarthikeyan G (2014) Advanced drug delivery systems in treating periodontal diseases—a review. IOSR J Dent Med Sci. 13(1):27–32. https://doi.org/10.9790/0853-13182732

Jain KK (2008) Drug delivery systems–An Overview BT–drug delivery systems. In: Jain KK, ed. Humana Press, pp 1–50. https://doi.org/10.1007/978-1-59745-210-6_1

Kami D, Gojo S (2014) Tuning cell fate: from insights to vertebrate regeneration. Organogenesis 10(2):231–240. https://doi.org/10.4161/org.28816

Kumar SR, Vijayalakshmi R (2006) Nanotechnology in dentistry. Indian J Dent Res 17(2):62–65. https://doi.org/10.4103/0970-9290.29890

Liang J, Peng X, Zhou X, Zou J, Cheng L (2020) Emerging applications of drug delivery systems in oral infectious diseases prevention and treatment. Molecules 25(3):516. https://doi.org/10.3390/molecules25030516

Mao AS, Mooney DJ (2015) Regenerative medicine: current therapies and future directions. Proc Natl Acad Sci U S A 112(47):14452–14459. https://doi.org/10.1073/pnas.1508520112

Mitchell MJ, Billingsley MM, Haley RM, Wechsler ME, Peppas NA, Langer R (2021) Engineering precision nanoparticles for drug delivery. Nat Rev Drug Discov 20(2):101–124. https://doi.org/10.1038/s41573-020-0090-8

Mota J, Yu N, Caridade SG, Luz GM, Gomes ME, Reis RL, Jansen JA, Walboomers XF, Mano JF (2012) Chitosan/bioactive glass nanoparticle composite membranes for periodontal regeneration. Acta Biomater 8:4173–4180. https://doi.org/10.1016/j.actbio.2012.06.040

Nair SC, Anoop KR (2012) Intraperiodontal pocket: An ideal route for local antimicrobial drug delivery. J Adv Pharm Technol Res. 3(1):9–15. https://doi.org/10.4103/2231-4040.93558

Nakagawa T, Yamada S, Oosuka Y et al (1991) Clinical and microbiological study of local minocycline delivery (Periocline) following scaling and root planing in recurrent periodontal pockets. Bull Tokyo Dent Coll 32(2):63–70

Nardo T, Chiono V, Carmagnola I, Fracchia L, Ceresa C, Tabrizian M, Ciardelli G (2021) Mussel-inspired antimicrobial coating on PTFE barrier membranes for guided tissue regeneration. Biomed Mater 16:035035. https://doi.org/10.1088/1748-605X/abf27e

Pavia M, Nobile GA, Angelillo IF (2003) Meta-analysis of local tetracycline in treating chronic periodontitis. J Periodontol 74(6):916–932. https://doi.org/10.1902/jop.2003.74.6.916

Peng L, Cheng XR, Zhuo RX, Lan J, Wang YN, Shi B, Li SQ (2009) Novel gene-activated matrix with embedded chitosan/plasmid DNA nanoparticles encoding PDGF for periodontal tissue engineering. J Biomed Mater Res Part A 90:564–576. https://doi.org/10.1002/jbm.a.32117

Position Paper (1999) The role of controlled drug delivery for periodontitis. J Periodontol 71(1):125–140. https://doi.org/10.1902/jop.2000.71.1.125

Schwach-Abdellaoui K, Vivien-Castioni N, Gurny R (2000) Local delivery of antimicrobial agents for the treatment of periodontal diseases. Eur J Pharm Biopharm 50(1):83–99. https://doi.org/10.1016/S0939-6411(00)00086-2

Sercombe L, Veerati T, Moheimani F, Wu SY, Sood AK, Hua S (2015) Advances and challenges of liposome assisted drug delivery. Front Pharmacol 6(2015):286. https://doi.org/10.3389/fphar.2015.00286

Southard GL, Dunn RL, Garrett S (1998) The drug delivery and biomaterial attributes of the ATRIGEL technology in the treatment of periodontal disease. Expert Opin Investig Drugs 7(9):1483–1491. https://doi.org/10.1517/13543784.7.9.1483

Thomas S, Baiju RM (2022) Nano technology and medicine: the interphase. In nano materials in dental medicine Ed. Thomas S, Baiju RM, pp 1–32. https://doi.org/10.1007/978-981-19-8718-2_1

Vacanti JP, Otte J-B, Wertheim JA (2014) Chapter 1–Introduction: regenerative medicine and solid organ transplantation from a historical perspective. In: Orlando G, Lerut J, Soker S, Stratta RJ, eds. Regenerative medicine applications in organ transplantation. Academic Press, pp 1–15. https://doi.org/10.1016/B978-0-12-398523-1.00001-X

Vargason AM, Anselmo AC, Mitragotri S (2021) The evolution of commercial drug delivery technologies. Nat Biomed Eng. 5(9):951–967. https://doi.org/10.1038/s41551-021-00698-w

Yun YH, Lee BK, Park K (2015) Controlled drug delivery: historical perspective for the next generation. J Control Release off J Control Release Soc. 219:2–7. https://doi.org/10.1016/j.jconrel.2015.10.005

Zhang Y, Jiang R, Lei L, Yang Y, Hu T (2022) Drug delivery systems for oral disease applications. J Appl Oral Sci 30:e20210349. https://doi.org/10.1590/1678-7757-2021-0349

Zhang C, Jahan SA, Zhang J, Bianchi MB, Volpe-Zanutto F, Baviskar SM et al (2023) Curcumin nanocrystals-in-nanofibres as a promising platform for the management of periodontal disease. Int J Pharm 648:123585. https://doi.org/10.1016/j.ijpharm.2023.123585

Zhao P, Chen W, Feng Z et al (2022) Electrospun nanofibers for periodontal treatment: a recent progress. Int J Nanomedicine 17:4137–4162. https://doi.org/10.2147/IJN.S370340

Chapter 5
Advancements and Applications of Polycaprolactone and Its Composites in Biomedical Engineering: A Comprehensive Review

Rubiya Mohammed and Raneesh Balakrishnan

Abstract Biodegradable polymers have gained extensive prominence within the medical realm, finding application as vital biomaterials across diverse domains, such as suture fabrication, drug delivery systems, tissue engineering constructs, and wound recuperation modalities. Polycaprolactone (PCL) stands out among these polymers due to its commendable biocompatibility and innate biodegradability, rendering it particularly suitable for these applications. The inherent physicochemical and mechanical attributes of pure PCL can be judiciously modulated utilizing blending with a spectrum of polymers, thus further enhancing its utility. This comprehensive review provides a detailed overview of PCL, encompassing its intrinsic characteristics, surface refinements, and diverse synthetic methodologies. This review also highlights the recent progress made in the biomedical field through the use of PCL and its derivatives, specifically emphasizing the latest advancements and emerging breakthroughs.

Keywords Biocompatible · Biodegradable · Polycaprolactone · Tissue engineering · Wound healing · Drug delivery

5.1 Introduction

The medical industry uses biodegradable polymers to aid the healthcare system. It has significant benefits, such as the facilitation of tissue regeneration, controlled drug delivery, reducing the need for additional surgeries to remove implants. However, challenges such as variable degradation rates, biocompatibility concerns, and higher

R. Mohammed · R. Balakrishnan (✉)
Department of Physics, Catholicate College, Pathanamthitta, Kerala, India
e-mail: raneesh.b@catholicatecollege.ac.in

R. Mohammed
Chemistry and Materials Science Program, Lakehead University, Thunder Bay, Canada

R. Balakrishnan et al. (eds.), *Biocompatible Nanocomposites*, Advanced Structured Materials 239, https://doi.org/10.1007/978-981-96-7450-3_5

research and development costs must be carefully addressed to maximize their benefits in healthcare (Nair and Laurencin 2007; Ulery et al. 2011). Biodegradable polymers are classified according to their origin as natural and synthetic polymers. Natural biopolymers include cellulose, starch, collagen, corn zein, etc., while synthetic consist of polyesters like poly (lactic acid) (PLA), poly (glycolic acid) (PGA), poly (trimethylene carbonate) (PTMC), poly (ε-caprolactone) (PCL), etc. (Rhim et al. 2013; Okamoto and John 2013; Vieira et al. 2011).

Biodegradable polymers are extensively utilized as biomaterials in numerous applications, such as suture fabrication, drug conveyance mechanisms, tissue engineering constructs, and wound recuperation modalities. Biocompatibility is the characteristic of a material that can interact with a biological system and not lead to tissue dysfunction or injury (inflammatory or necrotic response), ensuring a favorable response, like growth cell proliferation or cell migration and angiogenesis, when in contact with living tissues or organisms. Synthetic classes of polyesters, including PLA, PCL, PGA, etc., are very promising in medical fields because of these characteristics (Bikiaris et al. 2007; Middleton and Tipton 2000; Gupta and Kumar 2007; Williams 2008; Hanks et al. 1996; Kweon 2003). Polycaprolactone (PCL) stands out among these polymers due to its commendable biocompatibility and innate biodegradability, rendering it particularly suitable for these applications. PCL is a biodegradable aliphatic polymer that is hydrophobic in nature and semi-crystalline in structure. Its melting point varies from 59 to 64 °C, with a glass transition temperature of approximately −60 °C (Woodruff and Hutmacher 2010; Feng et al. 2015; Eshraghi and Das 2010; Azevedo et al. 2003). PCL is compatible with many other polymers and has adjustable biodegradability. The degradation rate of PCL can vary from several weeks to several years depending upon the environmental medium and physiochemical properties. PLA, PGA, etc., can form copolymers with PCL, enhancing its degradation rate (Woodruff and Hutmacher 2010; Heimowska et al. 2017; Sinha et al. 2004; Lam et al. 2007, 2009; Chen et al. 2000; Bartnikowski et al. 2019). For example, PCL has been copolymerized with DL-lactide for absorbable nerve guides. According to degradation studies reported in the literature, PCL degrades in two stages: (1) the hydrolytic cleavage of ester groups without the use of enzymes and (2) intracellular breakdown of the polymer when it is firmly crystalline and has a low molecular weight (Chen et al. 2000; Bartnikowski et al. 2019; Meek et al. 1996; Woodward et al. 1985; Geng and Discher 2005).

PCL is soluble in a wide range of solvents, such as, benzene, toluene, cyclohexanone, dichloromethane, chloroform, and carbon tetrachloride at room temperature. Insoluble in alcohol, petroleum ether, and diethyl ether, it has a low solubility in acetone, 2-butanone, ethyl acetate, dimethyl formamide, and acetonitrile (Azari et al. 2021; Bordes et al. 2010; Altun et al. 2022; Boucher 2020; He et al. 2016).

The ability to be combined with a large variety of other polymers and the high degree of thermal stability, which makes the melting process easy, made PCL more beneficial. Adjustable biodegradability and low manufacturing cost are its other attractions. In addition, PCL was approved by the Food and Drug Administration for use in humans. The development of biodegradable injectable implants for long-term contraceptive drug delivery using two different levonorgestrel (LNG) microsphere formulations, where (LNG) was encapsulated in a PCL microspheric matrix using the emulsion process by Manoukian et al., and the development of a rod-shaped implant for subretinal medication administration which is created by embedding the corticosteroid triamcinolone acetonide within a PCL matrix by Beeley et al. are some examples for biomedical applications (Manoukian 2018; Beeley et al. 2005).

The purpose of the present review is to provide a brief overview of PCL, its physiochemical and biological properties, surface alterations, and some techniques utilized to create PCL and PCL-based materials. It also describes how PCL and its derivatives have recently been used in biomedical applications.

5.2 Basic Physiochemical Properties of PCL

PCL exhibits a unique combination of thermal, mechanical, and surface properties that make it suitable for diverse biomedical applications. Table 5.1 summarizes the key physiochemical characteristics of PCL reported in the literature.

Table 5.1 Various physiochemical properties of PCL

Properties	Description	Reference
Melting point	59–64 °C	(Woodruff and Hutmacher 2010; Feng et al. 2015; Gunatillake et al. 2003)
Glass transition temperature	– 60 °C	Feng et al. 2015; Gunatillake et al. 2003)
Average molecular weight (M_n/g mol^{-1})	3000–80,000	Woodruff and Hutmacher 2010)
Density (ρ/g cm^{-3})	1.11	Nair and Laurencin 2007; Gunatillake et al. 2003; Mondal et al. 2016a)
Crystallinity (%)	69%	Polymer Data Handbook 2009)
Intrinsic viscosity (Z/ cm3 g -1)	0.9	Polymer Data Handbook 2009)
Surface tension (g) in mN/m	51	Nair and Laurencin 2007; Gunatillake et al. 2003; Mondal et al. 2016a)
Elongation at break (%)	> 500	Nair and Laurencin 2007; Gunatillake et al. 2003; Mondal et al. 2016a; Labet and Thielemans 2009)
Water permeability at 25°C (g/m^2/day)	177	Nair and Laurencin 2007; Gunatillake et al. 2003; Mondal et al. 2016a)
Tensile stress at break or max (MPa)	14	Nair and Laurencin 2007; Gunatillake et al. 2003; Mondal et al. 2016a)

5.3 Synthesis of Polycaprolactone

Polycaprolactone can be synthesized using two methods: (1) the condensation of 6-hydroxycaproic acid (6-hydroxyhexanoic acid) and (2) ring-opening polymerization (ROP) of caprolactone (ε-CL). Although the polycondensation method is less expensive than ROP, high molecular weight polymers are challenging to obtain (Labet and Thielemans 2009).

Polycondensation: This type of polymerization generally involves condensation between two bifunctional monomers. C. Braud et al. reported the synthesis of PCL oligomers by using the polycondensation of 6-hydroxyhexanoic acid without any catalyst. Under vacuum, the condensation reaction was carried out by heating for two hours at 80° C. For six hours, the temperature was steadily raised to 150 °C. The detailed preparation method of PCL by polycondensation is only reported in a few papers (Mahapatro et al. 2004; Braud 1998).

Ring-Opening Polymerization (ROP): Synthesis of polymers with high molecular weight can be obtained using ROP. Therefore, it is the most preferred method for the synthesis of PCL.

PCL is obtained by the ring-opening polymerization of ε-caprolactone monomers. Typically, the catalyst utilized determines which of various processes is responsible for the ROP of lactones (Labet and Thielemans 2009). Table 5.2 shows the different methods, such as anionic, cationic, and coordination insertion, involved in ROP. In addition to these, radical and zwitterionic initiations are also possible, but such methods are rarely used (Saegusa et al. 1978; Stridsberg et al. 2002).

The primary method for producing cyclic PCL is the ring expansion of lactones. Kricheldorf et al. developed this method by employing cyclic tin alkoxides as

Table 5.2 Different methods and their properties involved in ROP

Method	Properties	Reference
Anionic ROP	• Occurs due to a negatively charged initiator's nucleophilic attack on the carbonyl carbon • Larger lactones, like -caprolactone (-CL) and lactide, only react when an anion attacks the carbonyl carbon, splitting the acyl-oxygen molecule to generate an alkoxide as the rising species	Stridsberg et al. 2002)
Cationic ROP	• Creates a cationic species that is attacked by the monomer's carbonyl oxygen in a bimolecular nucleophilic substitution (SN2) process	Labet and Thielemans 2009; Stridsberg et al. 2002)
Coordination-Insertion ROP	• It is a pseudo-anionic ROP • The monomer is first coordinated to the active species and then placed into the metal oxide link to continue the propagation process	Labet and Thielemans 2009; Stridsberg et al. 2002; Dubois and Jérôme 1995; Mecerreyes et al. 1999)

the starting species during the ROP of caprolactone to create macrocyclic PCL (Kricheldorf 2004).

A wide range of catalysts are used for the ring-opening polymerization of caprolactone. Lithium diisopropylamide used to polymerize caprolactone in dioxane by Bhaw-Luximon et al. (2005) is a metal-based catalyst. Phenyl lithium (Deng et al. 1999), cyclopentadienyl sodium (Yuan et al. 1998), and tert-butoxyl potassium are examples of alkali-based catalysts. Because of their low toxicity and high activity, alkaline earth metal-based catalysts are regularly employed for biomedical applications. Among them, magnesium and calcium are interesting materials that can be used to synthesize polymers (Platel et al. 2008). Aluminium provides effective control over the polymerization reaction and is widely used for the ROP of lactones (Platel et al. 2008). Studies conducted by Dubois et al. used diethylaluminium alkoxide for the ROP of ε-CL (Dubois et al. 1992). Low molecular weight alcohols can be utilized to regulate the molecular weight of the polymer, and catalysts like stannous octoate are employed to catalyze the polymerization (Azimi et al. 2014).

The monomer-to-solvent ratio, the kinds and concentrations of catalysts, and the temperature and length of the polymerization process, among other variables, were all factors in the polymerization reactions. Each of these elements impacts the produced polymers. The process utilized affects the synthetic polymer's molecular weight and molecular weight distribution. The technique used also impacts the end group composition and chemical structure of the copolymers (Okada 2002). Changing these one or two factors can produce PCL with a suitable molecular weight. Its molecular weight and weight distribution influence the tensile toughness and behavior of the PCL's degradation (Tang et al. 2004). Depending on the intended medical use, toughness is a significant factor in the design of biomaterials. Soft functions like skin and cartilage regeneration must be smooth and flexible to ensure comfort and mobility. In contrast, biomaterials for tough applications like bone fillers and dental materials must withstand mechanical stress and resist wear. Choosing the right tensile strength is crucial for success in regenerative medicine and medical devices. (Santis et al. 2013; Zheng et al. 2014; Ghaee et al. 2019) (Fig. 5.1).

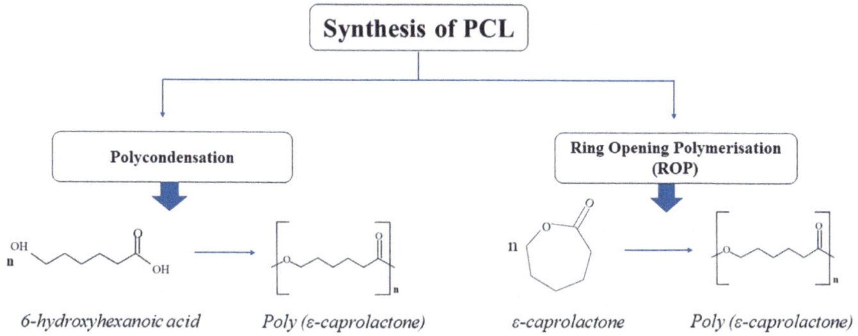

Fig. 5.1 Methods of PCL synthesis

5.4 PCL Films and Mats

Many industrial and scientific sectors, including optics, electronics, and biotechnology, are increasingly using thin polymer films. Solvent casting is a common method for creating polymeric films. This process involves dissolving the polymer in a solvent, casting it in a mold, and then removing the solvent to reveal a thin film. The selected solvent type affects the surface characteristics of these types of films (Tang et al. 2004). In many applications of biological engineering in which an active platform is needed to fill in for lost tissue or form a healing patch, biodegradable polymeric films are introduced. However, solution casting is typically impractical for creating commercial films. Solubilization, casting, and drying are a few of the phases in this film manufacturing, resulting in a lengthy processing time (Byun et al. 2010).

Published works have described two techniques for making PCL films: solvent casting of the polymer onto a petri dish and spin casting onto glass coverslips. These two techniques can produce PCL films with significant variations in pore size, porosity, surface composition and morphology, surface energy, and molecular structure. Polar solvents are often used to recover polymeric materials with surface contamination after surface treatment, which may result in significant alterations to the surface energy and contact angle hysteresis (Tang et al. 2004; Xiao 1995). PCL and PLA nanocomposite scaffolds have been combined for bone tissue engineering applications using solvent casting (Wang 2013). Martínez-Abad et al. used solvent-cast PCL films to incorporate cinnamaldehyde (CNMA) and allyl isothiocyanate (AITC) to study its antibacterial performance (Martínez-Abad et al. 2013). One method to improve cell adhesion and increase mechanical qualities is chemically modifying PCL film surfaces. Mojtaba Khorramnezhad and the team used the physical entrapment process to modify the surface of PCL films. Better cell adhesion, proliferation, and growth were accomplished due to gelatin surface modification increasing PCL films' hydrophilicity (Khorramnezhad et al. 2021).

Spin coating is another technique for applying homogeneous thin coatings to flat substrates. The extra solution is applied to the substrate, which is then rotated rapidly to distribute the fluid by centrifugal force. The film thickness can be changed by adjusting the rotational parameters of speed, duration, and solution concentration (Wang et al. 2013). The spin coating technique was used for the work by M. Afifi and co-workers to contrast the characteristics of ε-polycaprolactone (PCL) thin films containing graphene, carbonated hydroxyapatite, and alumina, each in the same amounts (Afifi et al. 2020).

There are many industrial domains where fiber mats made from polymer melt or solution are used. The surface morphology, porosity, and preparation techniques of mats make them distinct from films. Most mats are created using the electrospinning method. High electric fields are employed in the fabrication technique of electrospinning to create mats made of polymers and biopolymers. Electrospinning creates polymer fibers with diameters between nanometers and a few microns. By adjusting the polymer content, separation length of the needle from the collector, and applied voltage, it is possible to customize the porosity and structure of the formed mats. To study the effect of different solvents, Lee et al. prepared the poly (ε- caprolactone) (PCL) nonwoven mats using three solutions, i.e., N, N-dimethylformamide (DMF), methylene chloride (MC), and toluene. The properties of the PCL mixture made from these solvent systems, such as surface tension, dielectric constant, conductivity, and viscosity, were investigated. The fiber diameter at 13 weight percent was 5500 nm with a tight diameter distribution for the PCL solution dissolved only in MC. Electrospinning was significantly easier for solutions dissolved in the MC/DMF mixture, and the fiber diameter was reduced by 200 nm (Lee et al. 2003).

In addition to the literature, we fabricated PCL fibers in our laboratory using the electrospinning method for biomedical applications. During the electrospinning procedure, a high voltage of 18 kV was delivered to the needle tip and maintained a distance of 15 cm between the collector and the tip. A syringe pump was used to regulate the polymer solution's flow rate. We optimized the variables to obtain smooth bead-free fibers. The nanofibers were collected onto the aluminum foil covering the collector. The obtained scaffolds were dried at room temperature.

5.5 Surface Modification and Blending of PCL

The versatility of PCL stems from its efficient co-polymerization with various polymers, enabling the adjustment of its physical, chemical, and mechanical attributes. The inherent hydrophobicity of PCL limits its potential as a material for cell colonization. This issue can be solved by modifying the hydrophobic polymer surface of PCL. To reduce the hydrophobic nature, hydrophilic chemicals have been introduced onto polymeric scaffold surfaces using a variety of techniques. Surface characteristics, including shape, charge, chemical composition, roughness, and surface energy, all impact how the biomaterial interacts with biological molecules. There are numerous ways to change these surface features, including laser treatment, plasma treatment, grafting or coating with natural polymers, chemical depositions, etc. (Neděla et al. 2017; Tavakoli 2005; Michael et al. 2018).

Tiaw et al. modified the PCL surface by ablation experiments utilizing an excimer laser and femtosecond laser for membrane tissue engineering applications. While blind holes are made in the membrane using the excimer laser, the femtosecond laser produces clean, drilled-through holes. The results demonstrated that as the laser alters the membrane's surface, the morphological surface changes, and the water contact angle lowers. The fact that the water contact angle dropped indicates that the membrane's surface had become more hydrophilic than the untreated membrane (Tiaw et al. 2005). Plasma treatment generated much interest because of its many benefits, particularly its capacity to evenly alter the surface without changing the bulk characteristics. A PCL surface treated with active screen plasma was discovered to have altered chemical composition and structure in a study conducted by Xin Fu et al. (Fu et al. 2012). They performed active screen plasma treatments for 10 min under 2 mbar using a nitrogen (25%) and hydrogen (75%) gas mixture at

50 °C. This active screen plasma nitriding technique proves effective and practical in enhancing osteoblast cell adhesion and spreading on the surface of PCL. Another modification via layer-by-layer depositions of chitosan-*g*-PCL-*b*-PEG/heparin using the electrostatic self-assembly approach, which improved surface morphology and hydrophilicity on the PCL membrane was successfully created by Li Liu et al. (Liu et al. 2008). The persistent surface modification provided by covalent chemical conjugation methods comes at the cost of arduous multi-step processes, strict reaction conditions, and the possibility of harmful reaction by-products. Qin et al. reported the use of an azide-terminated amphiphilic graft polymer as a surface alteration method for electrospun polycaprolactone nanofibers with enhanced biocompatibility and regulated growth factor release for improved stem cell differentiation (Qin et al. 2022).

Gelatin (Gel), a naturally occurring biopolymer formed from collagen hydrolysis, is employed extensively in the pharmaceutical and medical sectors (Kalantary et al. 2020). Blended PCL/Gel scaffolds have been used for tissue regeneration since they have desired qualities such as improved hydrophilicity, porosity, tensile strength, and biocompatibility (Gil-Castell et al. 2020). These improvements can promote better cell adhesion, proliferation, and differentiation, facilitating a favorable tissue growth and regeneration environment. Improved tensile strength ensures structural integrity, and heightened biocompatibility minimizes adverse reactions, collectively promoting successful tissue regeneration with a higher chance of functional recovery. Studies by Zhang et al. demonstrated that PCL-gelatin nanofibrous membranes have high mechanical strength and biocompatibility qualities (Zhang et al. 2013). Gautam and group fabricated a tri-polymer PCL/gelatin/collagen type I scaffold for skin tissue engineering applications. In this study, grafting with 1.0 wt% collagen type I showed the highest proliferation among all collagen type I-modified scaffolds (Gautam et al. 2014). To create an implantable scaffold effective for linking the lengthy nerve gaps that will give results comparable to autografts, Gupta et al. evaluated PCL/gelatin scaffold for the qualities such as functional groups, mechanical characteristics, distribution of pore size development of Schwann cells, etc. (Gupta et al. 2009). Work by Hwang and co-workers provides a unique dual-channel regeneration method for biopolymer and bioactive molecule regeneration into PCL nanofiber. They electrospun PCL nanofibrous scaffold with cellulose acetate solutions and calcium lactate to improve the bio-physio-chemical characteristics of the composite fiber (Hwang et al. 2019).

PEG-PCL copolymer has been consistently used for medicinal applications. For example, Pang Z and the group developed a folic acid-modified PEG-PCL copolymer with ditelluride linkage as a targeted and redox-responsive cancer treatment delivery system. Based on PCL-PEG-PCL tri-block polyols and cellulose nano whisker (CNW), which acts as a cross-linker and a regulator of microphase separation of blocks, two groups of polyurethane (PU) nanocomposites were created as a promising option for tissue engineering by Noormohammadi et al. is another example (Noormohammadi et al. 2021; Pang et al. 2020).

5.6 PCL-Based Materials for Tissue Engineering

Tissue engineering (also known as regenerative medicine) attempts to generate human tissue from cells in a lab. Its ultimate goal is to heal disease by repairing or replacing tissues and organs that fail as a result of illness, genetic errors, congenital abnormalities, or major trauma. Tissue engineering depends on four key elements: the right cells to do the job, the right conditions to sustain the cells, the right biomolecules like growth factors, to make those cells effective and viable, and the appropriate mechanical and physical forces to affect the formation of cells. A new set of tissue replacement components and implementation techniques have been produced due to significant advancements in the interdisciplinary field of regenerative engineering. Using a combination of constructed extracellular matrix scaffolds, cells, and physiologically active compounds, it is now possible to fabricate tissues in the lab. This is made possible by scientific developments in biomaterials, biomimetic settings, stem cells, etc. Biomaterials are essential to tissue engineering. They are designed to build an architectural framework resembling a natural extracellular matrix to encourage cell development and tissue regeneration. The primary component of a scaffold ought to be harmless to cells, biocompatible, and degradable. The scaffold material serves as a mechanically viable substrate for cell adherence, proliferation, and differentiation. It should have adequate tensile strength to hold out against the load applied to the scaffold by the neighboring skeletal tissue. A popular polymer for making these scaffolds is PCL. The scaffold can provide the required mechanical support to the regenerating tissue and will gradually disintegrate once healing is complete (Mondal et al. 2016a; Helder et al. 2007).

5.6.1 Bone Tissue Engineering (BTE)

Bone is a dynamic tissue in which ongoing bone remodeling replaces old bone with new bone. Bone tissue engineering has become popular over the past few decades as a viable method to address the drawbacks of conventional procedures. The hydrophobic surfaces and absence of cell recognition sites on synthetic polymers are the key problems resulting in poor cell affinity (Tamboli et al. 2013). Due to its thermoplastic properties, PCL is both simple to handle and biocompatible, which has led to its widespread application in the construction of 3D scaffolds. To enhance the qualities of the scaffolds, researchers, therefore, mix PCL with different polyesters, inorganic materials, metal elements, or collagen (Park et al. 2021; Shim et al. 2012). Several research teams have created a composite scaffold with a well-organized design using PCL. The study by Fereshteh Sharifi et al. concentrated on the physical characteristics and cell viability of polycaprolactone/chitosan (PCL/CTS) and PCL/carboxymethyl chitosan (PCL/CMC)-based nanofibrous scaffolds. In contrast to PCL and PCL/CTS scaffolds, PCL/CMC scaffolds exhibited more hydrophilic surfaces. Studying how well the scaffolds performed biologically, it became clear that adding CMC with PCL fibers produced a greater effective scaffold for bone tissue regeneration applications (Sharifi et al. 2018). Applying nHAp to the surface of the nanofibrous gelatin-PCL scaffold allowed Sneh Gautam et al. to create the gelatin-PCL-nHAp nanostructured scaffold successfully. The nanocomposite scaffold had more osteoblast functionality and proliferation than the gelatin-PCL scaffold, and this suggested that the nHAp-treated scaffold is most suitable for BTE, as demonstrated by the MTT experiment and DNA quantification (Gautam et al. 2021). It is considered that PCL-HAp composite films work better for bone-to-tendon interface repair. By combining PCL and nanohydroxyapatite (nHAp) via heat pressing, Tong et al. (Tong et al. 2017) created a composite of this type. The elastic modulus of the films increased with the amount of nHA added. By combining the benefits of filament winding and the phase inversion/salt leaching process, a novel composite material for bone regeneration was created by Guarino V, consisting of a porous PCL matrix reinforced with continuous poly-L-lactide (PLLA) fibers. On the porous PCL/(PLLA) fibrous composite scaffolds, it has been demonstrated that for up to three weeks, stromal cells derived from bone marrow and trabecular osteoblasts multiply quickly, promoting the directed movement of bone cells along the fibers (Guarino et al. 2008).

PCL can have its characteristics changed by biocompatible chemical copolymers. Fu et al. used a PCL-PEG-PCL copolymer created using solvent casting from a chloroform/dimethylformamide mixture to create a film for cartilage tissue engineering. The films became more hydrophilic due to the PEG segmentation (Fu et al. 2016).

The scaffold morphologies are essential for tissue regeneration. It was presumed that these structures are inadequate to address the novel issues of quick and safe proliferation and cellular differentiation in all directions. This assumption was based on their cell biology studies, pharmaceutical protocol, biochemistry, and surface modification with biomolecules. The desire to replicate the 3D composition and structure of natural cellular environments has sped up the development of materials and production methods. Compared to the 3D printing method, the scaffolds made using solvent casting and electrospinning techniques exhibit inadequate control over the architecture and the pore size (Lee et al. 2008). Implant-related infections are significantly more likely to cause morbidity. Radhakrishnan and coworkers fabricated 3D-printed antimicrobial PCL scaffolds incorporated with Ag Nps, which exhibit the bactericidal effect with no inflammation at the point of contact. The PCL scaffold's stiffness and enzymatic stability were increased by adding AgNps (Radhakrishnan et al. 2021). Maxillofacial bone abnormalities are frequently filled and repaired using allogeneic and autologous bone. Even if immunological rejection can be prevented with allograft transplantation, the donor site will suffer secondary harm. Lei et al. created in situ tissue regeneration maxillo-facial bone scaffolds using fused deposition modeling (FDM) 3D printing technique, using various PCL and Mg_3PO_4 ratios. Their findings showed that the 20% Mg_3PO_4-containing PCL scaffolds have outstanding in vitro osteogenic differentiation, cytocompatibility, and mineralization properties (Lei et al. 2022). Different fiber orientation angles in 3D-printed PCL scaffolds modified with hydroxyapatite and polypropylene fumarate were created and evaluated on rabbits. A study found that customized scaffolds significantly improved the speed and quality of bone repair. For the restoration of bone tissue, cell-laid mineralized ECM in 3D-printed scaffolds has received much attention. By using three-dimensionally printed scaffolds using a mixture of poly(lactic-co-glycolic acid) (PLGA), polycaprolactone (PCL), and b-tricalcium phosphate (b-TCP), as well as mineralized ECM laid down by human nasal inferior turbinate tissue-derived mesenchymal stromal cells, Pati F and group were able to create bone graft substitutes (hTMSCs) (Pati et al. 2015; Buyuksungur et al. 2017). The ability of a 3D nanofibrous scaffold made of PCL to sustain and support bone marrow-derived human mesenchymal stem cell multilineage differentiation (hMSCs) was examined by Li et al. (Li et al. 2005). Aluminium oxide (Al_2O_3) and hydroxyapatite (HA) can be added to PCL nanocomposite scaffolds, which have been reported to increase biocompatibility. PCL scaffolds reinforced with Al_2O_3 retained good biocompatibility, as shown by tests on cell morphology and proliferation (Dong et al. 2012). Thomas et al. have studied the structural and mechanical properties of specifically oriented PCL nanofiber meshes, made at various rotation rates of collector during electrospinning, as potential bone tissue scaffolds. The narrowing of fiber diameters and the mechanical characteristics of both individual fibers and bulk fibrous meshes are significantly impacted by the deposition of fibers at faster rotation speed (rpm) (Thomas et al. 2006a) (Table 5.3).

5.6.2 *Skin Tissue Engineering*

Building on advances in biocompatible matrix technology and cell culture method established in the late twentieth century, skin tissue engineering has rapidly evolved. Reduced scar formation, enhanced wound healing, and reconstruction of the functional and structural elements of skin are the goals of skin tissue engineering (Clark et al. 2020; Greenwood 2016). Tissue-engineered skin alternatives are being created to address the requirement for an affordable and convenient synthetic or engineered skin transplant that can be used to repair acute and chronic wounds to get around the constraints associated with skin grafts. To revive the skin's function and structure, the ultimate goal of biomaterial design is to supply the required impulses to trigger and carry out cellular processes. Biocompatibility is a must-have for skin substitutes, meaning it must not be toxic, inflammatory, or immune-reactive, and it must not pose a threat of infection that could spread. The skin substitute must have the same biomechanical characteristics and barrier function as natural skin, as well as the ability to hasten wound healing. As the skin regenerates, the skin substitute should ideally encourage the creation of new tissue (Zeng et al. 2011; Supp and Boyce 2005).

Wound dressing and wound healing are the two main uses for electrospun nanofibrous scaffolds and patches. The nanofiber membrane's unusually tiny pores and high porosity can shield the wound from pathogens and prevent infection. Different polymers, both natural and artificial, are utilized in this context (Liu et al. 2021). For the efficient distribution of growth factors to the wound site for appropriate wound healing, PCL has been employed in various wound dressings (Raina et al. 2022).

Table 5.3 PCL-based materials in bone tissue engineering

Material	Filler	Filler content	Reference
PCL-PEG-PCL	(B–Sr–Mg) multi-doped HA composite	10, 20wt%	Yedekçi et al. 2022; Shen et al. 2019; Hokmabad et al. 2019; Pazarçeviren et al. 2017; Abedi et al. 2022; Zhang et al. 2017)
	MgO	0.5, 1, 2, 3 g	
	Elaeagnus angustifolia	0, 5, 10, 15%	
	Clinoptilolite	10, 20%	
	Fe_3O_4	10% w/w	
	Fe_3O_4, HA (to PCEC/Gel)	6, 6%w/w	
PCL/Gel	DCPD-modified GO	–	Setia Budi et al. 2022; Cakmak et al. 1962; Sattary et al. 2019; Shirani et al. 2019; Marins et al. 2019; Hamlekhan et al. 2010; Duymaz et al. 2019)
	Bacterial cellulose, HA	BC: 0.25 wt% HA: 0.25, 0.5 wt%	
	nHA, Vit D3	nHA: 15wt% Vit D3: 0.1 µl	
	Bioactive glass	5 wt%	
	Nb_2O_5 and HA	HA: 2wt% Nb2O5: 0, 3, 7, 10 wt%	
	Halomonas levan	0, 0.2, 1.2, 1.6 wt%	
	HAp	Gel: 10% w/v HAp: 40 wt%	
PCL/PVP and PVA/β-TCP	–	–	Sohrabi et al. 2022)
PCL/HA	Hydroxyapatite	0, 10, 20, 30 wt%	Thijssen et al. 2022; Moeini et al. 2017; Lei et al. 2020; Santis et al. 2015; Pedrosa et al. 2021; Guarino et al. 2014; Thomas et al. 2006b; Shitole et al. 2019; Ronca et al. 2016; Maleki-Ghaleh et al. 2021; Sousa et al. 2022; Biscaia et al. 2022)
	PEGDA	6 wt%	
	Hydroxyapatite	20 wt%	
	Si-HA	0.8 wt%	
	Fe-HA	20 wt%	
	Zn-HA	0, 1, 5, 10 wt%	
	Mg/CO_3-HA	13%	
	Hydroxyapatite	20 wt%	
	ZnO-nHA	nHA: 10 wt% ZnO: 1, 5, 10, 15, 30% w/w	
	Hydroxyapatite	10 wt%	
	HA, ZnHA, ZnHA-rGO	–	

(continued)

Table 5.3 (continued)

Material	Filler	Filler content	Reference
	Hydroxyapatite	20, 40 wt%	
PCL/β-TCP	MgO	10 wt%	Safiaghdam et al. 2023; Ozkan et al. 2009; Wu et al. 2023)
	–	–	
	–	–	
PCL/α-TCP	Heparin	α-TCP: 0.5, 1, 2 wt% heparin: 0.1, 1, 2 wt%	Alehosseini et al. 2018)
PCL/CNT	Multiwalled CNT	12.5, 25, 75, 150, 500 mg/ml	Mattioli-Belmonte et al. 2012; Pan et al. 2012; Tohidlou et al. 2019; Flores-Cedillo et al. 2016; Yang et al. 2019)
	Multiwalled CNT	0.25, 0.5, 1, 2%	
	Singlewalled CNT	0, 0.1, 0.2, 0.5%	
	BGP-multiwalled CNT	0.3 wt%	
	Multiwalled CNT-nHA	mwCNT: 1, 2 wt% nHA: 15, 30 wt%	
CS-g-PCL	Hydroxyapatite	1, 3, 5, 10 wt%	Shirzaei Sani et al. 2021; Amini et al. 2021; Sedghi et al. 2020)
	Magnetic bioactive glass Cisplatin	MBG: 10 wt% MBG/Cisplatin: 20 wt%	
	Mg-HA	5 wt%	
PCL	Graphene Graphene oxide Cissus quadrangularis	–	Kashte et al. 2019; Diba et al. 2012; Kotela et al. 2009; Jaidev et al. 2017; Wang et al. 2022; Nitya et al. 2012; Dong et al. 2021; Rezaei et al. 2023; Joseph et al. 2023; Joy et al. 2023)
	Forsterite	10, 20, 30, 40, 50 wt%	
	Wollastonite	0.5, 1, 2, 4, 6%	
	GO_Cu	1, 3, 5 g	
	Zn	1, 2, 3 wt%	
	Halloysite nanoclay	1, 2, 4, 6 wt%	
	Mg micro-particles	1, 3, 5, 7, 9 wt%	
	Eggshell micro-particles	3, 5, 7 wt%	
	Graphene oxide, strontium	–	
	ZrO_2	0.5, 1, 3, 5, 7 wt%	
PCL/CaP coating			Yang et al. 2008; Kim et al. 2022)

(continued)

Table 5.3 (continued)

Material	Filler	Filler content	Reference
	Graphene oxide	–	
PCL/PBAPCL	Hydroxyapatite	15 wt%	Chakrapani et al. 2017)
PCL/col	Hydroxyapatite	20 wt%	Phipps et al. 2011)
PCL	–	–	Yoshimoto et al. 2003)
PCL/HPG	Magnesium phosphate nHA	nHA and nMP: 10 wt%	Perumal et al. 2020)
PCL/PLLA	Hydroxyapatite	20 wt%	Bakhshandeh et al. 2011; Samadian et al. 2021; Siqueira et al. 2019; Samiei et al. 2016)
	Gel NFs Au NPs	40, 80, 160 ppm	
	Hydroxyapatite	3.5%	
	Simvastatin Hydroxyapatite	0.1, 0.5 w/w	
PCL/FHA	Fluoridated hydroxyapatite	0, 5, 10, 15 wt% 25 wt% F^-, 75 wt% OH^-	Johari et al. 2020)
PCL/MCC	Whitlockite nHA	1–10 wt%	Abbas et al. 2022)

Wound dressings based on PCL have been created both with and without the introduction of bioactive ingredients that promote wound healing or antibacterial agents. Liao et al. created PCL composite electrospun mats comprising tetracycline hydrochloride, dextran, and cellulose acetate (CA), testing its effectiveness against bacteria and blood coagulation. PCL was dissolved in DMF/THF (1:1) to make 10 wt% solutions. CA was dissolved in acetone/DMA to make 18 wt% solutions (2:1). Before the electrospinning, the two solutions—PCL and CA solutions—were combined in a variety of weight ratios (1:1, 1:2, and 1:3), and dextran and tetracycline hydrochloride were added as supplements. The PCL-CA-dextran-TCH mat displayed good bioactivity and efficient antibiotic activity against bacteria, according to the results (Liao et al. 2015). Studies by Bakhsheshi-Rad et al. loaded cephalexin (CEX), an efficient antibiotic, into the PCL-GelMA matrix. Their results revealed that the hydrophilic property and swelling ratios can both be improved by an increase in CEX concentration. As CEX increased, fibrous diameters decreased, improving nanofibrous tensile strength. Both gram-positive *Staphylococcus aureus* (*S. aureus*) and gram-negative *Escherichia coli* were significantly inhibited by PCL/GelMA loading with CEX's antibacterial activity (Bakhsheshi-Rad et al. 2019) and in a similar antimicrobial study by Oh et al. fabricated PCL/chitosan-caffeic acid nanocomposite fiber mat. The findings suggest that, in comparison with PCL and PCL/CH fibrous mats, PCL/CCA fibrous mats can increase normal human dermal fibroblast-neonatal (NHDF-neo) cell proliferation and dissemination for the healing process with greater antibacterial activity over *S. aureus* (Oh et al. 2016).

Since ancient times, plant-based herbal medications have been used to quicken the healing of wounds, and the scientific literature is abundant with accounts of how different stages of wound healing are impacted by medicinal herbs. To incorporate four individual plant extracts: *Indigofera aspalathoides* (*I. aspalathoides*), *Azadirachta indica* (*A. indica*), *Memecylon edule* (*M. edule*), and *Myristica andamanica* (*M. andamanica*) to PCL, Jin G and group employed electrospinning technique. According to the results, the HDF had a functioning metabolism on plant extracts that contained nanocomposites, particularly on PCL/ME. F-actin and collagen staining data provided additional evidence that supported the notion that PCL/ME was the ideal medium for skin tissue creation (Jin et al. 2013). Skin protects against dangerous physical and chemical substances, including tobacco smoke, poisonous metals, and ultraviolet (UV) radiation. These substances lead to the overproduction and build-up of incredibly hazardous reactive oxygen species (ROS) that are linked to skin aging, accelerated cancer growth, etc. Antioxidant activity of polycaprolactone-gelatin nanofiber incorporated with liposomes of epigallocatechin-3-gallate (EGCG), molecule from green tea was analyzed by Pires et al. and results showed that the hazardous ROS species produced by exposure to H_2O_2 or UV radiation are scavenged by the EGCG released from PCL/gelatin nanofibers, which also slows down the oxidation events linked to damage (Pires et al. 2019).

For skin tissue formation or injury healing brought on by illnesses or damage, tissue engineering technologies based on scaffolds are widely used. PCL-poly propylene succinate (PPSu) block copolymer incorporated with Ag Np exhibited antibacterial activity by reducing the adherence of the bacteria on the membrane and displaying non-toxicity toward human dermal fibroblast (HDF) (Afghah et al. 2020). One of the interesting methods currently presented in the biomedical applications of wound healing is the modification of nano-fibrous structures that resemble the natural extracellular matrix. Using the pulsed laser deposition (PLD) approach, different compositions of AgNPs were deposited on the surface of carbonated hydroxyapatite (CHAP), which is substituted partially via selenite ions (Se-CHAP/PCL) fibers. Investigation of the human osteoblast cell line (HFB4) proliferation on the exterior of AgNPs@Se-CHAP/PCL fibers showed that cells multiplied and distributed not only on the fibers' surface but also throughout the scaffold's length. Lengthening the coating time improved the antibacterial potential of the composite biomaterial. Multiplying diverse biological elements with polymers for electrospinning can create remarkable multirole scaffolds. Multirole scaffolds are created by integrating electrospinning with other advanced techniques. An example is the experiments by Motealleh et al., including chamomile in a combination of electrospun PCL and polystyrene (PS) for treating wounds. chamomile exhibited remarkable properties in promoting cell growth and enhancing cell proliferation. It provided a favorable environment for their viability and proliferation, according to the culture of mesenchymal stem cells on PCL/PS nanofibrous mats. The simultaneous use of chamomile drug and PS as a tissue cultured polystyrene (TCPS) with a nanofibrous structure can produce an increased rate of re-epithelialization and collagen fibers accumulation in the dermis and, finally an increased rate of granulation tissue formation after 14 days of post-treatment periods, according to in vivo studies on rat wounds for PCL/PS nanofibrous wound dressing samples with and without drug loading (Motealleh et al. 2014; Menazea et al. 2020). Table 5.4 shows some more PCL-based materials used in skin tissue engineering.

5.7 PCL-Based Drug Delivery Systems

Technologies that transfer medications into or throughout the body are drug delivery systems. Selective or preferential delivery of therapeutic molecules to the specific site at the required dosage is crucial for a successful drug delivery system. The constraints of conventional pharmaceuticals highlight the significance of developing novel drugs for treating a wide range of disorders. Medicinal substances are created using nanotechnology in biocompatible nanocomposites like micellar systems, conjugates, nanoparticles, and nanocapsules (Panyam and Labhasetwar 2003). The effectiveness of drug entrapment, particle size, surface characteristics, drug content, and drug release qualities can all be used to evaluate nanoparticle-based drug delivery systems. A large variety of biocompatible or biodegradable polymers have been investigated for the administration of various therapeutic compounds. Despite the existence of metallic and ceramic drug delivery systems, polymeric ones have gained significantly more prominence due to their adaptability. Because of its well-known biocompatibility, biodegradability, and a wide range of degrading characteristics, PCL can be regarded as one of the major polymer used in different medical benefits. The homogeneous distribution of the components in the matrix is caused by PCL's compatibility with a wide range of medications. In fact, a carrier's ability to distribute an ingredient over time depends on how slowly it degrades the chemical. Following in vitro and in vivo biocompatibility and effectiveness tests, the FDA approved a number of PCL-based drug delivery systems (Azari et al. 2021; Mondal et al. 2016b).

Karuppswamy et al. investigated the possibility of PCL nanofibers as a controlled therapeutic delivery system by incorporating it with different concentrations of antibiotic tetracycline hydrochloride ranging from 2 to 5%. The medication was first dissolved in methanol, and then chloroform and PCL were added. For drug-loaded membranes containing 2 and 3%, the initial burst release was less than 15%, but for 4 and 5%, it jumped to about 20%. The surface-loaded medication is released as soon as the membrane is submerged in PBS. One of the longest drug release periods of PCL membranes was reported after the initial burst release of tetracycline hydrochloride, lasting until the eighth day. The result also showed that tensile strength roughly decreases by 10% for every 1% increase in drug concentration (Karuppuswamy et al. 2015). The sample's inherent surface area often serves as a limiter for net drug loading and release. An electrospun fiber capsule that facilitates long-term distribution made by sintering the fiber was developed by Chaparro et al. sintering at various temperatures in conjunction with a steady reduction in porosity significantly changed the structure of as-spun capsules. After 56 and 57 °C exposures, sintering reduced the electrospun pore size by tenfold and 28-fold, respectively. The PCL capsules lacked

Table 5.4 PCL-based materials used in skin tissue engineering

Material	Filler	Filler content	Reference
PCL	Mg/Se-HAP/GO	110 mg	Ahmed et al. 2020; Alsenany et al. 2022; Sagart 2024; Ahmed et al. 2021a; Kouser et al. 2021; Hao et al. 2020; Kzar et al. 2023; Al-Wafi et al. 2023; Nejaddehbashi et al. 2020)
	Te/V-HAP	–	
	HA, Zeolite	HA—0.05% Zeolite—0.5 g	
	Ag-MNP	110 mg	
	Halloysite nanotubes	0, 0.01, 0.02, 0.03, 0.04, 0.05	
	Cerium oxide	5, 10, 15, 20, 25 wt%	
	TiO_2	–	
	CuO, V_2O_5	–	
	Silver sulfadiazine	0.3%	
PCL/Aloe vera/chitosan	–	–	Miguel et al. 2017; Yin and Xu 2020)
	AV, CS	AV—1 wt% CS—1, 2, 3 wt%	
PCL/PU	Graphene oxide	0.5–4%	Sadeghianmaryan et al. 2020; Khalili et al. 2022)
	TiO_2, curcumin	–	
PCL/Carboxymethyl cellulose	Co-HA	5, 10, 15 wt%	Wang and Shi 2022; Noppadol Sriputtha et al. 2022)
	CMC	0, 0.12, 0.39, 0.66, 0.93, 1.2 g	
PCL/Au@Se	–	–	Al-Mogbel et al. 2021)
PCL/Gelatin	Ag	–	Tra Thanh et al. 2018; Doostmohammadi et al. 2021; Pavliňáková et al. 2018; Karuppannan et al. 2021; Ravichandran and Radhakrishnan 2022; Hivechi et al. 2020; Heidari et al. 2017; Ramírez-Agudelo et al. 2018; Moazzami Goudarzi et al. 2020; Rather et al. 2018)
	Selenium, vitamin E	Se—12% Vit E—1.5%	
	Halloysite nanotubes	0.5, 1, 3, 6, 9 wt%	
	Copper oxide	1 wt%	
	Lupeol	–	
	Cellulose	1, 5 wt%	
	Graphene	0–2 wt%	
	nHA, Dox	nHA—9wt% Dox—2.8 wt%	
	Cellulose	1, 2, 3, 6 wt%	

(continued)

Table 5.4 (continued)

Material	Filler	Filler content	Reference
	Cerium oxide	25% v/v	
ACR-PCL-NPs	–	–	Nawaz et al. 2021)
AgNPs@Se-CHAP/PCL	Se-CHAP	–	Menazea et al. 2020)
PCL/Hydroxyapatite	Bismuth, strontium	–	Ahmed et al. 2021b; Ashraf et al. 2021; El-Naggar et al. 2021)
	Aluminium, vanadate	–	
	Thallium, vanadate	–	
PCL-Col/ZnO	NH_2-ZnO	0, 0.25, 0.5, 0.75% w/v	Li et al. 2021)
PCL/GS PCL/GYM	Gymnema sylvestre	0.5 wt%	Subramanian et al. 2021)
Gel/TiO_2 + PCL/Silk fibroin	–	–	Golipour et al. 2022)
PCL + Curcumin + NH_2-SBA-15	Curcumin, NH_2-SBA-15	–	Rathinavel et al. 2021)
PCL/Chitosan	Resveratrol, ferulic acid	1, 2, 5, 10 mg/ml	Poornima and Korrapati 2017)

all visible surface porosity at 58 and 59 °C. However, confined pores were visible in the 58 °C cross-section. Biphasic release of RhB and RB was seen from capsules heated to 56 and 57 °C; a more regulated discharge happened from capsules heated to 58 and 59 °C (Chaparro et al. 2019). Electrospun mats have a substantially higher degree of crystallinity than spin-cast films, which significantly reduces diffusivity. Dexamethasone, a medication used to treat illnesses of the posterior region of the eye and the inner ear, was incorporated into PCL mats by Hsu et al. The PCL polymer solution was supplemented with dexamethasone at a 10% polymer loading. These hybrid methods enabled dexamethasone to be released for an additional month (Hsu et al. 2016). Due to the potential to improve medication delivery and the ability to control release kinetics, the manufacture of antibiotic-loaded fibers has benefits over coated fibers and textiles. Gentamicin sulfate (GS) was added to gravity-spun PCL fibers by Chang et al. to demonstrate the possibility for regulated, local distribution of antibiotics from wound closure materials and tissue alternatives such as textile vascular grafts. Compared to drug-free fibers, the diameter of GS-loaded fibers made from homogenized suspensions rose by a maximum of 18%, with no discernible relationship to GS loading. On the other hand, GS-loaded fibers made from non-homogenized suspensions showed significant diameter increases (50%) that were indicative of the bigger size of the integrated particles. Over the course of

50 days in PBS, fibers made from non-homogenized suspensions released approximately 80% of the antibiotic load gradually. The GS released after 14 days prevented the growth of S. epidermidis in culture (Chang et al. 2008).

Nimesulide is a selective COX-2 isoform inhibitor and hydrophobic anti-inflammatory medication that has been approved for use in the symptomatic management of severe pain conditions. However, a distribution technique that avoids passing through the hepatic circulation is required due to the toxicity revealed in its hepatic metabolism. Simoes and team successfully created drug-loaded core-sheath fibers from a mixture of poly (methyl methacrylate) (PMMA) and polycaprolactone (PCL) utilizing a coaxial electrospinning technology were adopted to avoid passage through hepatic circulation. It was discovered that the production of the fibers was best with an electric field potential of 7–8 kV. The presence of nimesulide increased the core-sheath fibers' thermal stability compared to that of regular PMMA fibers. According to the results, nimesulide is entirely solubilized in the polymer matrix. The placement of the active within the drug-loaded, core-sheath fibers was discovered to have an important effect on the release of nimesulide (Simões et al. 2019). Ferulic acid (anti-inflammatory) and resveratrol (proangiogenic)-containing core–shell chitosan-PCL coaxial electrospun nanofibers were created in a related investigation. Resveratrol and ferulic acid were simultaneously delivered for up to 120 h in a sustained manner using drug-loaded coaxial electrospun nanofibers (240 nm thick), which also sped up the healing of wounds on rat skin (Poornima and Korrapati 2017). The results of numerous clinical studies suggest that anti-platelet therapy can lower the incidence of recurrent strokes. A medication called dipyridamole (DIP) is known to inhibit platelet aggregation. Repanas et al. created a polycaprolactone (PCL)-based drug delivery system with encapsulated dipyridamole (DIP). Most of the medication was kept inside the core of the electrospun fibers using the coaxial electrospinning technique. The outcomes demonstrated that the various attributes were affected by the concentration of DIP and PCL. Thinner, more hydrophobic fibers with lower encapsulation effectiveness and faster DIP dispersion through the polymeric matrix were produced at higher DIP concentrations (Repanas and Glasmacher 2015).

Due to the excellent therapeutic outcomes, tailored medicine delivery has attracted much attention recently. Yang et al. created a drug delivery implant with a regulated drug release function utilizing the FDM technique after preparing the drug-loaded PCL-CS filament by hot-melt extrusion (HME) without the use of a solvent. Water-soluble CS was added to the insoluble matrix to create release channels and provide medicines with efficient and regulated release. PCL was employed as a drug delivery vehicle to achieve a low processing temperature and avoid drug activity declines. The findings demonstrated that when the amount of medication and chitosan increased, the filament's tensile strength declined (Yang et al. 2022). Table 5.5 shows more PCL-based materials used in drug delivery systems.

Table 5.5 PCL-based materials used in drug delivery systems

Material	Filler	Filler content	Reference
PCL	L,L-diphenylalanine	2.5–50%	Chaparro et al. 2019; Liberato et al. 2016; Radisavljevic et al. 2018; Wang et al. 2016; Kamath et al. 2020; Luong-Van et al. 2006; Chou et al. 2021; Ke et al. 2016; Hivechi et al. 2019; Wang et al. 2019; Alhusein et al. 2013; Wong et al. 2012; Javaid et al. 2180)
	Cefazolin	20wt%	
	Ketoconazole	0.5, 1.2, 3, 6, 9% w/w	
	Ibuprofen	4 mg/ml	
	Heparin	0.05, 0.5 wt%	
	nanohydroxyapatite	–	
	Allantoin loaded PSNs	10, 20, 30 wt%	
	Cellulose nanocrystal	0, 0.5, 1.0, 1.5, 2.5, 4 wt%	
	ZnAl-LDH	2, 4, 8 wt%	
	–	–	
	Zein	–	
	–	–	
	Cefotaxime	3 mg	
PCL/Cellulose acetate	Vitamin D3	6, 12, 20%	Wsoo et al. 2021)
PCL-GSH-GLM	Glutathione	–	Khandaker et al. 2022)
PCL/CNT	Docetaxel	–	Niezabitowska et al. 2018)
PCL/Gelatin	Graphene oxide	2.5 wt%	Budi et al. 2022; Ahmadi et al. 2020; Mohamadi et al. 2022)
	Si-Mg-FA	0, 5, 10 wt%	
	Cindamycin HCl	–	
PCL/Chitosan	Doxycycline	10, 20, 30, 40, 50 wt%	Sahoo et al. 2010; Ghazalian et al. 2022; Yang et al. 2020)
	Tetracycline hydrochloride	–	
	Lidocaine hydrochloride Mupirocin	LID—1 wt% Mupirocin—2 wt%	
PCL/β-TCP	Lovastatin	100 μg	Tarafder et al. 2013)
PCL	Methotrexate	–	Dhanka et al. 2017)
PCL/Collagen	Dexamethasone Simvastatin Dexamethasone-simvastatin	0.2 wt%	Rather et al. 2020)
PCL/IBP	Ibuprofen	10%	Salmoria et al. 2017)

5.8 Challenges and Future Aspects

According to published research, PCL is employed in many ways to produce different scaffolds and devices. Remarkably, PCL-based devices and scaffolds have not yet seen widespread commercialization. Long-term research plans must address this and offer more applicable, helpful findings that enhance human health. The lack of cell recognition sites on its surface and its poor hydrophilicity have limited the ability of PCL to promote cellular adhesion, proliferation, and differentiation despite its high mechanical strength. Several surface modification techniques have been employed to reduce the hydrophobic nature of PCL by introducing hydrophilic groups. Also, PCL has a low glass transition temperature and melting point,limiting its use in high-temperature applications. Conducting further research to address these challenges is imperative for expanding the potential use of PCL. Innovative methods, such as integrating reinforcing agents and formulating composite materials, offer promise in enhancing the thermal stability of PCL and enabling its effective utilization in a broader range of high-temperature environments.

5.9 Conclusion

The extensive exploration of polycaprolactone (PCL) in various research studies underscores its potential as a versatile biomaterial. Its unique mechanical properties, compatibility with different polymers, biodegradability, and biocompatibility make it a promising candidate for various biomedical applications. The ease of modification and blending with other polymers further enhances its adaptability and suitability for tailored applications. Consequently, PCL continues to be a focal material in developing innovative solutions for tissue engineering, drug delivery systems, and other biomedical fields, promising significant contributions to advancing medical technology and patient care.

References

Abbas L et al (2022) Hydroxyapatite and whitlockite incorporated cellulose reinforced polycaprolactone (PCL): biomimetic nanocomposites for bone tissue engineering applications. J Polym Environ 31(5):2093–2113

Abedi F et al (2022) Synthesis and characterization of growth factor free nanoengineered bioactive scaffolds for bone tissue engineering. J Biol Eng 16:28

Afghah F et al (2020) 3D printing of silver-doped polycaprolactone-poly propylene succinate composite scaffolds for skin tissue engineering. Biomed Mater 15:035015

Afifi M, Ahmed MK, Fathi AM, Uskoković V (2020) Physical, electrochemical and biological evaluations of spin-coated ε-polycaprolactone thin films containing alumina/graphene/carbonated hydroxyapatite/titania for tissue engineering applications. Int J Pharm 585

Ahmadi T et al (2020) Fabrication and characterization of polycaprolactone fumarate/gelatin-based nanocomposite incorporated with silicon and magnesium co-doped fluorapatite nanoparticles using electrospinning method. Mater Sci Eng, C 106:110172

Ahmed MK, Mansour SF, Al-Wafi R, Menazea AA (2020) Composition and design of nanofibrous scaffolds of Mg/Se- hydroxyapatite/graphene oxide @ ε-polycaprolactone for wound healing applications. J Mater Res Tech 9:7472–7485

Ahmed MK et al (2021a) Nanofibrous ε-polycaprolactone scaffolds containing Ag-doped magnetite nanoparticles: physicochemical characterization and biological testing for wound dressing applications in vitro and in vivo. Bioact Mater 6:2070–2088

Ahmed MK, Mansour SF, Al-Wafi R, Abdel-Fattah E (2021b) Nanofibers scaffolds of co-doped Bi/Sr-hydroxyapatite encapsulated into polycaprolactone for biomedical applications. J Mater Res Tech 13:2297–2309

Alehosseini M, Golafshan N, Kharaziha M, Fathi M, Edris H (2018) Hemocompatible and bioactive heparin-loaded PCL-α-TCP fibrous membranes for bone tissue engineering. Macromol Biosci 18:1800020

Alhusein N, Blagbrough IS, De Bank PA (2013) Zein/polycaprolactone electrospun matrices for localised controlled delivery of tetracycline. Drug Deliv Transl Res 3:542–550

Al-Mogbel MS et al (2021) Conditions adjustment of polycaprolactone nanofibers scaffolds encapsulated with core shells of Au@Se via laser ablation for wound healing applications. Spectrochim Acta A Mol Biomol Spectrosc 259:119899

Alsenany N, Mansour SF, Eldera SS (2022) Controlled compositions of tellurium/vanadium co-doped into hydroxyapatite/ε-polycaprolactone for wound healing applications. New J Chem 46:14816–14825

Altun E, Ahmed J, Onur Aydogdu M, Harker A, Edirisinghe M (2022) The effect of solvent and pressure on polycaprolactone solutions for particle and fibre formation. Eur Polym J 173:111300

Al-Wafi R, Hammad MS, Mansour SF (2023) Development of antibacterial and morphological features of scaffolds based on polycaprolactone encapsulated with copper oxide/vanadium oxide for wound dressing applications. Ceram Int 49:26182–26190

Amini Z et al (2021) Magnetic bioactive glasses/Cisplatin loaded-chitosan (CS)-grafted- poly (ε-caprolactone) nanofibers against bone cancer treatment. Carbohydr Polym 258:117680

Ashraf S et al (2021) Nanofibers of polycaprolactone containing hydroxyapatite doped with aluminum/vanadate ions for wound healing applications. New J Chem 45:22610–22620

Azari A, Golchin A, Mahmoodinia Maymand M, Mansouri F, Ardeshirylajimi A (2021) Electrospun polycaprolactone nanofibers: current research and applications in biomedical application. Adv Pharm Bull 12:658–672

Azevedo MC, Reis RL, Claase MB, Grijpma DW, Feijen J (2003) Development and properties of polycaprolactone/hydroxyapatite composite biomaterials. J Mater Sci Mater Med 14:103–107

Azimi B, Nourpanah P, Rabiee M, Arbab S (2014) Poly (ε-Caprolactone) fiber: an overview. J Eng Fibers Fabr, 74–90

Bakhshandeh B, Soleimani M, Ghaemi N, Shabani I (2011) Effective combination of aligned nanocomposite nanofibers and human unrestricted somatic stem cells for bone tissue engineering. Acta Pharmacol Sin 32:626–636

Bakhsheshi-Rad HR et al (2019) Antibacterial activity and in vivo wound healing evaluation of polycaprolactone-gelatin methacryloyl-cephalexin electrospun nanofibrous. Mater Lett 256:126618

Bartnikowski M, Dargaville TR, Ivanovski S, Hutmacher DW (2019) Degradation mechanisms of polycaprolactone in the context of chemistry, geometry and environment. Prog Polym Sci 96:1–20

Beeley NRF et al (2005) Fabrication, implantation, elution, and retrieval of a steroid-loaded polycaprolactone subretinal implant. J Biomed Mater Res A 73:437–444

Bhaw-Luximon A, Jhurry D, Motala-Timol S, Lochee Y (2005) Polymerization of e-Caprolactone and its copolymerization with g-Butyrolactone using metal complexes, In Macromolecular symposia 231:60–68

Bikiaris DN, Papageorgiou GZ, Achilias DS, Pavlidou E, Stergiou A (2007) Miscibility and enzymatic degradation studies of poly(ε-caprolactone)/poly(propylene succinate) blends. Eur Polym J 43:2491–2503
Biscaia S et al (2022) 3D printed poly(ε-caprolactone)/hydroxyapatite scaffolds for bone tissue engineering: a comparative study on a composite preparation by melt blending or solvent casting techniques and the influence of bioceramic content on scaffold properties. Int J Mol Sci 23:2318
Bordes C et al (2010) Determination of poly(ε-caprolactone) solubility parameters: application to solvent substitution in a microencapsulation process. Int J Pharm 383:236–243
Boucher DS (2020) Solubility parameters and solvent affinities for polycaprolactone: a comparison of methods. J Appl Polym Sci 137:48908
Braud C et al (1998) Capillary zone electrophoresis in normal or reverse polarity separation modes for the analysis of hydroxy acid oligomers in neutral phosphate buffer. J Chromatogr B 706
Budi HS et al (2022) Alendronate reinforced polycaprolactone-gelatin-graphene oxide: a promising nanofibrous scaffolds with controlled drug release. Mater Today Commun 32:104108
Buyuksungur S et al (2017) 3D printed poly(ε-caprolactone) scaffolds modified with hydroxyapatite and poly(propylene fumarate) and their effects on the healing of rabbit femur defects. Biomater Sci 5:2144–2158
Byun Y, Kim YT, Whiteside S (2010) Characterization of an antioxidant polylactic acid (PLA) film prepared with α-tocopherol, BHT and polyethylene glycol using film cast extruder. J Food Eng 100:239–244
Cakmak AM et al (2020) 3D Printed Polycaprolactone/Gelatin/Bacterial Cellulose/hydroxyapatite composite scaffold for bone tissue engineering. Polymers (Basel) 12:1962
Chakrapani VY, Kumar TSS, Raj, Deepa K, Kumary TV (2017) Electrospun cytocompatible polycaprolactone blend composite with enhanced wettability for bone tissue engineering. J Nanosci Nanotechnol 17:2320–2328
Chang H-I, Lau Y-C, Yan C, Coombes AGA (2008) Controlled release of an antibiotic, gentamicin sulphate, from gravity spun polycaprolactone fibers. J Biomed Mater Res A 84A:230–237
Chaparro FJ, Presley KF, Coutinho da Silva MA, Lannutti JJ (2019) Sintered electrospun polycaprolactone for controlled model drug delivery. Mater Sci Eng: C 99:112–120
Charuchinda A, Molloy R, Siripitayananon J, Molloy N, Sriyai M (2003) Factors influencing the small-scale melt spinning of poly(ε-caprolactone) monofilament fibres. Polym Int 52:1175–1181
Chen DR, Bei JZ, Wang SG (2000) Polycaprolactone microparticles and their biodegradation. Polym Degrad Stab 67:455–459
Chou P-Y et al (2021) Fabrication of drug-eluting nano-hydroxylapatite filled polycaprolactone nanocomposites using solution-extrusion 3D printing technique. Polym (Basel) 13:318
Clark RAF, Musillo M, Stransky T (2020) Wound repair: basic biology to tissue engineering. In: Principles of tissue engineering. Elsevier, pp 1309–1329
De Santis R et al (2013) Advanced composites for hard-tissue engineering based on PCL/organic–inorganic hybrid fillers: from the design of 2D substrates to 3D rapid prototyped scaffolds. Polym Compos 34:1413–1417
de Santis R et al (2015) Towards the design of 3D fiber-deposited Poly(-caprolactone)/Iron-doped hydroxyapatite nanocomposite magnetic scaffolds for bone regeneration. J Biomed Nanotechnol 11:1236–1246
de Siqueira L et al (2019) Influence of PLLA/PCL/HA scaffold fiber orientation on mechanical properties and osteoblast behavior. Materials 12:3879
Deng X, Yuan M, Xiong C, Li X (1999) Polymerization of Lactides and Lactones. IV. Ring-opening polymerization of-caprolactone by rare earth phenyl compounds. J Appl Polym Sci 73
Dhanka M, Shetty C, Srivastava R (2017) Injectable methotrexate loaded polycaprolactone microspheres: physicochemical characterization, biocompatibility, and hemocompatibility evaluation. Mater Sci Eng, C 81:542–550
Diba M, Kharaziha M, Fathi MH, Gholipourmalekabadi M, Samadikuchaksaraei A (2012) Preparation and characterization of polycaprolactone/forsterite nanocomposite porous scaffolds designed for bone tissue regeneration. Compos Sci Technol 72:716–723

Dong Z et al (2012) Reinforcement of electrospun membranes using nanoscale Al2O3 whiskers for improved tissue scaffolds. J Biomed Mater Res A 100A:903–910

Dong Q et al (2021) 3D-printed Mg-incorporated PCL-based scaffolds: a promising approach for bone healing. Mater Sci Eng, C 129:112372

Doostmohammadi M et al (2021) Bioactive anti-oxidative polycaprolactone/gelatin electrospun nanofibers containing selenium nanoparticles/vitamin E for wound dressing applications. J Biomater Appl 36:193–209

Dubois P, Deghe P, JBrbme R, Teyssi P (1992) Macromolecular Engineering of Polylactones and Polylactides. 8. Ring-Opening Polymerization of ECaprolactone Initiated by Primary Amines and Trialkylaluminum. Macromolecules 25

Dubois P, Jérôme R (1995) Recent advances in ring-opening polymerization of lactones and related compounds. J Macromol Sci, Part C 35:379–418

Duymaz BT et al (2019) 3D bio-printing of levan/polycaprolactone/gelatin blends for bone tissue engineering: characterization of the cellular behavior. Eur Polym J 119:426–437

El-Naggar ME et al (2021) Thallium/vanadate co-substitutions through hydroxyapatite/ polycaprolactone nanofibrous scaffolds for biomedical domains. Mater Chem Phys 271:124879

Eshraghi S, Das S (2010) Mechanical and microstructural properties of polycaprolactone scaffolds with one-dimensional, two-dimensional, and three-dimensional orthogonally oriented porous architectures produced by selective laser sintering. Acta Biomater 6:2467–2476

Feng S et al (2015) Study on the condensed state physics of poly(ε-caprolactone) nano-aggregates in aqueous dispersions. J Colloid Interface Sci 450:264–271

Flores-Cedillo ML et al (2016) Multiwall carbon nanotubes/polycaprolactone scaffolds seeded with human dental pulp stem cells for bone tissue regeneration. J Mater Sci Mater Med 27:35

Fu X, Sammons RL, Bertóti I, Jenkins MJ, Dong H (2012) Active screen plasma surface modification of polycaprolactone to improve cell attachment. J Biomed Mater Res B Appl Biomater 100B:314–320

Fu N et al (2016) PCL-PEG-PCL film promotes cartilage regeneration in vivo. Cell Prolif 49:729–739

Gautam S, Chou C-F, Dinda AK, Potdar PD, Mishra NC (2014) Surface modification of nanofibrous polycaprolactone/gelatin composite scaffold by collagen type I grafting for skin tissue engineering. Mater Sci Eng, C 34:402–409

Gautam S et al (2021) Gelatin-polycaprolactone-nanohydroxyapatite electrospun nanocomposite scaffold for bone tissue engineering. Mater Sci Eng, C 119:111588

Geng Y, Discher DE (2005) Hydrolytic degradation of Poly(ethylene oxide)- *block*- polycaprolactone worm micelles. J Am Chem Soc 127:12780–12781

Ghaee A, Bagheri-Khoulenjani S, Amir Afshar H, Bogheiri H (2019) Biomimetic nanocomposite scaffolds based on surface modified PCL-nanofibers containing curcumin embedded in chitosan/ gelatin for skin regeneration. Compos B Eng 177:107339

Ghazalian M, Afshar S, Rostami A, Rashedi S, Bahrami SH (2022) Fabrication and characterization of chitosan-polycaprolactone core-shell nanofibers containing tetracycline hydrochloride. Colloids Surf A Physicochem Eng Asp 636:128163

Gil-Castell O et al (2020) Polycaprolactone/gelatin-based scaffolds with tailored performance: in vitro and in vivo validation. Mater Sci Eng, C 107:110296

Golipour H, Ezzatzadeh E, Sadeghianmaryan A (2022) Investigation of co-electrospun gelatin: <scp> TiO_2 </scp> /polycaprolactone:silk fibroin scaffolds for wound healing applications. J Appl Polym Sci 139

Greenwood JE (2016) Hybrid biomaterials for skin tissue engineering. Ski Tissue Eng Regen Med 185–210

Guarino V et al (2008) Polylactic acid fibre-reinforced polycaprolactone scaffolds for bone tissue engineering. Biomaterials 29:3662–3670

Guarino V et al (2014) MgCHA particles dispersion in porous PCL scaffolds: *in vitro* mineralization and *in vivo* bone formation. J Tissue Eng Regen Med 8:291–303

Gunatillake PA, Adhikari R, Gadegaard N (2003) Biodegradable synthetic polymers for tissue engineering. Eur Cells Mater 5:1–16
Gupta AP, Kumar V (2007) New emerging trends in synthetic biodegradable polymers—Polylactide: a critique. Eur Polym J 43:4053–4074
Gupta D et al (2009) Aligned and random nanofibrous substrate for the in vitro culture of Schwann cells for neural tissue engineering. Acta Biomater 5:2560–2569
Hamlekhan A, Mozafari M, Nezafati N, Azami M, Hadipour H (2010) A proposed fabrication method of novel PCL-GEL-HAp nanocomposite scaffolds for bone tissue engineering applications. Adv Compos Lett 19:123–130
Hanks CT, Wataha JC, Sun Z (1996) In vitro models of biocompatibility: a review. Dent Mater 12:186–193
Hao D, Zhang G, Gong Y, Ma Z (2020) Development and biological evaluation of cerium oxide loaded polycaprolactone dressing on cutaneous wound healing in nursing care. Mater Lett 265:127401
He Y, Wildman RD, Tuck CJ, Christie SDR, Edmondson S (2016) An Investigation of the behavior of solvent based polycaprolactone ink for material jetting. Sci Rep 6:20852
Heidari M, Bahrami H, Ranjbar-Mohammadi M (2017) Fabrication, optimization and characterization of electrospun poly(caprolactone)/gelatin/graphene nanofibrous mats. Mater Sci Eng, C 78:218–229
Heimowska A, Morawska M, Bocho-Janiszewska A (2017) Biodegradation of poly(ϵ-caprolactone) in natural water environments. Pol J Chem Technol 19:120–126
Helder MN, Knippenberg M, Klein-Nulend J, Wuisman PIJM (2007) Stem cells from adipose tissue allow challenging new concepts for regenerative medicine. Tissue Eng 13:1799–1808
Hivechi A, Bahrami SH, Siegel RA (2019) Drug release and biodegradability of electrospun cellulose nanocrystal reinforced polycaprolactone. Mater Sci Eng, C 94:929–937
Hivechi A, Bahrami SH, Siegel RA, Milan PB, Amoupour M (2020) In vitro and in vivo studies of biaxially electrospun poly(caprolactone)/gelatin nanofibers, reinforced with cellulose nanocrystals, for wound healing applications. Cellulose 27:5179–5196
Hokmabad VR et al (2019) Effect of incorporating Elaeagnus angustifolia extract in PCL-PEG-PCL nanofibers for bone tissue engineering. Front Chem Sci Eng 13:108–119
Hoskins JN, Grayson SM (2009) Synthesis and degradation behavior of cyclic poly(ε-caprolactone). Macromolecules 42:6406–6413
Hsu K-H et al (2016) Hybrid electrospun polycaprolactone mats consisting of nanofibers and microbeads for extended release of dexamethasone. Pharm Res 33:1509–1516
Hwang TI, Kim JI, Joshi MK, Park CH, Kim CS (2019) Simultaneous regeneration of calcium lactate and cellulose into PCL nanofiber for biomedical application. Carbohydr Polym 212:21–29
Jaidev LR, Kumar S, Chatterjee K (2017) Multi-biofunctional polymer graphene composite for bone tissue regeneration that elutes copper ions to impart angiogenic, osteogenic and bactericidal properties. Colloids Surf B Biointerfaces 159:293–302
Javaid S et al (2021) Cefotaxime loaded polycaprolactone based polymeric nanoparticles with antifouling properties for In-Vitro drug release applications. Polym (Basel) 13:2180
Jin G et al (2013) Tissue engineered plant extracts as nanofibrous wound dressing. Biomaterials 34:724–734
Johari N, Fathi M, Fereshteh Z, Kargozar S, Samadikuchaksaraei A (2020) The electrospun poly(ε-caprolactone)/fluoridated hydroxyapatite nanocomposite for bone tissue engineering. Polym Adv Technol 31:1019–1026
Joseph B et al (2023) Processing and evaluation of the structure-properties of electrospun PCL/zirconium nanoparticle scaffolds. Mater Today Commun 34:104961
Joy A et al (2023) Design of biocompatible polycaprolactone-based nanocomposite loaded with graphene oxide/strontium nanohybrid for biomedical applications. Appl Nanosci 13:4471–4484
Kalantary S, Golbabaei F, Latifi M, Shokrgozar MA, Yaseri M (2020) Feasibility of using vitamin e-loaded poly(ε -caprolactone)/Gelatin Nanofibrous mat to prevent oxidative stress in skin. J Nanosci Nanotechnol 20:3554–3562

Kamath SM et al (2020) Fabrication of tri-layered electrospun polycaprolactone mats with improved sustained drug release profile. Sci Rep 10:18179

Karuppannan SK et al (2021) Copper oxide nanoparticles infused electrospun polycaprolactone/gelatin scaffold as an antibacterial wound dressing. Mater Lett 294:129787

Karuppuswamy P, Reddy Venugopal J, Navaneethan B, Luwang Laiva A, Ramakrishna S (2015) Polycaprolactone nanofibers for the controlled release of tetracycline hydrochloride. Mater Lett 141:180–186

Kashte S, Arbade G, Sharma RK, Kadam S (2019) Herbally painted biofunctional scaffolds with improved osteoinductivity for bone tissue engineering. J Biomim, Biomater Biomed Eng 41:49–68

Ke M et al (2016) Allantoin-loaded porous silica nanoparticles/polycaprolactone nanofiber composites: fabrication, characterization, and drug release properties. RSC Adv 6:4593–4600

Khalili M et al (2022) Preparation and characterization of bi-layered polycaprolactone/polyurethane nanofibrous scaffold loaded with titanium oxide and curcumin for wound dressing applications. Appl Phys A 128:497

Khandaker M et al (2022) Glutathione immobilized polycaprolactone nanofiber mesh as a dermal drug delivery mechanism for wound healing in a diabetic patient. Processes 10:512

Khorramnezhad M, AkbariB, Akbari M, Kharaziha M (2021) Effect of surface modification on physical and cellular properties of PCL thin film. Colloids Surf B Biointerfaces 200

Kim SE et al (2022) Evaluation of calcium phosphate–coated polycaprolactone/graphene oxide scaffold with macro- and microporous structure for bone tissue engineering. In Vitro Models 1:261–272

Kotela I, Podporska J, Soltysiak E, Konsztowicz KJ, Blazewicz M (2009) Polymer nanocomposites for bone tissue substitutes. Ceram Int 35:2475–2480

Kouser S et al (2021) Chitosan functionalized halloysite nanotube/poly (caprolactone) nanocomposites for wound healing application. Appl Surf Sci Adv 6:100158

Kricheldorf HR (2004) Biodegradable polymers with variable architectures via ring-expansion polymerization. J Polym Sci, Part A: Polym Chem 42:4723–4742

Kweon H (2003) A novel degradable polycaprolactone networks for tissue engineering. Biomaterials 24:801–808

Kzar HH et al (2023) The biomedical potential of polycaprolactone nanofibrous scaffold containing titanium oxide for wound healing applications. Int J Microstruct Mater Prop 16:278

Labet M, Thielemans W (2009) Synthesis of polycaprolactone: a review. Chem Soc Rev 38:3484–3504

Lam CX, Teoh SH, Hutmacher DW (2007) Comparison of the degradation of polycaprolactone and polycaprolactone–(β-tricalcium phosphate) scaffolds in alkaline medium. Polym Int 56:718–728

Lam CXF, Hutmacher DW, Schantz J-T, Woodruff MA, Teoh SH (2009) Evaluation of polycaprolactone scaffold degradation for 6 months *in vitro* and *in vivo*. J Biomed Mater Res A 90A:906–919

Lee J, Cuddihy MJ, Kotov NA (2008) Three-dimensional cell culture matrices: state of the art. Tissue Eng Part B Rev 14:61–86

Lee KH, Kim HY, Khil MS, Ra YM, Lee DR (2003) Characterization of nano-structured Poly(1-Caprolactone) nonwoven mats via electrospinning. polymer 44:1287–1294

Lei T et al (2020) Silicon-incorporated nanohydroxyapatite-reinforced poly(ε-caprolactone) film to enhance osteogenesis for bone tissue engineering applications. Colloids Surf B Biointerfaces 187:110714

Lei B, Gao X, Zhang R, Yi X, Zhou Q (2022) In situ magnesium phosphate/polycaprolactone 3D-printed scaffold induce bone regeneration in rabbit maxillofacial bone defect model. Mater des 215:110477

Li W-J, Tuli R, Huang X, Laquerriere P, Tuan RS (2005) Multilineage differentiation of human mesenchymal stem cells in a three-dimensional nanofibrous scaffold. Biomaterials 26:5158–5166

Li P et al (2021) Electrospun scaffold of collagen and polycaprolactone containing ZnO quantum dots for skin wound regeneration. J Bionic Eng 18:1378–1390

Liao N et al (2015) Electrospun bioactive poly (ε-caprolactone)–cellulose acetate–dextran antibacterial composite mats for wound dressing applications. Colloids Surf A Physicochem Eng Asp 469:194–201

Liberato MS et al (2016) Polycaprolactone fibers with self-assembled peptide micro/nanotubes: a practical route towards enhanced mechanical strength and drug delivery applications. J Mater Chem B 4:1405–1413

Lin B, Waymouth RM (2017) Urea anions: Simple, fast, and selective catalysts for ring-opening polymerizations. J Am Chem Soc 139:1645–1652

Liu L et al (2008) Surface modification of polycaprolactone membrane via layer-by-layer deposition for promoting blood compatibility. J Biomed Mater Res B Appl Biomater 87B:244–250

Liu X, Xu H, Zhang M, Yu D-G (2021) Electrospun medicated nanofibers for wound healing: review. Membranes (Basel) 11:770

Luong-Van E et al (2006) Controlled release of heparin from poly(ε-caprolactone) electrospun fibers. Biomaterials 27:2042–2050

Mahapatro A, Kumar A, Gross RA (2004) Mild, solvent-free ω-hydroxy acid polycondensations catalyzed by Candida antarctica lipase B. Biomacromol 5:62–68

Maleki-Ghaleh H et al (2021) Effect of zinc-doped hydroxyapatite/graphene nanocomposite on the physicochemical properties and osteogenesis differentiation of 3D-printed polycaprolactone scaffolds for bone tissue engineering. Chem Eng J 426:131321

Manoukian OS et al (2018) Biodegradable polymeric injectable implants for long-term delivery of contraceptive drugs. J Appl Polym Sci 135

Marins NH et al (2019) Niobium pentoxide and hydroxyapatite particle loaded electrospun polycaprolactone/gelatin membranes for bone tissue engineering. Colloids Surf B Biointerfaces 182:110386

Martínez-Abad A, Sánchez G, Fuster V, Lagaron JM, Ocio MJ (2013) Antibacterial performance of solvent cast polycaprolactone (PCL) films containing essential oils. Food Control 34:214–220

Mattioli-Belmonte M et al (2012) Tuning polycaprolactone–carbon nanotube composites for bone tissue engineering scaffolds. Mater Sci Eng C 32:152–159

Mecerreyes D, Jerome R, Dubois P (1999) Novel macromolecular architectures based on aliphatic polyesters: relevance of the 'Coordination-Insertion' ring opening polymerization, In: Hilborn, J.G., et al. Macromolecular Architectures. Advances in Polymer Science, vol 147. Springer, Berlin, Heidelberg

Meek MF, Den Dunnen W, Schakenraad JM, Robinson PH (1996) Evaluation of functional nerve recovery after reconstruction with a Poly(DL-Lactide-ε-Caprolactone) nerve guide, filled with modified denatured muscle tissue

Menazea AA, Abdelbadie SA, Ahmed MK (2020) Manipulation of AgNPs coated on selenium/carbonated hydroxyapatite/ε-polycaprolactone nano-fibrous via pulsed laser deposition for wound healing applications. Appl Surf Sci 508:145299

Michael FM, Khalid M, Walvekar R, Siddiqui H, Balaji AB (2018) Surface modification techniques of biodegradable and biocompatible polymers. In: Biodegradable and biocompatible polymer composites. Elsevier, pp 33–54

Middleton JC, Tipton AJ (2000) Synthetic biodegradable polymers as orthopedic devices. Biomaterials 21:2335–2346

Miguel S, Ribeiro M, Coutinho P, Correia I (2017) Electrospun polycaprolactone/aloe vera_chitosan nanofibrous asymmetric membranes aimed for wound healing applications. Polymers (Basel) 9:183

Moazzami Goudarzi Z, Behzad T, Ghasemi-Mobarakeh L, Kharaziha M, Enayati MS (2020) Structural and mechanical properties of fibrous poly (caprolactone)/gelatin nanocomposite incorporated with cellulose nanofibers. Polym Bull 77:717–740

Moeini S, Mohammadi MR, Simchi A (2017) In-situ solvothermal processing of polycaprolactone/hydroxyapatite nanocomposites with enhanced mechanical and biological performance for bone tissue engineering. Bioact Mater 2:146–155

Mohamadi P et al (2022) Electrospinning of poly(Caprolactone)/gelatin/clindamycin nanocomposites as an antibacterial wound dressing. Mater Sci Forum 1063:71–81

Mondal D, Griffith M, Venkatraman SS (2016b) Polycaprolactone-based biomaterials for tissue engineering and drug delivery: current scenario and challenges. Int J Polym Mater Polym Biomater 65:255–265

Mondal D, Griffith M, Venkatraman SS (2016) Polycaprolactone-based biomaterials for tissue engineering and drug delivery: current scenario and challenges. Int J Polym Mater Polym Biomater 65:255–265

Motealleh B et al (2014) Morphology, drug release, antibacterial, cell proliferation, and histology studies of chamomile-loaded wound dressing mats based on electrospun nanofibrous poly(ε-caprolactone)/polystyrene blends. J Biomed Mater Res B Appl Biomater 102:977–987

Nair LS, Laurencin CT (2007) Biodegradable polymers as biomaterials. Prog Polym Sci (Oxf) 32:762–798

Nawaz T et al (2021) Development and optimization of acriflavine-loaded polycaprolactone nanoparticles using box–behnken design for burn wound healing applications. Polymers (Basel) 14:101

Neděla O, Slepička P, Švorčík V (2017) Surface modification of polymer substrates for biomedical applications. Materials 10:1115

Nejaddehbashi F et al (2020) Incorporation of silver sulfadiazine into an electrospun composite of polycaprolactone as an antibacterial scaffold for wound healing in rats. Cell J 21:379–390

Niezabitowska E et al (2018) Facile production of nanocomposites of carbon nanotubes and polycaprolactone with high aspect ratios with potential applications in drug delivery. RSC Adv 8:16444–16454

Nitya G, Nair GT, Mony U, Chennazhi KP, Nair SV (2012) In vitro evaluation of electrospun PCL/nanoclay composite scaffold for bone tissue engineering. J Mater Sci Mater Med 23:1749–1761

Noormohammadi F, Nourany M, Mir Mohamad Sadeghi G, Wang P-Y, Shahsavarani H (2021) The role of cellulose nanowhiskers in controlling phase segregation, crystallization and thermal stimuli responsiveness in PCL-PEGx-PCL block copolymer-based PU for human tissue engineering applications. Carbohydr Polym 252:117219

Sriputtha N, Wiwatwongwana F, Promma N (2022) Evaluation of polycaprolactone/carboxymethylcellulose scaffolds using hyperelastic model. Int J Adv Technol Eng Explor 9

Oh G-W et al (2016) Fabrication, characterization and determination of biological activities of poly(ε-caprolactone)/chitosan-caffeic acid composite fibrous mat for wound dressing application. Int J Biol Macromol 93:1549–1558

Okada M (2002) Chemical Syntheses of Biodegradable Polymers. Prog Polym Sci 27:87–133

Okamoto M, John B (2013) Synthetic biopolymer nanocomposites for tissue engineering scaffolds. Prog Polym Sci 38:1487–1503

Ozkan S, Kalyon DM, Yu X (2009) Functionally graded β-TCP/PCL nanocomposite scaffolds: *In vitro* evaluation with human fetal osteoblast cells for bone tissue engineering. J Biomed Mater Res A 9999A

Pan L, Pei X, He R, Wan Q, Wang J (2012) Multiwall carbon nanotubes/polycaprolactone composites for bone tissue engineering application. Colloids Surf B Biointerfaces 93:226–234

Pang Z, Zhou J, Sun C (2020) Ditelluride-bridged PEG-PCL copolymer as folic acid-targeted and redox-responsive nanoparticles for enhanced cancer therapy. Front Chem 8:156

Panyam J, Labhasetwar V (2003) Biodegradable nanoparticles for drug and gene delivery to cells and tissue. Adv Drug Deliv Rev 55:329–347

Park SA et al (2021) Three-dimensionally printed polycaprolactone/beta-tricalcium phosphate scaffold was more effective as an <scp>rhBMP</scp> -2 carrier for new bone formation than polycaprolactone alone. J Biomed Mater Res A 109:840–848

Pati F et al (2015) Ornamenting 3D printed scaffolds with cell-laid extracellular matrix for bone tissue regeneration. Biomaterials 37:230–241

Pavliňáková V, Fohlerová Z, Pavliňák D, Khunová V, Vojtová L (2018) Effect of halloysite nanotube structure on physical, chemical, structural and biological properties of elastic polycaprolactone/gelatin nanofibers for wound healing applications. Mater Sci Eng, C 91:94–102

Pazarçeviren E, Erdemli Ö, Keskin D, Tezcaner A (2017) Clinoptilolite/PCL–PEG–PCL composite scaffolds for bone tissue engineering applications. J Biomater Appl 31:1148–1168

Pedrosa MCG et al (2021) Structure and biological compatibility of polycaprolactone/zinc-hydroxyapatite electrospun nanofibers for tissue regeneration. J Bioact Compat Polym 36:314–333

Perera WPTD et al (2022) Albumin grafted coaxial electrosparyed polycaprolactone-zinc oxide nanoparticle for sustained release and activity enhanced antibacterial drug delivery. RSC Adv 12:1718–1727

Perumal G, Sivakumar PM, Nandkumar AM, Doble M (2020) Synthesis of magnesium phosphate nanoflakes and its PCL composite electrospun nanofiber scaffolds for bone tissue regeneration. Mater Sci Eng, C 109:110527

Phipps MC et al (2011) Mesenchymal stem cell responses to bone-mimetic electrospun matrices composed of polycaprolactone, collagen I and nanoparticulate hydroxyapatite. PLoS ONE 6:e16813

Pires F et al (2019) Polycaprolactone/gelatin nanofiber membranes containing EGCG-loaded liposomes and their potential use for skin regeneration. ACS Appl Bio Mater 2:4790–4800

Platel RH, Hodgson LM, Williams CK (2008) Biocompatible initiators for lactide polymerization. Polym Rev 48:11–63

Polymer Data Handbook (2009), 2nd ed. J Am Chem Soc 131:16330–16330

Poornima B, Korrapati PS (2017) Fabrication of chitosan-polycaprolactone composite nanofibrous scaffold for simultaneous delivery of ferulic acid and resveratrol. Carbohydr Polym 157:1741–1749

Qin X et al (2022) Surface modification of polycaprolactone scaffold with improved biocompatibility and controlled growth factor release for enhanced stem cell differentiation. Front Bioeng Biotechnol 9

Radhakrishnan S et al (2021) Fabrication of 3D printed antimicrobial polycaprolactone scaffolds for tissue engineering applications. Mater Sci Eng, C 118:111525

Radisavljevic A et al (2018) Cefazolin-loaded polycaprolactone fibers produced via different electrospinning methods: characterization, drug release and antibacterial effect. Eur J Pharm Sci 124:26–36

Raina N, Pahwa R, Khosla JK, Gupta PN, Gupta M (2022) Polycaprolactone-based materials in wound healing applications. Polym Bull 79:7041–7063

Ramírez-Agudelo R et al (2018) Hybrid nanofibers based on poly-caprolactone/gelatin/hydroxyapatite nanoparticles-loaded doxycycline: effective anti-tumoral and antibacterial activity. Mater Sci Eng, C 83:25–34

Rather HA et al (2018) Antioxidative study of cerium oxide nanoparticle functionalised PCL-Gelatin electrospun fibers for wound healing application. Bioact Mater 3:201–211

Rather HA, Patel R, Yadav UCS, Vasita R (2020) Dual drug-delivering polycaprolactone-collagen scaffold to induce early osteogenic differentiation and coupled angiogenesis. Biomed Mater 15:045008

Rathinavel S, Korrapati PS, Kalaiselvi P, Dharmalingam S (2021) Mesoporous silica incorporated PCL/Curcumin nanofiber for wound healing application. Eur J Pharm Sci 167:106021

Ravichandran S, Radhakrishnan J (2022) Anticancer efficacy of lupeol incorporated electrospun Polycaprolactone/gelatin nanocomposite nanofibrous mats. Nanotechnology 33:295104

Repanas A, Glasmacher B (2015) Dipyridamole embedded in Polycaprolactone fibers prepared by coaxial electrospinning as a novel drug delivery system. J Drug Deliv Sci Technol 29:132–142

Rezaei H et al (2023) Fabrication and characterization of three-dimensional polycaprolactone/ sodium alginate and egg whites and eggshells hybrid scaffold in bone tissue engineering. J Polym Eng 43:47–52
Rhim JW, Park HM, Ha CS (2013) Bio-nanocomposites for food packaging applications. Prog Polym Sci 38:1629–1652
Ronca D et al (2016) Bone tissue engineering: 3D PCL-based nanocomposite scaffolds with tailored properties. Procedia CIRP 49:51–54
Sadeghianmaryan A et al (2020) Electrospinning of scaffolds from the polycaprolactone/ polyurethane composite with graphene oxide for skin tissue engineering. Appl Biochem Biotechnol 191:567–578
Saegusa T, Kobayashi S, Hayashi K (1978) Polymerization via Zwitterion. 17.1 Alternating Copolymerization of 2-Phenylimino-1,3-Dioxolane with P-Propiolactone, Macromolecules, 11:360–361
Safiaghdam H et al (2023) 3D-printed MgO nanoparticle loaded polycaprolactone β -tricalcium phosphate composite scaffold for bone tissue engineering applications: In-vitro and in-vivo evaluation. J Biomed Mater Res 111:322–339
Sagart A, Jahandideh A, Asghari A, Akbarzadeh A, Mortazavi P (2024) The Comparative effects of platelet-rich plasma and polycaprolactone-hydroxyapatite zeolite nanocomposites on wound healing after tooth extraction. Crescent J Med Biol Sci 11:183–189
Sahoo S, Sasmal A, Sahoo D, Nayak P (2010) Synthesis and characterization of chitosan-polycaprolactone blended with organoclay for control release of doxycycline. J Appl Polym Sci 118:3167–3175
Salmoria GV, Sibilia F, Henschel VG, Fare S, Tanzi MC (2017) Structure and properties of poly-caprolactone/ibuprofen rods prepared by melt extrusion for implantable drug delivery. Polym Bull 74:4973–4987
Samadian H, Khastar H, Ehterami A, Salehi M (2021) Bioengineered 3D nanocomposite based on gold nanoparticles and gelatin nanofibers for bone regeneration: in vitro and in vivo study. Sci Rep 11:13877
Samiei M et al (2016) Osteogenic/Odontogenic bioengineering with co-administration of simvas-tatin and hydroxyapatite on poly caprolactone based nanofibrous scaffold. Adv Pharm Bull 6:353–365
Sattary M, Rafienia M, Kazemi M, Salehi H, Mahmoudzadeh M (2019) Promoting effect of nano hydroxyapatite and vitamin D3 on the osteogenic differentiation of human adipose-derived stem cells in polycaprolactone/gelatin scaffold for bone tissue engineering. Mater Sci Eng, C 97:141–155
Sedghi R, Shaabani A, Sayyari N (2020) Electrospun triazole-based chitosan nanofibers as a novel scaffolds for bone tissue repair and regeneration. Carbohydr Polym 230:115707
Setia Budi H et al (2022) Preparation of antibacterial Gel/PCL nanofibers reinforced by dicalcium phosphate-modified graphene oxide with control release of clindamycin for possible application in bone tissue engineering. Inorg Chem Commun 139:109336
Sharifi F et al (2018) Polycaprolactone/carboxymethyl chitosan nanofibrous scaffolds for bone tissue engineering application. Int J Biol Macromol 115:243–248
Shen J et al (2019) 3D-printed nanocomposite scaffolds with tunable magnesium ionic microenvi-ronment induce in situ bone tissue regeneration. Appl Mater Today 16:493–507
Shim J-H, Lee J-S, Kim JY, Cho D-W (2012) Bioprinting of a mechanically enhanced three-dimensional dual cell-laden construct for osteochondral tissue engineering using a multi-head tissue/organ building system. J Micromech Microeng 22:085014
Shirani K, Nourbakhsh MS, Rafienia M (2019) Electrospun polycaprolactone/gelatin/bioactive glass nanoscaffold for bone tissue engineering. Int J Polym Mater Polym Biomater 68:607–615
Shirzaei Sani I, Rezaei M, Baradar Khoshfetrat A, Razzaghi D (2021) Preparation and characteri-zation of polycaprolactone/chitosan-g-polycaprolactone/hydroxyapatite electrospun nanocom-posite scaffolds for bone tissue engineering. Int J Biol Macromol 182:1638–1649

Shitole AA et al (2019) Electrospun polycaprolactone/hydroxyapatite/ZnO nanofibers as potential biomaterials for bone tissue regeneration. J Mater Sci Mater Med 30:51
Simões MCR, Cragg SM, Barbu E, de Sousa FB (2019) The potential of electrospun poly(methyl methacrylate)/polycaprolactone core–sheath fibers for drug delivery applications. J Mater Sci 54:5712–5725
Sinha VR, Bansal K, Kaushik R, Kumria R, Trehan A (2004) Poly-ε-caprolactone microspheres and nanospheres: an overview. Int J Pharm 278:1–23
Sohrabi M, Abbasi M, Ansar MM, Soltani Tehrani B (2022) Evaluation of electrospun nanofibers fabricated using PCL/PVP and PVA/β-TCP as potential scaffolds for bone tissue engineering. Polymer Bulletin 79:8397–8413
Sousa AC et al (2022) Assessment of 3D-printed polycaprolactone, hydroxyapatite nanoparticles and diacrylate poly(ethylene glycol) scaffolds for bone regeneration. Pharmaceutics 14:2643
Stridsberg KM, Ryner M, Albertsson A-C (2002) Controlled ring-opening polymerization: polymers with designed macromolecular architecture. In: Degradable Aliphatic Polyesters. Advances in Polymer Science, vol 157. Springer, Berlin, Heidelberg
Subramanian S et al (2021) Effect of gamma sterilization on Gymnema sylvestre leaf extract fused Polycaprolactone nanofiber for effective wound dressing applications. Mater Lett 300:130145
Supp DM, Boyce ST (2005) Engineered skin substitutes: practices and potentials. Clin Dermatol 23:403–412
Tamboli V, Mishra GP, Mitra AK (2013) Novel pentablock copolymer (PLA–PCL–PEG–PCL–PLA)-based nanoparticles for controlled drug delivery: effect of copolymer compositions on the crystallinity of copolymers and in vitro drug release profile from nanoparticles. Colloid Polym Sci 291:1235–1245
Tang ZG et al (2004) Surface properties and biocompatibility of solvent-cast poly [ε-caprolactone] films. Biomaterials 25:4741–4748
Tarafder S, Nansen K, Bose S (2013) Lovastatin release from polycaprolactone coated β-tricalcium phosphate: effects of pH, concentration and drug–polymer interactions. Mater Sci Eng, C 33:3121–3128
Tavakoli M (2005) Surface modification of polymers to enhance biocompatibility. Surf Interfaces Biomater 719–744
Thijssen Q et al (2022) Tough photo-cross-linked PCL-hydroxyapatite composites for bone tissue engineering. Biomacromol 23:1366–1375
Thomas V et al (2006a) Mechano-morphological studies of aligned nanofibrous scaffolds of polycaprolactone fabricated by electrospinning. J Biomater Sci Polym Ed 17:969–984
Thomas V et al (2006b) Electrospun bioactive nanocomposite scaffolds of polycaprolactone and nanohydroxyapatite for bone tissue engineering. J Nanosci Nanotechnol 6:487–493
Tiaw KS et al (2005) Laser surface modification of poly(ε-caprolactone) (PCL) membrane for tissue engineering applications. Biomaterials 26:763–769
Tohidlou H, Shafiei SS, Abbasi S, Asadi-Eydivand M, Fathi-Roudsari M (2019) Amine-functionalized single-walled carbon nanotube/polycaprolactone electrospun scaffold for bone tissue engineering: in vitro study. Fibers and Polymers 20:1869–1882
Tong SY, Wang Z, Lim PN, Wang W, Thian ES (2017) Uniformly-dispersed nanohydroxapatite-reinforced poly(ε-caprolactone) composite films for tendon tissue engineering application. Mater Sci Eng, C 70:1149–1155
Tra Thanh N et al (2018) Optimization and characterization of electrospun polycaprolactone coated with gelatin-silver nanoparticles for wound healing application. Mater Sci Eng: C 91:318–329
Ulery BD, Nair LS, Laurencin CT (2011) Biomedical applications of biodegradable polymers. J Polym Sci B Polym Phys 49:832–864
Vieira MGA, da Silva MA, dos Santos LO, Beppu MM (2011) Natural-based plasticizers and biopolymer films: a review. Eur Polym J 47:254–263
Wang L et al. Biodegradable Poly-Epsilon-Caprolactone (PCL) for tissue engineering applications: a review. Rev Adv Mater Sci 34

Wang Y, Shi D (2022) In vitro and in vivo evaluations of nanofibrous nanocomposite based on carboxymethyl cellulose/polycaprolactone/cobalt-doped hydroxyapatite as the wound dressing materials. Arab J Chem 15:104270

Wang B, Zheng H, Chang M-W, Ahmad Z, Li J-S (2016) Hollow polycaprolactone composite fibers for controlled magnetic responsive antifungal drug release. Colloids Surf B Biointerfaces 145:757–767

Wang H, Wu J, Zheng L, Cheng X (2019) Preparation and properties of ZnAl layered double hydroxide/polycaprolactone nanocomposites for use in drug delivery. Polym-Plast Technol Mater 58:1027–1035

Wang S et al (2022) 3D-Printed PCL/Zn scaffolds for bone regeneration with a dose-dependent effect on osteogenesis and osteoclastogenesis. Mater Today Bio 13:100202

Williams DF (2008) On the mechanisms of biocompatibility. Biomaterials 29:2941–2953

Wong BS, Teoh S-H, Kang L (2012) Polycaprolactone scaffold as targeted drug delivery system and cell attachment scaffold for postsurgical care of limb salvage. Drug Deliv Transl Res 2:272–283

Woodruff MA, Hutmacher DW (2010) The return of a forgotten polymer—Polycaprolactone in the 21st century. Prog Polym Sci (Oxf) 35:1217–1256

Woodward SC, BrewerPS, Moatamed F, Schindler A, Pitt CG (1985) The Intracellular Degradation of Poly(E-Caprolactone)

Wsoo MA et al (2021) Vitamin D3-loaded electrospun cellulose acetate/polycaprolactone nanofibers: characterization, in-vitro drug release and cytotoxicity studies. Int J Biol Macromol 181:82–98

Wu H et al (2023) Dynamic degradation patterns of porous polycaprolactone/β-tricalcium phosphate composites orchestrate macrophage responses and immunoregulatory bone regeneration. Bioact Mater 21:595–611

Xiao G (1995) Solvent induced changes on corona discharge-treated polyolefin surfaces probed by contact angle measurement. J Colloid Interface Sci 171(1):200–204

Yang F, Wolke JGC, Jansen JA (2008) Biomimetic calcium phosphate coating on electrospun poly(ε-caprolactone) scaffolds for bone tissue engineering. Chem Eng J 137:154–161

Yang H et al (2019) In vitro evaluation of a novel multiwalled carbon nanotube/nanohydroxyapatite/ polycaprolactone composite for bone tissue engineering. J Mater Res 34:532–544

Yang S et al (2020) Multifunctional chitosan/polycaprolactone nanofiber scaffolds with varied dual-drug release for wound-healing applications. ACS Biomater Sci Eng 6:4666–4676

Yang Y et al (2022) 3D-printed polycaprolactone-chitosan based drug delivery implants for personalized administration. Mater des 214:110394

Yedekçi B, Tezcaner A, Yılmaz B, Demir T, Evis Z (2022) 3D porous PCL-PEG-PCL / strontium, magnesium and boron multi-doped hydroxyapatite composite scaffolds for bone tissue engineering. J Mech Behav Biomed Mater 125:104941

Yin J, Xu L (2020) Batch preparation of electrospun polycaprolactone/chitosan/aloe vera blended nanofiber membranes for novel wound dressing. Int J Biol Macromol 160:352–363

Yoshimoto H, Shin YM, Terai H, Vacanti JP (2003) A biodegradable nanofiber scaffold by electrospinning and its potential for bone tissue engineering. Biomaterials 24:2077–2082

Yuan M, Xiong C, Deng X (1998) Ring-opening polymerization of 1-caprolactone initiated by cyclopentadienyl sodium. J Appl Polym Sci 67

Zeng Q et al (2011) Skin tissue engineering. In: Comprehensive biomaterials. Elsevier, pp467–499

Zhang H et al (2017) Magnetic nanoparticle-loaded electrospun polymeric nanofibers for tissue engineering. Mater Sci Eng, C 73:537–543

Zhang et al (2013) Engineering of epidermis skin grafts using electrospun nanofibrous gelatin/ polycaprolactone membranes. Int J Nanomedicine 2077

Zheng R et al (2014) The influence of Gelatin/PCL ratio and 3-D construct shape of electrospun membranes on cartilage regeneration. Biomaterials 35:152–164

Chapter 6
Stimulus-Responsive Biocompatible Nanocomposites in Medicine and 4D Printing

Hossein Rouhbakhsh and Roya sedghi

Abstract Small molecule drugs often exhibit poor solubility, inefficient cellular uptake, and rapid clearance from the body, limiting their therapeutic efficacy. Additionally, these drugs typically lack the ability to selectively target diseased tissues, leading to undesirable side effects. Stimuli-responsive nanocarriers, particularly those based on biodegradable and biocompatible biopolymers, offer a promising solution to overcome these challenges in drug delivery. Biopolymer-based nanocomposites can be engineered to exhibit multi-stimuli responsiveness and tunable properties, enabling prolonged circulation, targeted accumulation, enhanced penetration, and controlled drug release at the site of disease. While endogenous stimuli, such as changes in pH, redox potential, or enzyme activity, provide some benefits, the variability between patients and limited tissue penetration can be limiting. Leveraging the combined effects of exogenous and endogenous stimuli through multi-stimuli-responsive carriers presents a more effective approach for precise drug delivery and improved therapeutic outcomes. This can be achieved through the design of biopolymer-based nanocomposites that respond to external triggers, such as light, temperature, or magnetic fields, in addition to internal physiological cues. Furthermore, the integration of stimuli-responsive biopolymer-based nanocomposites with 4D printing technology opens up new possibilities for advanced medical devices. 4D printing refers to the fabrication of smart materials that can change shape, properties, or function over time in response to environmental stimuli. By incorporating stimuli-responsive biopolymer-based nanocomposites into 4D-printed medical devices, it becomes possible to create customized, dynamic, and adaptive structures that can interact with the body and deliver therapeutics in a spatiotemporally controlled manner. This approach has the potential to revolutionize the field of personalized medicine and targeted drug delivery. This chapter provides a comprehensive overview of the use of stimuli-responsive biopolymer-based nanocomposites in medicine and

H. Rouhbakhsh · R. sedghi (✉)
Department of Polymer and Materials Chemistry, Faculty of Chemistry and Petroleum Sciences, Shahid Beheshti University, G.C, 1983969411 Tehran, Iran
e-mail: r_sedghi@sbu.ac.ir

H. Rouhbakhsh
e-mail: h_rouhbakhsh@sbu.ac.ir

R. Balakrishnan et al. (eds.), *Biocompatible Nanocomposites*, Advanced Structured Materials 239, https://doi.org/10.1007/978-981-96-7450-3_6

their potential applications in areas such as 4D printing for advanced medical devices. The discussion includes the design principles, fabrication methods, and specific examples of how these smart materials can be leveraged to improve drug delivery and therapeutic outcomes.

Keywords Biocompatible · Nanoparticles · 4D printing · Quantum dots · Stimuli · Medical devices

6.1 Introduction

Many small molecule drugs exhibit poor solubility in bodily fluids, which limits their absorption in the gastrointestinal tract. Solubility is a critical factor because, for a drug to be absorbed effectively, it must first dissolve in the aqueous environment of the gastrointestinal tract. Drugs with low solubility often fail to reach a sufficiently high concentration in the bloodstream to produce a therapeutic effect. The mechanisms behind poor solubility are primarily governed by the drug's lipophilicity (tendency to dissolve in fats or oils rather than water) and crystallinity (whether the drug forms solid crystalline structures that are difficult to dissolve). The Nernst–Brunner equation and Higuchi's equation describe how factors such as particle size, polymorphism, and formulation (e.g., using solubilizing agents, nanoparticles, or liposomes) can influence solubility. Additionally, salt formation, amorphous formulations, and the use of solid dispersion techniques can be used to enhance the solubility of poorly soluble compounds by improving their dissolution rate. Once the drug is in the bloodstream, it must cross cellular membranes to reach its target site. This process is facilitated by several mechanisms, such as passive diffusion, active transport, and endocytosis. Small molecules often struggle to efficiently cross cell membranes due to their lipophilicity, molecular size, or the presence of efflux transporters (e.g., P-glycoprotein) that pump the drug out of the cell before it can exert its effect the lipid bilayer of the cell membrane acts as a barrier, and drugs with high hydrophobicity tend to partition into the membrane, potentially limiting their ability to pass into the cytoplasm. To improve cellular uptake, modifications can be made to the drug's molecular structure to enhance its hydrophilicity or to exploit specific receptor-mediated transport mechanisms, where the drug is actively transported into cells through specific protein channels or endocytosis. Nanoparticle-based delivery systems (e.g., liposomes or micelles) and prodrug strategies can also be used to enhance cellular uptake by overcoming these barriers. Moreover, drugs that are substrates for efflux transporters (e.g., ABCB1/P-glycoprotein) can be actively pumped out of cells, reducing their intracellular concentration and therapeutic efficacy. In such cases, strategies that inhibit these transporters or design drugs to avoid recognition by them are being explored. Even when drugs succeed in entering the bloodstream and reaching target cells, they often face rapid clearance from the body, leading to low bioavailability. Bioavailability refers to the fraction of the drug that reaches systemic circulation in an active form after administration. Factors contributing to rapid drug clearance include metabolism in the liver (primarily via cytochrome P450 enzymes), renal excretion by the kidneys, and biliary secretion through the liver and intestines The kidneys are a major route for the elimination of small molecules, especially water-soluble drugs, via glomerular filtration and tubular secretion. Drugs with low molecular weight and high water solubility are more likely to be rapidly eliminated through renal excretion. Drugs are typically metabolized in the liver by enzymes like CYP450 enzymes, which can either deactivate the drug or convert it into more hydrophilic metabolites, making it easier for the body to eliminate the compound To overcome these challenges, strategies such as drug modifications to evade rapid clearance, extended-release formulations, and the use of biodegradable drug delivery systems have been proposed. These strategies can slow down metabolism and renal clearance, allowing the drug to remain in circulation for a longer period, increasing its bioavailability and therapeutic effect. PEGylation (attachment of polyethylene glycol chains) is one example of a technique that can reduce renal clearance and improve the pharmacokinetics of biologic drugs (Blum et al. 2015; Vlieghe et al. 2010). Perhaps most significantly, these drugs often lack the ability to target specific areas of disease. This indiscriminate targeting can lead to severe side effects, as exemplified by the cytotoxicity (cell killing) effects of many common anticancer drugs that damage healthy tissues alongside cancerous ones. In living organisms, intricate processes rely on a series of signals that trigger specific responses at different scales and times. Nature often achieves this by directly manipulating materials with stimuli, like activating an enzyme through a trigger molecule. Similarly, effective drug delivery systems need to selectively target specific tissues and molecules at the right time, while avoiding side effects and the body's natural defenses (Blum et al. 2015; Widakowich et al. 2007). While both synthetic and biopolymers are explored for stimuli-responsive nanocarriers, biopolymers are gaining significant traction due to their inherent biodegradability, biocompatibility, natural abundance, and unique chemical structures that offer distinct advantages (Xiao et al. 2023a; Sedighi et al. 2023). Biopolymer-based nanocomposites hold particular promise for overcoming challenges in drug delivery, such as poor water solubility, imprecise targeting, uneven distribution, and high toxicity, thereby offering exciting possibilities for effective treatment of a wider range of disease (Santos et al. 2021; Parhi 2020) (Fig. 6.1). The diverse microenvironments of diseases necessitate the development of biopolymer-based nanocomposites with multi-stimuli responsiveness and tunable structures (size, surface charge, shape, stability). These advanced carriers can achieve prolonged circulation time, targeted accumulation at the diseased site, enhanced penetration, and efficient cellular uptake, ultimately leading to controlled drug release (Parhi 2020; Xiao et al. 2023b). While endogenous stimuli for drug delivery offer some promise, their variability between patients and limited tissue penetration can hinder effectiveness. Exogenous stimuli, although offering more control, require expensive equipment. A more promising approach utilizes multi-stimuli-responsive carriers that leverage the combined effects of different stimuli for precise drug release at the target site (Arjama et al. 2022; Jia et al. 2021).

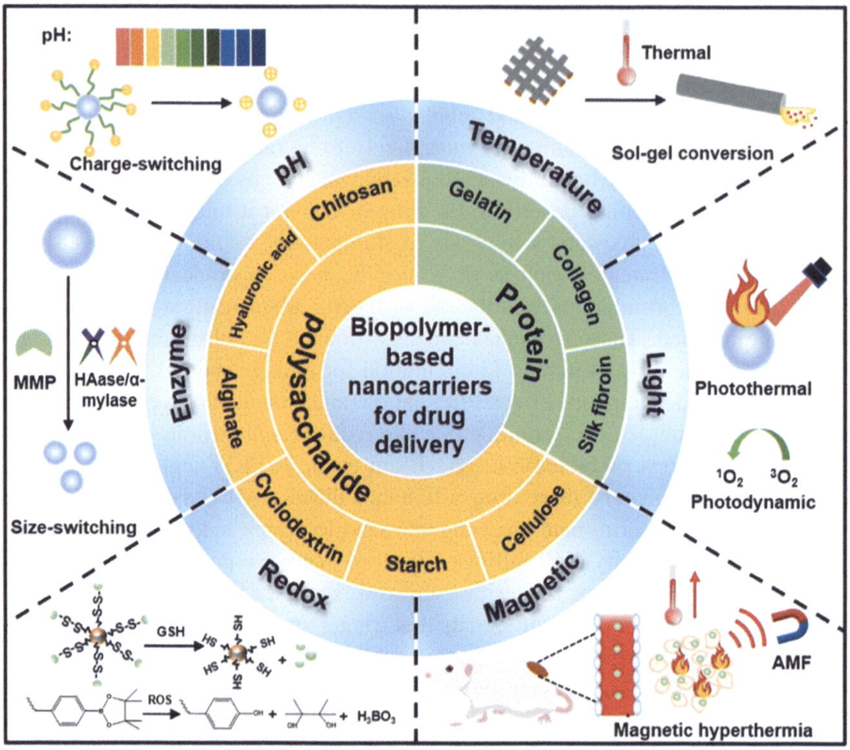

Fig. 6.1 Classification of biopolymer-based nanocarriers according to their stimulus-responsive behavior (Xiao et al. 2023a). Copy Right 2023 Elsevier

6.2 Nanomaterials Used in Medicine

6.2.1 Organic Nanoparticles

6.2.1.1 Polymer

Polymeric nanoparticles (NPs) are an effective platform for drug delivery, owing to their adaptability and compatibility with biological systems. These microscopic carriers can be crafted from various natural or synthetic materials, offering a wide range of possible structures and functionalities. Scientists can precisely control multiple features of these NPs during the formulation process, making them highly customizable for specific drug delivery needs. Polymeric NPs are typically synthesized using techniques like emulsification or nanoprecipitation, and the chosen method influences the final characteristics of the nanoparticles. Furthermore, these NPs offer various strategies for incorporating therapeutic cargo. Drugs can be incorporated into the core, dispersed throughout the polymeric matrix, covalently bonded to the polymer, or adsorbed onto the surface of the nanoparticles. This versatility in design and drug loading strategies enhances the potential of polymeric NPs for targeted drug delivery applications (Khanna et al. 2023; Mitchell et al. 2021). Polymeric NPs excel in delivering a wide range of drugs, from small molecules to proteins and vaccines, due to their tunable properties and ability to encapsulate both hydrophobic and hydrophilic cargo. This allows for co-delivery of multiple drugs and precise control over release for optimized therapeutic effect (Ref. 30–38). Polymeric nanoparticles come in various shapes, including nanocapsules (hollow spheres) and nanospheres (solid particles). Further distinctions include polymersomes, micelles, and dendrimers (Mitchell et al. 2021; Zhang et al. 2020a).

6.2.1.2 Polymersomes

Polymersomes, in particular, are artificial vesicles with enhanced stability and cargo retention compared to liposomes. This makes them efficient carriers for delivering drugs directly into cells (Khanna et al. 2023; Mitchell et al. 2021). Common building blocks for polymeric nanoparticles include copolymers like poly(ethylene glycol) (PEG) and poly(dimethylsiloxane) (PDMS). Notably, polymeric micelles, self-assembled from responsive block copolymers, form spheres with a water-loving core to shield drug cargo and a water-repelling shell to improve circulation time in the body. These versatile micelles can carry various drugs, from small molecules to proteins, and have shown promise in delivering cancer treatments in clinical trials (Liu et al. 2020; Lee et al. 2018).

6.2.1.3 Dendrimers

Dendrimers, star-shaped polymers with intricate 3D structures, offer highly tunable size, shape, and surface properties. Dendrimers, typically sized between 1 and 15 nm, possess a well-defined structure with controlled molecular weight, charge, and flexible branches (Mitchell et al. 2021; Kulkarni et al. 2019). These unique features allow them to act as versatile carriers. Functional groups on their exterior can be attached to biomolecules or imaging agents, while drugs are encapsulated within. Dendrimers are particularly well-suited for delivering nucleic acids and small molecules, and positively charged dendrimers like poly(ethylenimine) (PEI) and poly(amidoamine) (PAMAM) are commonly explored for these applications. Notably, several dendrimer-based products are undergoing clinical trials for various purposes, including theranostics, transfection, topical gels, and contrast agents (Xu et al. 2014; Palmerston Mendes et al. 2017; Menjoge et al. 2010).

6.2.1.4 Liposomes

Liposomes hold a significant place in drug delivery history, being the first nanoparticles approved for therapeutic use in 1965. These spherical vesicles, composed of phospholipids, form a bilayer structure with a hydrophilic core and a hydrophobic shell. This unique structure allows liposomes to encapsulate a wide range of drugs, both hydrophilic and hydrophobic, protecting them from degradation. Additionally, liposomes offer several advantages, including low inherent toxicity, minimal immune response, and biological inertness (reduced interaction with the body's natural processes). These characteristics contribute to improved drug efficacy and bioavailability. Notably, liposomal formulations of anticancer drugs like doxorubicin and paclitaxel (Doxil®, Myocet®) have demonstrated success in treating metastatic breast cancer (Khanna et al. 2023; Samad et al. 2007; Zhang et al. 2008; Wang et al. 2017).

6.2.1.5 Nanoemulsions

Nanoemulsions, which are oil droplets dispersed in water with sizes typically ranging from 10 to 1000 nm, represent a promising category of drug carriers. These emulsions can be designed in various forms, with the oil-in-water configuration being the most prevalent (Khanna et al. 2023; Jaiswal et al. 2015). Notably, researchers have investigated altering the membrane of nanoemulsions to enhance their functionality. For example, nanoemulsions loaded with spirulina and paclitaxel, featuring modified membranes, have demonstrated improved anticancer effects by influencing immune regulation pathways. Moreover, nanoemulsions have been utilized for the delivery of drugs like rapamycin, temozolomide, and bevacizumab to treat advanced melanoma. In comparison with liposomes and other nanoparticles, nanoemulsions present benefits such as enhanced stability, improved transparency, and biodegradability (Mitchell et al. 2021; Du et al. 2020). Nevertheless, their clinical utility is currently constrained by the requirement for specialized equipment and high-pressure/temperature procedures (Khanna et al. 2023).

6.2.1.6 Solid Lipid Nanoparticles (SLN)

Solid lipid nanoparticles (SLNs) are another type of three-dimensional nanocarrier, typically sized between 1 and 100 nm. These colloidal particles share similarities with micelles, with a hydrophobic core that encapsulates drugs. They are composed of an emulsifying agent, a single layer of phospholipids (fats), and water (Khanna et al. 2023; Üner and Yener 2007). Additionally, some SLNs incorporate polyethylene glycol (PEG) for enhanced stability and function. Similar to micelles, SLNs offer protection for the encapsulated drug and potentially reduce its toxicity. Studies have shown promise in using SLNs to deliver drugs like mitoxantrone, doxorubicin (DOX), and idarubicin (IDA). Notably, SLNs have demonstrated effectiveness in overcoming

P-glycoprotein (Pgp)-mediated multidrug resistance in leukemia cells (Ma et al. 2009).

Solid lipid nanoparticles (SLNs) modified with epidermal growth factor (EGF) on their surface demonstrated increased cytotoxicity against liver cancer cells (HepG2 and SMMC 7701) compared to unmodified SLNs. This targeted delivery approach, achieved through ligand-receptor interaction, led to enhanced antitumor efficacy and significantly reduced tumor weight in vivo. These findings suggest that EGF-conjugated SLNs hold promise for improved liver cancer treatment (Zhang et al. 2015).

6.2.1.7 Carbon Nanoparticles

Carbon nanoparticles (CNPs) are a unique class of drug carriers based on elemental carbon. Their biocompatible nature and diverse properties, including hydrophobicity for drug encapsulation (π-π stacking), make them attractive for therapeutic applications. CNPs come in various forms, such as graphene sheets and nanotubes, each offering distinct advantages. For example, reduced graphene oxide (rGO) and graphene oxide (GO) can absorb high amounts of near-IR light due to their high surface area and thermal conductivity and can be used as effective thermal stimuli in tumor destruction. Studies have shown promise in using GO to deliver doxorubicin for enhanced anticancer activity (Khanna et al. 2023).

Fullerenes, a type of carbon nanoparticle with diverse shapes like spheres and tubes, possess unique properties like extended π-conjugation. This property allows them to generate reactive oxygen species upon light exposure, making them promising candidates for photodynamic therapy in cancer treatment, especially when modified with PEG for improved functionality (Chen et al. 2012; Mroz et al. 2007; Tabata et al. 1997) (Ref. [55–57). Carbon nanotubes (CNTs), cylindrical structures similar to rolled-up graphene, can come in single-walled or multi-walled forms. They interact with immune cells to suppress tumor growth and hold promise for delivering DNA or drugs like doxorubicin (DOX) for targeted cancer therapy. Notably, fluorescent single-walled CNTs combined with antibodies have shown success in combating colon cancer (Heister et al. 2009).

6.3 Inorganic Nanoparticles

6.3.1 Metallic Nanoparticles

Metallic nanoparticles are employed in imaging and targeted drug delivery systems because of their exceptional magnetic, optical, and photothermal properties. Gold, silver, copper, and iron-derived nanoparticles are frequently employed in the realm of cancer treatment. Gold nanoparticles, in particular, are extensively utilized owing to the ease with which their surface attributes and dimensions can be controlled (Khanna et al. 2023; Zhao et al. 2004). The intrinsic properties of the materials composing inorganic nanoparticles impart distinct physical, electrical, magnetic, and optical characteristics. For example, gold nanoparticles (AuNPs) display photothermal behavior that is dependent on their size and shape, attributed to the oscillation of surface electrons. This, along with their ease of functionalization, makes AuNPs versatile tools for drug deliver y(Yang et al. 2019; Wang et al. 2019).

6.3.2 Magnetic Nanoparticles

Augustine et al. 2017) Magnetic nanoparticles, often coated with fatty acids or polymers for biocompatibility, are valuable for magnetic resonance imaging (MRI) and delivering drugs complexed with metals or metal oxides (Khanna et al. 2023; Basoglu et al. 2018). Iron oxide nanoparticles, particularly magnetite (Fe3O4) and maghemite (Fe2O3), are a leading material for FDA-approved inorganic nanomedicines due to their superparamagnetic properties (valuable for MRI and magnetic targeting). These versatile nanoparticles also hold promise for drug and gene delivery, alongside other inorganic options like calcium phosphate and mesoporous silica NPs**.** Inorganic nanoparticles (NPs) are particularly notable for their promise in diagnostics, imaging, and photothermal therapies, owing to their distinctive magnetic, radioactive, or plasmonic properties. Their good biocompatibility and stability make them well-suited for specialized applications where organic materials fall short. However, low solubility and potential toxicity, particularly in formulations containing heavy metals, limit their widespread clinical use (Khanna et al. 2023; Fiorillo et al. 2015).

6.3.3 Quantum Dots (QDs)

Quantum dots)QDs(, known for their semiconducting properties, offer exceptional advantages for biological imaging due to their high photostability, broad absorption spectra, and narrow emission bands. These nanoparticles consist of a core material, often encased by another semiconductor like ZnS. This layered structure, sometimes referred to as a "bilayer" QD, can be composed of various elements and further protected by an additional outer shell (Khanna et al. 2023; Jamieson et al. 2007; Ghasemi et al. 2009). Quantum dots can be categorized into various types, including nanodiamond, carbon, and graphene QDs. Their applications extend beyond bioimaging, with ongoing research exploring their potential in cancer therapy. Graphene QDs are particularly attractive due to their biocompatibility and efficient elimination from the body. Studies have shown promising results, such as the use of QD-DOX conjugates for effective prostate cancer treatment (Bagalkot et al. 2007). Researchers have developed a promising nanocomposite material (Bi@RP-PEG-DOX) for cancer treatment. This composite combines two-dimensional bismuthene sheets, zero-dimensional red phosphorus quantum dots, and the chemotherapeutic drug DOX. The unique design allows for both photodynamic and photothermal therapeutic effects. Additionally, the nanocomposite responds to the acidic environment of tumors and near-infrared II laser irradiation, triggering the release of DOX. Furthermore, its bismuth component offers excellent contrast properties for computed tomography (CT) scans, enabling potential CT-guided therapy. However, optimizing the production process for the quantum dots remains a key challenge (Mehrali et al. 2017).

The general classification of nanoparticles along with their advantages and disadvantages is briefly shown in Fig. 6.2.

6.4 Stimuli-Responsive Nanocomposites for Drug Delivery

While traditional drug delivery methods offer extended drug action and reduced dosing compared to conventional options, the emergence of stimuli-responsive biomaterials promises even greater advancements in biomedicine (Xiao et al. 2023a; Cui et al. 2019). This progress is driven by the remarkable ability to design these materials with multiple built-in responses to different stimuli. Additionally, the well-understood principles of dynamic interactions allow these materials to react to a wide range of external and internal cues. Based on the type of stimulus, nanocomposites that respond to stimuli are divided into two categories: internal and external, each of which is reviewed separately in this section (Lavrador et al. 2021).

6.5 Internal Stimuli

6.5.1 pH-Responsive

pH is a commonly altered physiological parameter in conditions such as tumors and inflammation. pH-responsive systems can take advantage of this variation by dynamically adjusting their structural characteristics, thereby enabling the targeted delivery of bioactive molecules in environments with pathologically acidic pH. Beyond its disruption in disease states, pH variation is a natural characteristic of specific anatomical locations. For instance, the gastrointestinal tract exhibits a pH gradient, ranging from the slightly acidic saliva in the mouth to the highly acidic gastric juices in the stomach and finally to the microbe-rich intestines with a more neutral pH. Notably, intracellular endosomal compartments maintain a naturally acidic environment. This inherent acidity can be leveraged for targeted cargo release from nanocomposite hydrogels. When released nanoparticles are internalized by neighboring cells, the acidic endosomal environment triggers the release of their encapsulated cargo polymers designed to respond to pH variations incorporate pendant

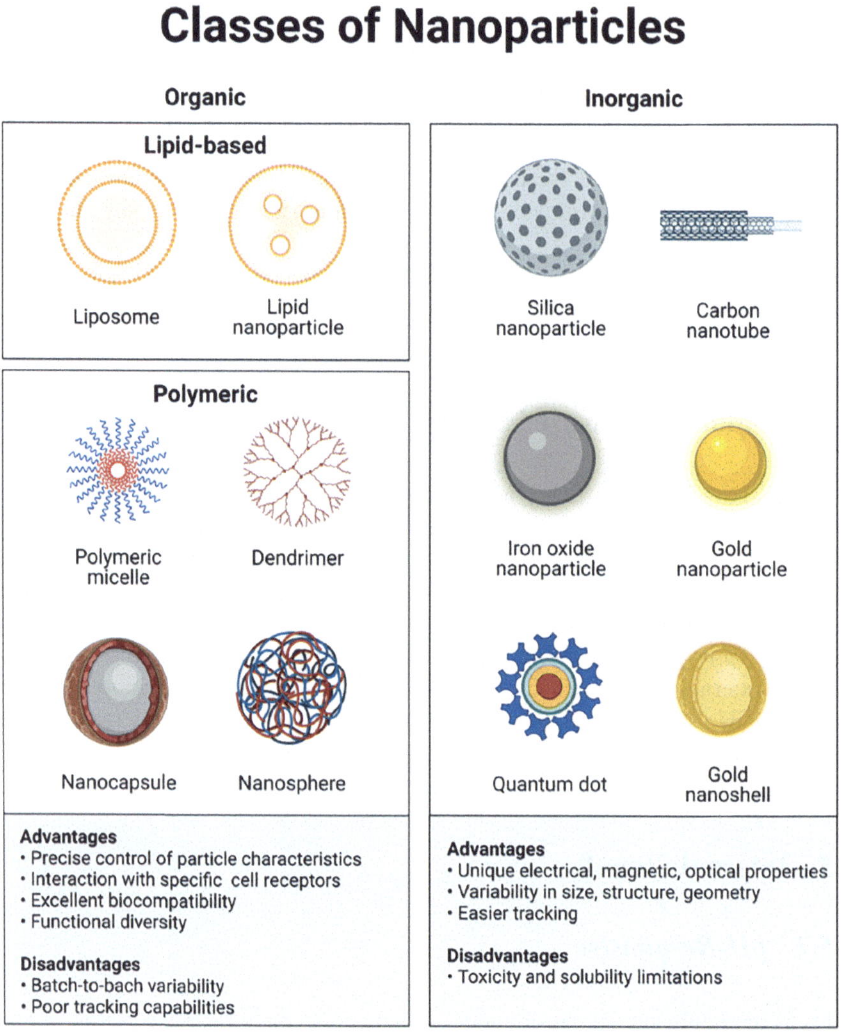

Fig. 6.2 Classification of types of nanoparticles along with advantages and disadvantages. Picture adapted from Vasconcelos and Santos (2023)

acidic or basic groups. These groups can either absorb or release protons in response to shifts in the surrounding pH of the environment. This responsiveness is triggered by conformational changes within the soluble polymer structure (Lavrador et al. 2021).

There are two main strategies for developing pH-responsive drug delivery systems (Lavrador et al. 2021):

Utilizing acid-labile linkers: This strategy involves covalently attaching the drug to the carrier molecule using special linker molecules that are sensitive to acidic environments. When exposed to an acidic environment, these linkers break down, releasing the drug.

Employing polyelectrolytes with ionizable groups: This approach utilizes polymers containing ionizable groups that can change their charge based on the surrounding pH. This change in charge can trigger the release of the encapsulated drug.

Researchers have utilized various polymers to design pH-responsive drug delivery systems. Some prominent examples include chitosan, albumin, and gelatin.

Introducing pH-sensitive bonds (pH-labile linkages) into these hybrid platforms offers another strategy to enhance design flexibility and achieve a wider range of desired properties in nanocomposite hydrogels. A recent breakthrough involved the development of injectable, self-healing, and highly pH-responsive nanocomposite hydrogels (Wu et al. 2020). These hydrogels utilize dynamic covalent bonds formed between amine-functionalized silica nanoparticles and pendant aldehyde groups within zwitterionic polymer. In the synthesized hydrogel, the significant number of open Schiff junctions made it sensitive to small pH changes, so that when the pH changed from 6.8 to 6.6, it showed significant drug release in the cancer environment. Also, due to the presence of nanoparticles in the hydrogel structure, the physical–mechanical properties increased significantly (Fig. 3a) (Wu et al. 2020).

In a study by Zhao et al., researchers constructed a novel, charge-reversible drug delivery system based on graphene oxide nanoparticles (GON). This system employed electrostatic interactions for two key steps: first, the self-assembly of GON with chitosan, and second, the coating of the complex with dimethylmaleic anhydride-modified chitosan (CS-DMMA). In this system, the chitosan shell protects the drug doxorubicin from various physiological factors. When the drug carrier reaches the tumor site, due to the acidic conditions around the tumor, the chitosan shell is separated and facilitates the release of the drug. The high amount of amino groups present in the structure of chitosan causes acidic hydrogens to be absorbed to a high extent and causes drug release in the low pH range. Hydrolysis of CS-DMMA enriches the amino groups and creates a positive surface charge at tumor pH, this positive charge increases the cellular uptake of the drug delivery system by cancer cells. This new GON-based system with its surface charge-modifiable shell is promising as a smart drug delivery platform for targeted cancer therapy. Its ability to control drug release and enhance cellular uptake indicates its potential to improve therapeutic efficacy (Fig. 3b) (Zhao et al. 2018).

6.5.2 *Redox-Responsive*

A biological measure of the equilibrium between electron-gaining and electron-losing processes within cells, reduction–oxidation potential (redox potential) is susceptible to alterations in the cellular environment. In inflammatory conditions, hypoxia (low oxygen availability), and disease states like cancer, it is known to

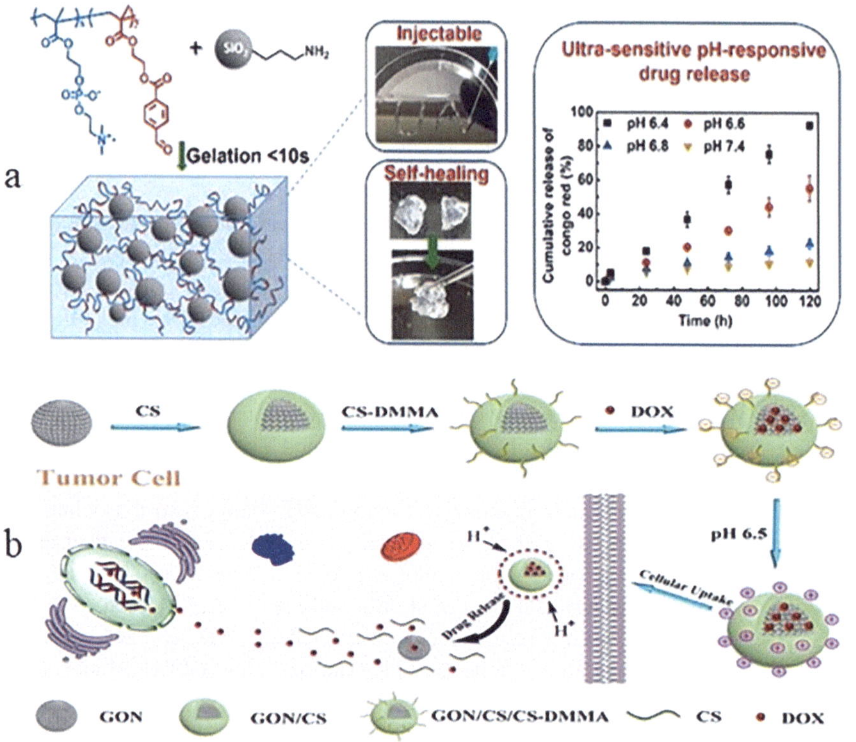

Fig. 6.3 A **a** Injectable and Self-Healing Nanocomposite Hydrogels with Ultrasensitive pH-Responsiveness for Controlled Drug Delivery. Reprinted with permission from Wu et al. (2020). Zhao et al. (2018) Copyright 2020 American Chemical Society. **b** pH-Sensitive Graphene Oxide Nanoparticle/Chitosan Hybrids for Efficient Intracellular Doxorubicin Delivery. Reprinted with permission from. Copyright 2018 American Chemical Society

be changed (Lavrador et al. 2021). Redox potential is interestingly determined by high glutathione concentrations in the major cellular region (cytosol) and endocytic organelles compared to the external environment. This is analogous to the varied pH levels across various compartments inside a cell. Drugs are released selectively within cells by taking advantage of this particular distinction (Lavrador et al. 2021; Wu et al. 2011).

The development of redox-responsive nanocomposite often relies on incorporating specific functional groups. These groups can be categorized into three main types (Lavrador et al. 2021; Tao and He 2018):

Reduction-cleavable linkers: These linkers, such as disulfide, diselenide, or thiol-maleimide bonds, are susceptible to breaking apart upon encountering a reducing environment.

Reactive oxygen species (ROS)-sensitive moieties: These moieties, including diselenide, phenylboronic ester, thioketal, or thioether groups, can respond to the presence of ROS, which are molecules involved in cellular stress.

Metallic nanocomponents: Certain metallic nanoparticles can also be incorporated to influence the redox responsiveness of the hydrogels.

In their study, Liang et al. developed a sophisticated hybrid hydrogel formed through a Michael-type addition reaction between PEG polymers and liposomal nanoparticles. These hybrid hydrogels undergo degradation upon exposure to thiol-containing molecules like glutathione (GSH), owing to the presence of thioether succinimide linkages between the PEG polymers and liposomal nanoparticles. The hierarchical structure of the network allowed for the simultaneous encapsulation and release of various chemotherapy-related drug molecules, including doxorubicin (DOX) and cytochrome c (Fig. 6.4) (Liang and Kiick 2016). The drug release mechanism is that the bonds sensitive to glutathione are broken by this antioxidant and the liposomes in the hydrogel network are released in response to this antioxidant. After releasing the drug molecules from the environment with higher concentration (inside the liposome) to the environment with lower concentration (outside the liposome) under Fickian diffusion, the release of the drug is carried out in a controlled manner.

In a study conducted by Zhang et al., a programmable drug delivery system was developed utilizing a composite nanoparticle based on chondroitin sulfate (CS) to improve and increase the effectiveness of chemotherapy in drug-resistant cancer. This nanoparticle targeted CD44 specifically in tumors, rapidly released drug molecules, continuously delivered cargo to drug-resistant tumor cells, and sensitized cancer cells to antitumor drugs by reducing the levels of Bcl-XL, an antiapoptotic protein. The programmable drug delivery system, employing a chondroitin sulfate (CS)-based composite nanoparticle, exhibited specific targeting of CD44 in tumors, quick release of drug molecules, sustained release in drug-resistant tumor cells, and enhanced the sensitivity of cancer cells to antitumor drugs by decreasing the levels of Bcl-XL protein. The nanoparticle showcased improved and sensitized chemotherapy for

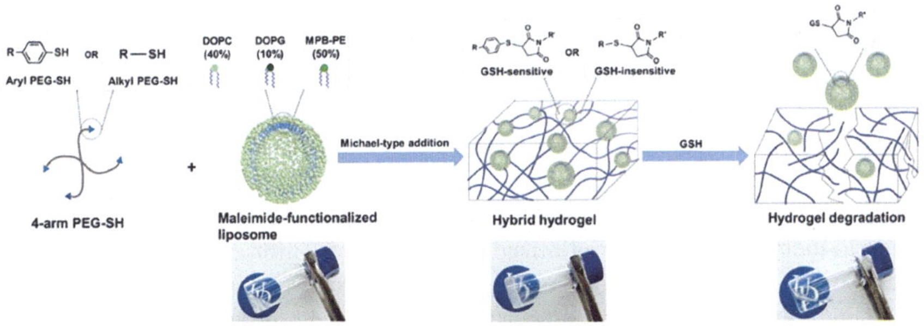

Fig. 6.4 Schematic formation and degradation of the liposome crosslinked hybrid hydrogels. Reprinted with permission from Liang and Kiick (2016). Copyright 2016 American Chemical Society

drug-resistant cancer, offering a promising strategy for precise drug delivery and regulated release in the field of biomedical applications (Fig. 5a) (Zhang et al. 2019).

Zhu et al. developed an innovative delivery system for substance P (SP) using zeolite imidazolate framework-8 (ZIF-8) nanoparticles to facilitate wound healing. ZIF-8 nanoparticles were synthesized via a biomimetic mineralization approach, after which they were loaded with SP and incorporated into a Ca^{2+}-crosslinked alginate/pectin hydrogel. The inclusion of SP-loaded ZIF-8 nanoparticles into the hydrogel enhanced the controlled release of SP in response to reactive oxygen species (ROS), which are pivotal in the wound healing process. The ROS-triggered release mechanism of SP from the hydrogel dressing has the potential to accelerate wound healing, as demonstrated in their study (Zhu et al. 2020).

In a separate study conducted by Yi et al., multifunctional nanogels were formulated to exhibit enhanced thermal responsiveness in order to facilitate the precise administration of therapeutic substances. These nanogels are produced through the in situ radical polymerization of N-isopropylacrylamide (NIPAM) and acrylic acid (AA) in the presence of β-cyclodextrin. That β -cyclodextrin, having hydrophilic and hydrophobic groups, causes it to form a complex with NIPAM, and the hydrophilic and hydrophobic interactions between NIPAM and cyclodextrin increase the sensitivity of the nanogel to temperature. The resultant nanogels exhibit heightened sensitivity toward alterations in temperature, thereby enabling the swift liberation of encapsulated doxorubicin (DOX) within specific localized environments. Furthermore, the rapid uptake of nanogels by cancerous cells coupled with an enhanced cytotoxic effect against cancer highlights their promise for targeted drug delivery applications that extend beyond anticancer medications (Fig. 5b) (Yi et al. 2017).

6.5.3 Enzyme-Responsive

Enzymes are ubiquitously present in the human body and are integral to nearly all vital biological processes necessary for organismal survival. Like other internal signals, the presence of endogenous enzymes, either localized in specific tissues or overexpressed in association with certain diseases, offers significant potential for the development of advanced, controlled drug delivery systems. Furthermore, enzyme-responsive systems, characterized by their specificity in both location and function, have the potential to outperform traditional stimuli such as pH, temperature, and light, which may lack efficiency and specificity due to factors like pH gradients in cellular compartments. Enzymes, on the other hand, offer enhanced selectivity owing to their evolutionarily optimized substrate recognition capabilities. Enzyme-responsive nanocomposite hydrogels can be developed by either employing naturally occurring polymers with inherent sensitivity (such as fibrin, collagen, gelatin, and hyaluronic acid) or by modifying non-responsive biomaterials through the integration of enzyme-sensitive linkers (Lavrador et al. 2021; Hu et al. 2012; Ooi et al. 2017).

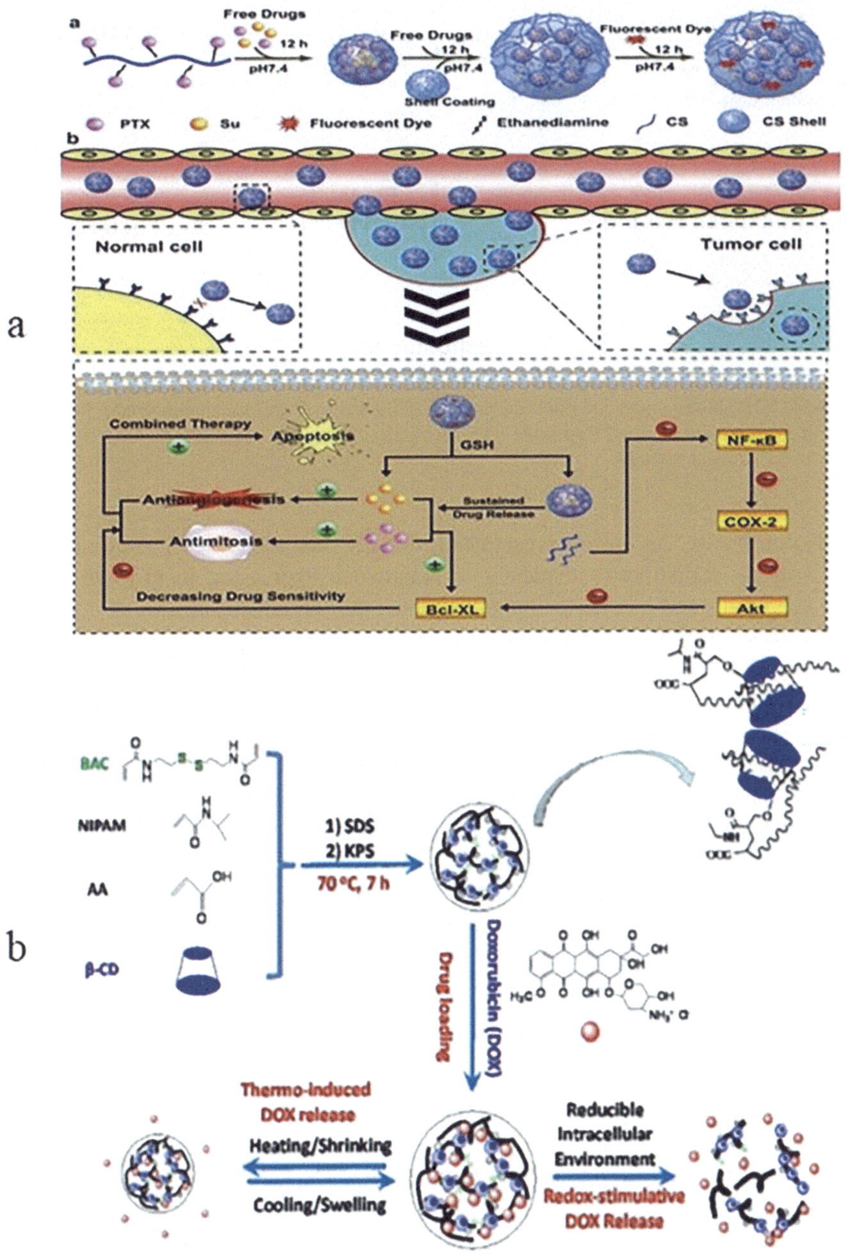

Fig. 6.5 Other drug deliver system-based redox-responsive nanocarrier. **a** Programmable Drug Delivery System for Enhancing and Sensitizing Chemotherapy to Drug-Resistant Cancer. Reprinted with permission from Zhang et al. (2019). Copyright 2019 American Chemical Society. **b** Stimulative nanogels-based β-cyclodextrin for therapeutic delivery. Reproduced with permission (Yi et al. 2017). Copy Right 2017 Elsevier

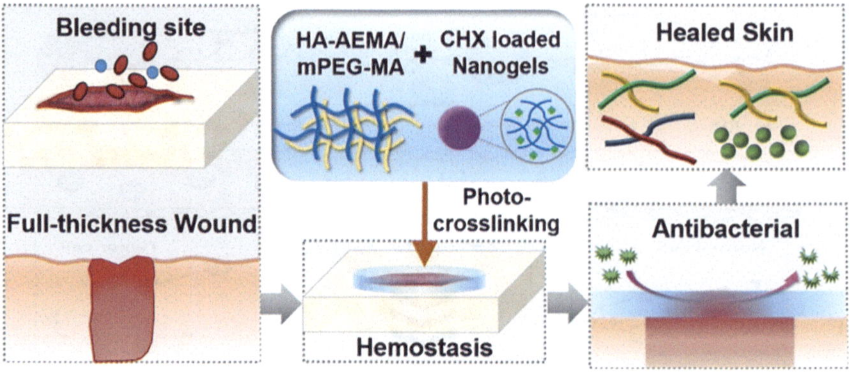

Fig. 6.6 A Hyaluronic acid and polyethylene glycol hybrid hydrogel with antibacterial properties for enhanced wound healing. Reprinted with permission from Zhu et al. (2018). Copyright 2018 American Chemical Society

Hyaluronidase, an enzyme responsible for the degradation of hyaluronic acid, demonstrates a distinctive tendency of heightened expression in invasive malignant neoplasms and is released by pathogenic microorganisms at sites of infection. Thus, nanocomposites that incorporate hyaluronic acid as structural units possess the inherent ability to serve as platforms sensitive to enzymes, making them applicable to various fields within biomedicine (Lavrador et al. 2018; Buhren et al. 2016). Zhu et al. formulated hybrid hydrogels that exhibit swift hemostatic effects and enduring antibacterial characteristics for the purpose of wound healing. This was achieved through the utilization of aminoethyl methacrylate hyaluronic acid (HA-AEMA) and methacrylated polyethylene glycol (mPEG) as the principal constituents. The hybrid hydrogels showcased prompt control of hemorrhaging, resistance against infections, and the capacity to sustain a moist milieu, all of which are vital aspects in the management of wounds (Zhu et al. 2018) (Fig. 6.6).

6.5.4 Glucose-Responsive

Glucose, serving as the predominant energy source for the majority of tissues, is commonly found in the bloodstream at levels that adhere to physiological norms, under the influence of a precise feedback control system. The human pancreas, in this specific scenario, functions as an inherent glucose-responsive platform, wherein β-cells monitor changes in blood glucose concentrations and secrete endogenous insulin correspondingly (Xiao et al. 2023a; Xu et al. 2023). Nevertheless, this delicate equilibrium can be substantially disturbed in conditions like diabetes mellitus, a disease characterized by a significant elevation (more than fourfold) in blood glucose levels, which occurs due to either insufficient insulin production or the onset of insulin resistance (Lavrador et al. 2021; Katsarou et al. 2017). In cases of Type 1 diabetes, individuals rely heavily on the regular administration of increased insulin doses to reduce and maintain proper blood glucose levels. The pioneering work by Gu's team has paved the way for the development of self-regulating hybrid platforms that respond to glucose fluctuations, expanding on this core principle. Specifically, through the use of pH-responsive hydrogel systems or hypoxia-sensitive nanovesicles in experimental diabetic models, Gu's team successfully demonstrated the effective regulation of both hyperglycemia and hypoglycemia in response to glucose challenges, by administering nanocomposite hydrogels as an alternative to conventional insulin therapies (Fig. 6.7) (Yu et al. 2015). Another strategy involves utilizing specific inorganic nanomaterials that respond to environmental changes triggered by the activity of glucose oxidase, an enzyme essential in glucose metabolism. For example, some research has investigated the use of insulin-loaded mesoporous silica nanoparticles functionalized with cyclodextrin-modified glucose oxidase, as well as mesoporous bioactive glass nanospheres integrated with ZnO quantum dots, which function as gatekeepers responsive to pH variations (Lavrador et al. 2021; Oroval et al. 2017). Additionally, Zeoliticimidazole framework-8 nanocrystals exhibit great promise as nanomaterials capable of detecting pH variations induced by glucose oxidase activity and subsequently triggering the release of insulin (Zhang et al. 2020b; Chen et al. 2018a). Beyond pH considerations, the potential harm posed by hydrogen peroxide, a byproduct of glucose oxidase activity, can be mitigated through the inclusion of an additional enzyme such as catalase, which helps convert hydrogen peroxide back into oxygen, thereby preventing destabilization of glucose oxidase. Alternatively, metal-based nanomaterials like MnO2 or ceria nanoparticles, known for their ability to scavenge hydrogen peroxide effectively, have been incorporated into hydrogel matrices to create glucose-responsive hybrid platforms with improved enzymatic stability over extended periods (Gordijo et al. 2010). Moreover, hydrogels containing platinum nanoparticles and polyaniline, which possess inherent electro-oxidation properties toward hydrogen peroxide, show significant promise as highly sensitive platforms for detecting glucose levels. Despite their considerable potential, the majority of glucose-responsive nanocomposite hydrogels developed to date rely on enzymes, which raises concerns regarding the feasibility of translating these platforms to clinical settings, particularly in relation to cost, enzymatic durability, and potential immune reactions triggered by the administration of enzyme-rich constructs (Lavrador et al. 2021).

6.5.5 Electro-Responsive

Stimuli-responsive nanocomposite designed with electrical conductivity have the potential to enhance the inherent electrochemical communication among neighboring cells or nearby tissues, thereby increasing tissue biofunctionality in cases where electrical conduction is impeded by injury. Electroactive tissues, such as bone, cartilage, cardiac, nerve, and skeletal muscles, depend significantly on bioelectrical signals for their function. The development of nanocomposite hydrogels designed

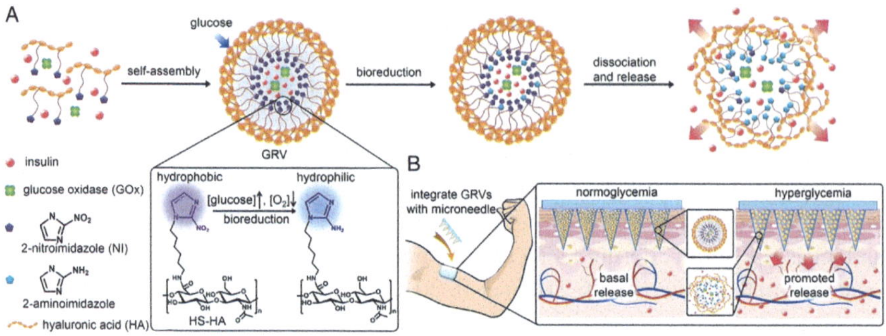

Fig. 6.7 Schematic of the glucose-responsive insulin delivery system using hypoxia-sensitive vesicle-loading MN-array patches. Adapted from Yu et al. (2015)

for these tissues has garnered considerable attention, as incorporating conductive nanomaterials into extracellular matrix (ECM)-mimicking networks can significantly improve both mechanical strength and biological functionality (Lavrador et al. 2021; Mehrali et al. 2017). When it comes to electro-responsive nanocomposite, the creation of three-dimensional electroactive frameworks typically involves incorporating conductive inorganic nanomaterials and select conductive biomaterials, given that the majority of polymers lack inherent conductivity (Lavrador et al. 2021). In the field of electrically responsive nanomaterials, various metallic (e.g., gold and silicon) and carbon-based (e.g., graphene) nanoparticles have been employed, primarily in the forms of nanosheets, nanotubes, and nanorods (Thoniyot et al. 2015). Notably, gold and silicon nanowires have been effectively utilized in the assembly of electroactive cardiac three-dimensional microtissues featuring biomimetic cardiomyocyte characteristics and maturation of contractile apparatus, demonstrating the capacity to induce synchronized contractions and potentially reduce post-transplantation arrhythmias. In the realm of conductive polymeric biomaterials, polypyrrole and polyaniline stand out as well-established options owing to their favorable cytocompatibility and customizable attributes. Nevertheless, there exist novel materials like poly(p-phenylene vinylene), polythiophene, and poly(3,4-ethylenedioxythiophene) that are gaining recognition (Richards et al. 2016; Dvir et al. 2011). Recent research has explored the incorporation of polypyrrole nanorods into photopolymerizable polyacrylamide/chitosan interpenetrating networks, resulting in the development of durable electroactive nanocomposite hydrogels. These hydrogels are capable of enabling electrostimulated dexamethasone delivery while promoting the proliferation of C2C12 mouse myoblasts, even in the absence of external electrical stimuli. Furthermore, evaluations of its regenerative potential in a rat wound model with full-skin defects revealed enhanced wound closure compared to controls lacking nanorods. The healing of epidermal tissue was further improved through the incorporation of epidermal growth factor within the nanocomposite hydrogel. On the other hand, nanocomposite hydrogels incorporating polyaniline have also demonstrated successful applications in diverse biomedical contexts. Notably, the integration of polyaniline nanowires into photocrosslinkable gelatin hydrogels led to the creation of electroactive platforms compatible with digital projection stereolithography, facilitating the bioprinting of complex structures containing osteoblast-like cells (Sawyer et al. 2017).

The integration of these conductive nanomaterials and polymers into hydrogel-based nanocomposites has enabled the development of robust, electroactive platforms that can be used for various biomedical applications, such as wound healing, bone regeneration, and cardiac tissue engineering. The electrical stimulation provided by these nanocomposites can enhance cellular responses, promote the formation of functional tissue structures, and even facilitate the delivery of therapeutic agents in a spatiotemporally controlled manner.

However, the translation of electro-responsive nanocomposites into clinical practice faces several challenges. Ensuring the long-term stability, biocompatibility, and seamless integration of these materials with host tissues is crucial. Additionally, optimizing the electrical stimulation parameters, such as the frequency, intensity, and duration, to elicit the desired biological responses remains an area of active research. Overcoming these challenges will be key to unlocking the full potential of electro-responsive nanocomposites in regenerative medicine and personalized healthcare.

6.6 External Stimuli

6.6.1 Thermo-Responsive

Temperature, being a gentle and easily controlled stimulus, is extensively studied in the development of stimuli-responsive platforms. Traditionally, these platforms utilize thermo-responsive polymers like methylcellulose derivatives, poly(N-isopropylacrylamide), poly(2-oxazoline), or poloxamers, which inherently undergo rapid phase or volume transitions in response to temperature fluctuations near their upper and lower critical solution temperatures (UCST and LCST) (Lavrador et al. 2021). Alternatively, advanced genetic engineering approaches can be used to create biomimetic synthetic proteins with tailored temperature-responsive behaviors. As a result, hydrogels exhibiting LCST properties can be formed by heating, while those with UCST characteristics can be developed through cooling (Lavrador et al. 2021). The human body's typical temperature of around 37 °C provides an opportunity to design hybrid platforms that respond to internal body heat by modulating their UCST/LCST threshold values. This enables the creation of systems that can either be assembled at room temperature and dissolve upon application or consist of liquid precursors that self-assemble into hydrogels upon injection into tissues. These biomaterial-based systems, which offer self-dissolving platforms or facilitate in situ hydrogel formation, have been widely investigated in the literature (Chen et al. 2018b; Dimatteo et al. 2018). In the study conducted by the researchers, thermo-responsive polymeric nanospheres were synthesized using chitosan oligosaccharide (COS) conjugated with pluronic F127 grafted with carboxyl groups to deliver both kartogenin (KGN) and diclofenac (DCF) within a single system. The nanospheres were prepared through the conjugation of COS with pluronic F127 grafted with carboxyl groups (F127ACOOH) via EDC chemistry. The release profiles of the two drugs were independently regulated by temperature fluctuations. KGN was covalently crosslinked to the outer surface of the nanospheres, while DCF was encapsulated in the core. The nanospheres exhibited an immediate release of DCF and a sustained release of KGN, both of which were temperature-responsive. The system showed promising results in inhibiting the progression of osteoarthritis in treated rats, with enhanced therapeutic effects observed under cold treatment (Kang et al. 2016). Qin et al. developed a drug delivery system activated by near-infrared (NIR) light, using an amphiphilic chitosan derivative-dispersed single-wall carbon nanotube (CNT) formulation encapsulated within a thermo/pH-sensitive nanogel (CS/PNIPAAm@CNT). This system was designed to enhance drug delivery to targeted tumors while minimizing adverse effects. The particles exhibited a notable drug carrying capacity and demonstrated accelerated drug release at elevated temperatures and decreased pH levels. The application of NIR irradiation additionally boosted drug liberation and cytotoxicity against cancerous cells. The investigation proposes that the developed nanocomposite represents a promising therapeutic carrier for precise cancer treatment (Fig. 8a) (Qin et al. 2015).

In a separate study, researchers developed a nanocomposite system known as "lipogels," which integrates fibrin hydrogels, near-infrared (NIR)-responsive nanomaterials, and thermosensitive liposomes (TSL) for controlled drug release. The lipogels are designed to release their liposomal cargo upon NIR irradiation, which elevates the temperature in the irradiated areas. The release rate of the drug can be precisely controlled by adjusting the concentration of NIR nanotransducers, the intensity of the electromagnetic energy, and the irradiation protocol. This remotely controlled drug delivery system holds significant promise for therapies requiring precise modulation of drug dosage within a defined spatiotemporal framework (Fig. 8b) (Martín-Saavedra et al. 2017).

6.6.2 *Light-Responsive*

Light has emerged as a powerful external trigger, offering distinct advantages for the development of advanced biomedical platforms that allow precise spatiotemporal control over biomaterial behavior and dynamic modification of their properties (Rapp and DeForest 2021). Moreover, light-responsive platforms present a wide range of opportunities for the remote-controlled delivery of therapeutics or the gradual degradation of implantable devices, all while ensuring safety and non-invasive procedures (Lavrador et al. 2021; Li et al. 2019). The response mechanisms of light-responsive materials can be classified into several categories: (1) alteration

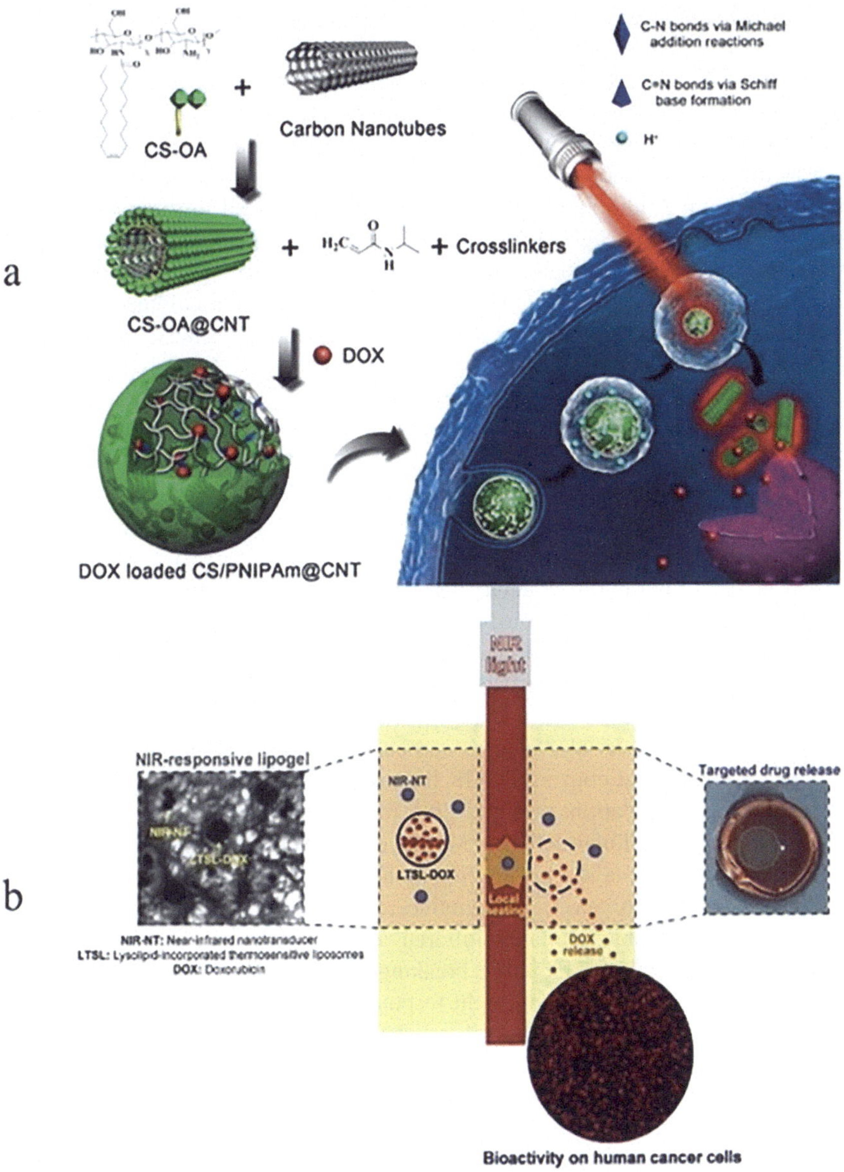

Fig. 6.8 Temperature-responsive drug nanocarriers. **a** Thermo/pH-sensitive nanogel (CS/PNIPAAm@CNT) for anticancer drug delivery. Reprinted with permission from Qin et al. (2015). Copy Right 2015 Elsevier. **b** Lipogels responsive to near-infrared light for the triggered release of therapeutic agents. Reprinted with permission from Martín-Saavedra et al. (2017). Copy Right 2017 Elsevier

of molecular structure through photoisomerization upon light exposure; (2) degradation of the polymer backbone; (3) structural damage to the material induced by the photothermal effect; and (4) degradation via photochemical effects. Among these, photoisomerization involves changes in chemical bonds with restricted rotation, particularly double bonds. This process typically entails a reversible alteration, such as the cis–trans isomerization of the trans double bond in azobenzene when exposed to light. Additionally, photoisomerization may lead to the breakdown of the chromophore due to light-induced structural changes, resulting in an irreversible transformation. Photodegradation of the polymer backbone can trigger the release of encapsulated drug molecules, while local destruction of the polymer matrix due to the photothermal effect also facilitates the release of embedded molecules. In some cases, light exposure can induce cleavage of the o-nitrobenzyl group within the polymer, thereby enabling the controlled release of the drug (Ma et al. 2023). The two primary categories of transducing nanomaterials in light-responsive hybrid platforms are photothermal nanoparticles, which convert light into heat, and upconversion nanoparticles, which convert near-infrared (NIR) light into ultraviolet (UV) light. For instance, photothermal nanoparticles—such as metal-based nanoparticles, carbon nanotubes, graphene oxide, or polydopamine—serve as transducers that convert incident light into heat at targeted locations. By integrating photoabsorbing nanomaterials with thermo-responsive elements, systems that initially respond solely to temperature changes can be transformed into light-responsive hybrid platforms. This conversion is particularly advantageous, as it allows for the use of low-energy NIR light, which offers greater tissue penetration depth compared to UV light. Carbon-based nanomaterials, such as carbon nanotubes and graphene oxide, are particularly effective at converting NIR light into heat, enabling remote-controlled nanocomposite degradation, pulsatile drug release, and the thermal disruption of endosomes for intracellular delivery (Merino et al. 2015; Gangrade et al. 2020).

The chromophore plays a pivotal role in the light-responsive mechanism. These chemical structures are specifically engineered to interact with different types of light, such as UV, visible light, near-infrared, and even laser light. UV light, with its high energy, is particularly effective at breaking chemical bonds, making it a common choice as the primary stimulus in light-responsive systems. Leveraging the advantages of UV light, a variety of photocrosslinkable materials have been developed to create light-responsive systems, particularly for applications in tissue engineering (Ma et al. 2023). The scientists utilized photocrosslinkable gelatin methacrylate (GelMA) for the creation of hydrogels. Within these hydrogels, periodontal ligament stem cells (PDLSCs) were enclosed (Ma et al. 2017). These specially developed hydrogel systems have the capability to enhance the regeneration of functional tissue and can serve as an effective platform for investigating the extracellular matrix. In order to address wound management and repair, the researchers synthesized an in situ imine crosslinking-based photoresponsive chitosan hydrogel (Ma et al. 2020). Nevertheless, there exist certain disadvantages associated with the utilization of UV light. Insufficient tissue permeability poses a constraint on the effectiveness of UV light, restricting its penetration to superficial tissue layers and hindering access to deeper tissue regions. Moreover, UV light is known to be exceedingly toxic, leading to the generation of free radicals upon irradiation (Ma et al. 2023).

Yin et al. effectively developed bioreducible hyaluronic acid-graphene oxide (HSG) nanosheets for the purpose of photothermally controlled drug delivery. The HSG nanosheets displayed improved stability, exhibited tumor targeting capabilities via HA receptors, and showed redox-sensitivity to enable controlled drug release. These nanosheets demonstrated a superior photothermal response in comparison with graphene oxide (GO), facilitating photothermally controlled endolysosomal escape and enabling cytoplasm-selective drug delivery. The rapid release of medications from HSG nanosheets was accomplished through redox-triggered mechanisms, thereby amplifying the pharmacological impact of the administered drug. Both in vitro and in vivo experiments substantiated the effectiveness of HSG nanosheets in targeted drug delivery and illustrated their potential in enhancing therapeutic outcomes (Yin et al. 2017).

Wang et al. proposed a new targeted nano-immunostimulant (ZGS-Si-Pc@HA) that employs persistent luminescence nanoparticles to elicit prolonged immune activation in the context of cancer immunotherapy. This nano-immunostimulant effectively enhances the infiltration of T lymphocytes into tumors, improves the efficacy of checkpoint blockade immunotherapy, and elicits an immunological memory response, providing a dependable approach to combatting cancer metastasis and relapse. The composite material mentioned above integrates persistent luminescence nanoparticles (ZGS), a photosensitizer (Si-Pc), and hyaluronic acid (HA) to achieve sustained immune activation in the realm of cancer immunotherapy. In essence, ZGS-Si-Pc@HA presents a reliable strategy for addressing cancer metastasis and recurrence by amplifying cancer immunotherapy and bolstering the immune response (Wang et al. 2021).

6.6.3 Mechanical-Responsive

The term "mechanical response of bio-based materials" pertains to materials derived from biological sources that exhibit the ability to react to external mechanical stimuli and display specific functionalities. These external stimuli can be categorized into three main types: compression, tension, and shear, each corresponding to a variety of specific materials. In a recent study, scientists fabricated a hydrogel that responds to mechanical stimuli and explored its potential for controlled drug release. By utilizing sodium alginate as a bio-based material, researchers combined it with specific molecules to create the hydrogel. Two distinct hydrogels were synthesized—one containing only free molecules and the other containing both free molecules and those attached to the network structure of the hydrogel. Experimental findings revealed that upon compression, the free molecules were released from the hydrogel. Conversely, when anchored molecules were present, the periodic compression led to a controlled release of the drug. A plausible explanation for this phenomenon is proposed: in the absence of compression, the molecules remained attached within the hydrogel;

however, upon the application of a compressive force, the anchoring structure was disrupted, resulting in a pulsatile release pattern. Furthermore, a compression-responsive hydrogel was developed by incorporating Cyclodextrin with sodium alginate (Ma et al. 2023; Tan et al. 2015). The hydrogel demonstrated enhanced aqueous solubility of hydrocortisone acetate and did not significantly impede the proliferation of cultured mouse fibroblasts. The release of hydrocortisone acetate from the hydrogel became increasingly responsive to mechanical compression. As the concentration of mono-6-deoxy-6-ethylenediamine-β-cyclodextrin increased, the mechanical sensitivity of the drug release also intensified, while the swelling rate and rigidity of the hydrogel decreased (Ma et al. 2023).

6.6.4 Ultrasound-Responsive

Ultrasound is widely used in medicine for imaging internal tissues, aiding in the diagnosis and treatment of various conditions due to its non-invasive and safe nature. Like other external stimuli, ultrasound waves can be utilized to remotely trigger drug release, leading to the development of theranostic platforms capable of penetrating deep tissues. Ultrasound-responsive systems offer the potential for targeted delivery of bioactive therapeutics through mechanisms such as pulsatile or burst release profiles, as well as sustained release over prolonged periods following in vivo administration. The acoustic energy generated by ultrasound stimulation can disrupt the self-assembly equilibrium of specific nanocarriers, such as liposomes, polymeric micelles, and nanocapsules, resulting in their disintegration. Hydrogels are often employed in the formulation of bio-based ultrasound-responsive materials due to their excellent mechanical strength, appropriate viscoelasticity, and exceptional biocompatibility. For example, researchers developed a cellulose-based hydrogel film that responded to ultrasound irradiation, triggering drug release. The results showed that ultrasound exposure disrupted the hydrogen bonds between cellulose and the drug molecules in the hydrogel, thereby facilitating drug release. The drug release efficiency of the hydrogel subjected to ultrasound treatment was found to be six times greater than that of the hydrogel without ultrasound exposure. Additionally, as the ultrasonic power increased, the amount of drug released also rose. Similarly, researchers developed a chitin hydrogel containing gallic acid, which, upon ultrasound irradiation, experienced disruption of the hydrogen bonds between chitin and gallic acid, accelerating drug release. Beyond hydrogen bond disruption, ultrasound can also facilitate the breakdown of ionic bonds within the material. For example, scientists designed a sodium alginate hydrogel loaded with chemotherapeutic agents. Through experimentation, it was demonstrated that pulsed chemotherapeutic drug delivery could be effectively controlled by modulating ultrasound activation. The therapeutic efficacy with ultrasound treatment was significantly enhanced compared to the group without ultrasound exposure. The underlying ultrasound response mechanism in this system involves the rupture of the calcium ion crosslinking structure within the hydrogel, resulting in the breakdown of the grid structure and the subsequent release of small chemotherapeutic molecules. After the ultrasound is deactivated, the free calcium ions in the hydrogel spontaneously re-establish crosslinking, thereby repairing the grid structure and encapsulating the drug molecules once again (Ma et al. 2023). In addition to sonodisruptable nanocarriers, silica-based nanomaterials have been extensively explored for ultrasound-mediated theranostic applications, leveraging enhanced acoustic cavitation and the ability to incorporate sonosensitizers within their mesoporous structures. Leading the way in the use of alginate hydrogels for on-demand ultrasound-responsive drug delivery, Mooney's research team has further advanced this platform to improve its applicability in tissue engineering. The team developed ultrasound-responsive nanocomposite hydrogels by integrating gold nanoparticles conjugated with bone morphogenetic protein-2 (BMP-2) into ionically crosslinked alginate matrices (Lavrador et al. 2021; Ma et al. 2023).

6.7 4D Printing in Medicine

Three-dimensional (3D) printing, a cutting-edge additive manufacturing (AM) technology, is increasingly used in clinical applications such as organ and tissue creation, drug delivery systems, and regenerative medicine. This technology combines imaging techniques like computed tomography (CT) and magnetic resonance imaging (MRI) to gather personalized human data, which is then converted into computer-aided design (CAD) models for the production of 3D-printed objects. The key advantage of 3D printing lies in its capacity to create customized biomedical solutions by fabricating patient-specific devices. Although 3D-printed products exhibit exceptional physiological properties, biomedical devices produced through this method are often static and not optimized for dynamic environments. The microenvironment of native tissues plays an essential role in supporting tissue growth and regulating biological functions, which presents a challenge for static 3D-printed structures. To enhance functionality, bioactive materials need to be adaptable to dynamic environments while meeting stringent requirements for biodegradability and biocompatibility. The technology of four-dimensional (4D) printing has been innovated as a solution to address this challenge (Arif et al. 2022). The concept of four-dimensional (4D) printing was first introduced by Tibbits as a dynamic material that can change its shape over time. Unlike 3D printing, where the output is static, the additional dimension in 4D printing refers to "time," highlighting the ability of advanced materials to evolve in appearance or function over time in response to specific environmental stimuli. Advanced 4D printing materials, also known as 4D bioprinting compounds, active origami, or shape-morphing systems, follow similar techniques to 3D printing, where polymer/nanocomposite layers are deposited successively to create three-dimensional objects. What sets 4D bioprinting apart is its ability to incorporate temporal transformations, allowing materials to be

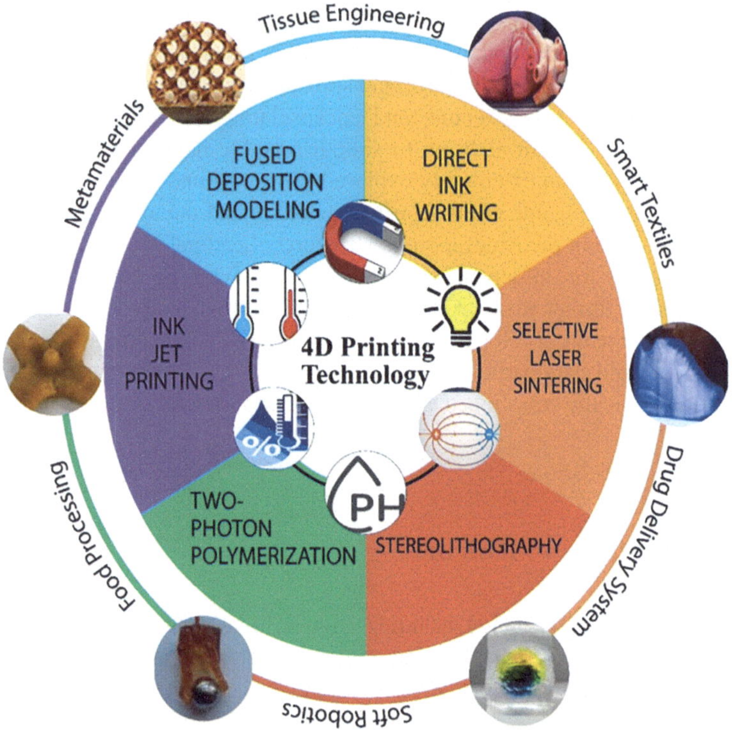

Fig. 6.9 4D printing from manufacturing to application. Reprinted with permission from Khalid et al. (2022). Copy Right 2022 Elsevier

programmed to respond to external triggers and change shape accordingly based on environmental factors (Zhou et al. 2020).

4D printing allows advanced materials to undergo transformations into new dimensions in response to various external stimuli, including moisture, temperature, light, magnetic fields, electric fields, biomolecules, pH, cell traction forces, and enzymes (Fig. 6.9) (Khalid et al. 2022).

6.7.1 Technologies Employed in 4D Printing

Traditional 3D printing technologies have paved the way for well-established 4D printing techniques, such as light- and extrusion-based methods. Stereolithography (SLA) is one of the most widely used printing technologies in the biomedical field. It utilizes visible or ultraviolet (UV) light from a laser to crosslink photosensitive ink materials into a desired three-dimensional (3D) structure. The small surface area of the emitted light allows for precise control and high-resolution printing by projecting it onto the liquid ink material's surface to drive the photocrosslinking reaction. A notable advantage of the SLA method is that it does not require the extrusion of printing materials through a nozzle or syringe, which eliminates common issues such as nozzle clogging and the potential damage to bioink-resident cells due to shear stress during the bioprinting process. However, extended exposure to UV radiation may cause cell damage, particularly for encapsulated cells. Additionally, the SLA method has limitations in its broader biomedical applications, including the inability to use bioplastic materials, high costs, and the non-uniform mechanical properties of SLA-printed structures (Agarwal et al. 2021).

Using light as a crosslinking source, digital light processing (DLP) printing has emerged as a promising technique for rapidly fabricating complex 3D and 4D objects with micro- to nanoscale structural features. DLP-based printing offers a wide range of material options and is known for achieving excellent shape-to-design fidelity and high printing resolution, allowing for the creation of intricate structures with remarkable precision (Zhao et al. 2004). In contrast to the SLA technique, DLP cures the liquid resin in a reservoir in just seconds per layer by using a digital light pattern instead of a scanning laser beam. Surface-projection-based printing, such as DLP, enables significantly faster printing speeds compared to SLA, often up to several times quicker. Both SLA and DLP printing techniques hold great promise for advancing the production of commercial and laboratory-made materials that can be crosslinked rapidly with high resolution, including photocrosslinkable 4D shape-memory polymers (SMPs) or liquid crystalline polymers (LCPs) (Wang et al. 2022).

Two-photon polymerization (TPP), also known as direct laser writing, enables the direct fabrication of 3D nanostructures with spatial resolution that exceeds the diffraction limit of the material by scanning a tightly focused femtosecond laser beam within a liquid photopolymer, leveraging the nonlinear threshold effect. However, the relatively short working distance of the high-numerical-aperture focal lens used in TPP devices limits the maximum height of the structures, which typically remain within the millimeter scale (Wang et al. 2022). Extrusion-based additive manufacturing systems, such as direct ink writing (DIW) and fused deposition modeling (FDM), are among the most widely recognized and extensively studied 3D printing methods due to their simplicity, widespread availability, and cost-effectiveness. These extrusion-based techniques offer a higher degree of versatility in terms of compatible printing materials compared to other 3D printing technologies. The DIW methodology functions on the principle of extruding a viscoelastic liquid ink, thereby enabling a broad spectrum of material options (including ceramics, hydrogels, plastic biomaterials, and cells) and facilitating controlled deposition and maintenance of ink shape post-printing (Agarwal et al. 2021). On the contrary, FDM platforms predominantly depend on intense heating to trigger the phase change of substances (from solid to semi-molten and back to solid) during the printing process, a technique that is compatible with a wide range of thermoplastic materials. Nevertheless, the applicability of the FDM method for 4D printing in medical settings necessitates further exploration, given the current absence of suitable Smart Responsive Materials (SRMs) that can be transformed into FDM printable filaments. Another significant limitation of the extrusion-based methods is the comparatively sluggish printing speed at which items can be manufactured (Agarwal et al. 2021).

6.7.2 Biomedical Applications of 4D Printing

6.7.2.1 Tissue Engineering

The creation of dynamic tissue scaffolds involves the application and positioning of a 4D ink solution, filaments of (bio)materials, or a combination of multiple (bio)materials, which can either encapsulate cells and growth factors during the printing process or load cells after printing. There are two primary types of printed tissue scaffolds: cell-laden scaffolds and cell-seeded scaffolds. Cell-laden scaffolds are created by simultaneously depositing bioinks that contain biocompatible materials and cells. Commonly used materials for 4D printing in tissue engineering include gelatin, collagen, GelMA, alginate, hyaluronic acid (HA), decellularized extracellular matrices (dECMs), and their (methyl)acrylate derivatives. On the other hand, cell-seeded scaffolds are easier to produce, as the structures are printed, cleaned, and then supplemented with cells and/or growth factors. This method allows for a broader selection of printable materials, including synthetic plastic polymers. In tissue engineering, these 4D materials serve as artificial extracellular matrices (ECMs), providing both biophysical and biochemical signals to cells, thereby mimicking the natural cues of the native ECM and promoting cellular growth and differentiation (Wang et al. 2022).

6.7.2.2 Bone Tissue Engineering

The application of 3D printing in tissue engineering was initially explored in the field of bone repair. The development of 4D printing for bone scaffolds offers significant advantages, particularly in terms of their ability to be easily implanted through minimally invasive procedures and their precise adaptation to irregularly shaped bone defects. In their study, You et al. introduced a dual stimulus-responsive shape-memory polymer (SMP) that can undergo both swelling-induced macroscopic shape changes and heat-induced microscopic adjustments. They utilized a DLP printer to produce dynamic cell culturing, static micropillar, and static flat surfaces. The findings indicated a continuous cytoskeleton spreading on the dynamic surface, while the nuclear shape initially appeared slightly elliptical and later transformed into anisotropy post micropattern alteration. The images illustrated the effect of the dynamic surface on mechanical cues, offering additional support for its positive role in modulating cell behavior. By integrating shape-memory polymer (SMP) and hydrogel layers, this 4D-printed membrane was able to control multiscale structural deformations over time, which could potentially enhance bone repair processes in vivo (Wang et al. 2022).

Dias et al. effectively developed chitosan-based surfaces that are pH-sensitive and biodegradable. These surfaces are integrated with poly(L-lactic acid) (PLLA) and bioactive glass (BG) films to regulate the biomineralization process. Their research demonstrated that the transformation of chitosan at its critical pH range influenced the formation of apatite. Specifically, the transition from acidic to neutral pH levels led to an increase in surface roughness, thereby facilitating apatite formation. Consequently, the manipulation of pH variations enables the successful creation of apatite (Dias et al. 2008).

Hada et al. developed porous scaffolds for mandibular bone repair in bone tissue engineering (BTE) using a hybrid HA/PCL-based shape-memory polymer (SMP). The study showed that these SMP-based scaffolds could be compressed into temporary deformed shapes and later returned to their original porous structure upon exposure to body temperature (Hada et al. 2022). Furthermore, these scaffolds hold significant promise for application in 4D printing for optimal bone tissue repair under dynamic in vivo conditions.

Lee et al. effectively produced 4D high cell density structures by utilizing shape-altering hydrogels composed of oxidized and methacrylated alginate (OMA) and methacrylated gelatin (GelMA). The sequential photocrosslinking of OMA and GelMA enabled the creation of dual-layered hydrogels with adjustable geometric transformations independent of external triggers. These structures exhibited the capability to undergo shape modifications and facilitated the differentiation of stem cells into chondrogenic and osteogenic lineages. The results indicate that employing shape-altering hydrogels holds potential in regenerative medicine applications for constructing intricate tissues with regulated temporal alterations in shape (Lee et al. 2021).

6.7.2.3 Cartilage Tissue Engineering

The bioprinting technology is significantly advancing the development of scaffolds for cartilage tissue engineering (CTE), enabling the regeneration of cartilage at damaged sites. This technology has also been applied to create curved tracheal implants for cartilage repair. For instance, the incorporation of uni-aligned nanocellulose fibrils into acrylamide matrices results in longitudinal deformation upon swelling, leading to the formation of complex curvatures after swelling (Arif et al. 2022). Another notable example is the work by Kim et al., who employed 4D printing to fabricate a trachea implant using photo-crosslinked silk fibroin-based hydrogel. Using digital light processing (DLP) technology, the researchers produced a cell-laden hydrogel structure that was implanted into rabbits to assess its in-vivo response. The results showed significant cartilage regeneration within the rabbit trachea, demonstrating the potential of 4D-printed implants in promoting tissue repair and regeneration in complex anatomical structures like the trachea (Kim et al. 2020). Betsch et al. advanced the fabrication of synthetic fibrocartilage by dynamically modifying the matrix orientation during the printing process. They achieved this by directing the alignment of collagen fibers, which is a key factor in replicating the native mechanical properties of cartilage. The researchers successfully engineered complex, multi-layered cartilage scaffolds using bio-inks loaded with magnetic chondrocytes. This approach allowed for precise control over the spatial distribution and orientation of cells within the scaffold, enhancing the functionality and potential for cartilage tissue engineering applications. The use of magnetic chondrocytes facilitated the fine-tuning of cell alignment and distribution during printing, offering new possibilities for creating scaffolds that mimic the intricate structure and mechanical properties of native cartilage (Betsch et al. 2018).

6.7.2.4 Medical Devices

Medical devices are instruments with the potential to be used in illness, injury, or disability diagnosis, prevention, monitoring, prognosis, treatment, or relief. The medical device industry has seen a substantial transformation because of 4D printing technology. Specifically, the integration of soft robotics and 4D printing can have a synergistic impact and almost limitless possibilities in this field. The goal of soft robotics is to build softrobots that, by virtue of the inherent elastomeric qualities of soft materials, can execute complicated tasks with little effort. This section will cover the latest advancements in soft robotics, biomedical devices, including 4D-printed stents (Agarwal et al. 2021). Lin et al. pioneered a novel approach for fabricating an absorbable left atrial appendage occluder (LAAO) using 4D printing techniques, designed to reduce stroke risk associated with atrial fibrillation. The LAAO was manufactured from a poly(lactic acid) (PLA) matrix combined with Fe_3O_4 magnetic nanocomposites, utilizing fused deposition modeling (FDM). The choice of material and geometric configuration was carefully optimized to align with the mechanical properties of the left atrial appendage (LAA) tissue, based on its stress–strain curve. The occluders were demonstrated to be durable, biodegradable, and biocompatible, with the remarkable ability to undergo a 4D transformation when exposed to an external magnetic field. This magnetic responsiveness allowed the LAAO to conform to the anatomical shape of the LAA during transcatheter deployment. In a proof-of-concept experiment, the LAAO was successfully deployed in a freshly isolated swine heart, showcasing the feasibility and potential clinical application of this 4D-printed device for transcatheter LAA closure, ultimately improving stroke prevention in patients with atrial fibrillation (Lin et al. 2021).

The research direction of 4D printing in medical devices, particularly when integrated with soft robotics, presents a promising frontier in the medical and biomedical fields. The key themes of this emerging research are focused on advancing the adaptability, functionality, and performance of medical devices to address complex clinical challenges in ways that were previously unattainable.

In summary, the future of 4D printing in medical devices is poised to revolutionize healthcare by enabling the development of highly adaptable, customizable, and efficient devices. Through the integration of soft robotics, responsive materials, and advanced sensing technologies, 4D printing could lead to breakthroughs in personalized medicine, minimally invasive surgeries, and dynamic therapeutic interventions.

As research continues, the field holds immense promise for improving patient care and treatment outcomes.

6.8 Conclusion

The intersection of nanotechnology, biomaterials science, and additive manufacturing has initiated a new era in the field of medicine and regenerative treatments. Leading this transformative movement are stimulus-sensitive biocompatible nanocomposites. These innovative substances blend the distinct characteristics of nanomaterials with the functionalities of biocompatible polymers, resulting in a novel category of intelligent materials capable of reacting to external stimuli like light, temperature, pH, or magnetic fields. This adaptability facilitates the creation of biocompatible frameworks, drug delivery systems, and implantable devices with unparalleled precision and capabilities. Nevertheless, the absence of feedback mechanisms that can effectively respond to multiple stimuli simultaneously and adapt to the body's physiological conditions remains a significant challenge. Additionally, these systems encounter obstacles such as intricate design requirements, long-term durability issues, and relatively high expenses. Scholars are actively seeking solutions to surmount these hurdles.

References

Agarwal T, Hann SY, Chiesa I, Cui H, Celikkin N, Micalizzi S et al (2021) 4D printing in biomedical applications: emerging trends and technologies. J Mater Chem B 9(37):7608–7632

Arif ZU, Khalid MY, Zolfagharian A, Bodaghi M (2022) 4D bioprinting of smart polymers for biomedical applications: recent progress, challenges, and future perspectives. React Funct Polym 179:105374

Arjama M, Mehnath S, Jeyaraj M (2022) Self-assembled hydrogel nanocube for stimuli responsive drug delivery and tumor ablation by phototherapy against breast cancer. Int J Biol Macromol 213:435–446

Augustine S, Singh J, Srivastava M, Sharma M, Das A, Malhotra BD (2017) Recent advances in carbon based nanosystems for cancer theranostics. Biomater Sci 5(5):901–952

Bagalkot V, Zhang L, Levy-Nissenbaum E, Jon S, Kantoff PW, Langer R et al (2007) Quantum dot– aptamer conjugates for synchronous cancer imaging, therapy, and sensing of drug delivery based on bi-fluorescence resonance energy transfer. Nano Lett 7(10):3065–3070

Basoglu H, Goncu B, Akbas F (2018) Magnetic nanoparticle-mediated gene therapy to induce Fas apoptosis pathway in breast cancer. Cancer Gene Ther 25(5):141–147

Betsch M, Cristian C, Lin YY, Blaeser A, Schöneberg J, Vogt M et al (2018) Incorporating 4D into bioprinting: real-time magnetically directed collagen fiber alignment for generating complex multilayered tissues. Adv Healthc Mater 7(21):1800894

Blum AP, Kammeyer JK, Rush AM, Callmann CE, Hahn ME, Gianneschi NC (2015) Stimuli-responsive nanomaterials for biomedical applications. J Am Chem Soc 137(6):2140–2154

Buhren BA, Schrumpf H, Hoff N-P, Bölke E, Hilton S, Gerber PA (2016) Hyaluronidase: from clinical applications to molecular and cellular mechanisms. Eur J Med Res 21:1–7

Chen Z, Mao R, Liu Y (2012) Fullerenes for cancer diagnosis and therapy: preparation, biological and clinical perspectives. Curr Drug Metab 13(8):1035–1045

Chen W-H, Luo G-F, Vázquez-González M, Cazelles R, Sohn YS, Nechushtai R et al (2018a) Glucose-responsive metal–organic-framework nanoparticles act as "smart" sense-and-treat carriers. ACS Nano 12(8):7538–7545

Chen Z, Wang J, Sun W, Archibong E, Kahkoska AR, Zhang X et al (2018b) Synthetic beta cells for fusion-mediated dynamic insulin secretion. Nat Chem Biol 14(1):86–93

Cui T, Yan Z, Qin H, Sun Y, Ren J, Qu X (2019) A Sequential target-responsive nanocarrier with enhanced tumor penetration and neighboring effect in vivo. Small 15(42):1903323

Dias CI, Mano JF, Alves NM (2008) PH-Responsive biomineralization onto chitosan grafted biodegradable substrates. J Mater Chem 18(21):2493–2499

Dimatteo R, Darling NJ, Segura T (2018) In situ forming injectable hydrogels for drug delivery and wound repair. Adv Drug Deliv Rev 127:167–184

Dos Santos AM, Carvalho SG, Meneguin AB, Sábio RM, Gremião MPD, Chorilli M (2021) Oral delivery of micro/nanoparticulate systems based on natural polysaccharides for intestinal diseases therapy: challenges, advances and future perspectives. J Control Release 334:353–366

Du M, Yang Z, Lu W, Wang B, Wang Q, Chen Z et al (2020) Design and development of spirulina polysaccharide-loaded nanoemulsions with improved the antitumor effects of paclitaxel. J Microencapsul 37(6):403–412

Dvir T, Timko BP, Brigham MD, Naik SR, Karajanagi SS, Levy O et al (2011) Nanowired three-dimensional cardiac patches. Nat Nanotechnol 6(11):720–725

Fiorillo M, Verre AF, Iliut M, Peiris-Pagés M, Ozsvari B, Gandara R et al (2015) Graphene oxide selectively targets cancer stem cells, across multiple tumor types: implications for non-toxic cancer treatment, via "differentiation-based nano-therapy." Oncotarget 6(6):3553

Gangrade A, Gawali B, Jadi PK, Naidu VG, Mandal BB (2020) Photo-electro active nanocomposite silk hydrogel for spatiotemporal controlled release of chemotherapeutics: an in vivo approach toward suppressing solid tumor growth. ACS Appl Mater Interfaces 12(25):27905–27916

Ghasemi Y, Peymani P, Afifi S (2009) Quantum dot: magic nanoparticle for imaging, detection and targeting. Acta Biomed 80(2):156–165

Gordijo CR, Shuhendler AJ, Wu XY (2010) Glucose-responsive bioinorganic nanohybrid membrane for self-regulated insulin release. Adv Func Mater 20(9):1404–1412

Hada V, Hashmi SAR, Mili M, Gorhe N, Sagiri SS, Pal K, Chawdhary R, Khan M, Naik A, Prashant N, Srivastava AK (2022). Polycaprolactone-based shape memory polymeric nanocomposites for biomedical applications. Food Med Environ Appl Nanomater 413–433

Heister E, Neves V, Tîlmaciu C, Lipert K, Beltrán VS, Coley HM et al (2009) Triple functionalisation of single-walled carbon nanotubes with doxorubicin, a monoclonal antibody, and a fluorescent marker for targeted cancer therapy. Carbon 47(9):2152–2160

Hu J, Zhang G, Liu S (2012) Enzyme-responsive polymeric assemblies, nanoparticles and hydrogels. Chem Soc Rev 41(18):5933–5949

Jaiswal M, Dudhe R, Sharma P (2015) Nanoemulsion: an advanced mode of drug delivery system. 3 Biotech 5:123–127

Jamieson T, Bakhshi R, Petrova D, Pocock R, Imani M, Seifalian AM (2007) Biological applications of quantum dots. Biomaterials 28(31):4717–4732

Jia R, Teng L, Gao L, Su T, Fu L, Qiu Z et al (2021) Advances in multiple stimuli-responsive drug-delivery systems for cancer therapy. Int J Nanomedicine 1525–1551

Kang M-L, Kim J-E, Im G-I (2016) Thermoresponsive nanospheres with independent dual drug release profiles for the treatment of osteoarthritis. Acta Biomater 39:65–78

Katsarou A, Gudbjörnsdottir S, Rawshani A, Dabelea D, Bonifacio E, Anderson BJ et al (2017) Type 1 diabetes mellitus. Nat Rev Dis Primers 3(1):1–17

Khalid MY, Arif ZU, Noroozi R, Zolfagharian A, Bodaghi M (2022) 4D printing of shape memory polymer composites: a review on fabrication techniques, applications, and future perspectives. J Manuf Process 81:759–797

Khanna N, Chatterji T, Singh S, Pandey S (2023) Application of stimuli responsive nanocomposites in drug delivery and theranostics to counter cancer proliferation and drug resistance. J Drug Deliv Sci Technol 104958

Kim SH, Seo YB, Yeon YK, Lee YJ, Park HS, Sultan MT et al (2020) 4D-bioprinted silk hydrogels for tissue engineering. Biomaterials 260:120281

Kulkarni JA, Witzigmann D, Leung J, Tam YYC, Cullis PR (2019) On the role of helper lipids in lipid nanoparticle formulations of siRNA. Nanoscale 11(45):21733–21739

Lavrador P, Gaspar VM, Mano JF (2018) Stimuli-responsive nanocarriers for delivery of bone therapeutics–barriers and progresses. J Control Release 273:51–67

Lavrador P, Esteves MR, Gaspar VM, Mano JF (2021) Stimuli-responsive nanocomposite hydrogels for biomedical applications. Adv Func Mater 31(8):2005941

Lee S-W, Kim Y-M, Cho CH, Kim YT, Kim SM, Hur SY et al (2018) An open-label, randomized, parallel, phase II trial to evaluate the efficacy and safety of a cremophor-free polymeric micelle formulation of paclitaxel as first-line treatment for ovarian cancer: a Korean Gynecologic Oncology Group study (KGOG-3021). Cancer Res Treat: Off J Korean Cancer Assoc 50(1):195

Lee YB, Jeon O, Lee SJ, Ding A, Wells D, Alsberg E (2021) Induction of four-dimensional spatiotemporal geometric transformations in high cell density tissues via shape-changing hydrogels. Adv Funct Mater 2010104:1–9

Li L, Scheiger JM, Levkin PA (2019) Design and applications of photoresponsive hydrogels. Adv Mater 31(26):1807333

Liang Y, Kiick KL (2016) Liposome-cross-linked hybrid hydrogels for glutathione-triggered delivery of multiple cargo molecules. Biomacromol 17(2):601–614

Lin C, Liu L, Liu Y, Leng J (2021) 4D printing of bioinspired absorbable left atrial appendage occluders: a proof-of-concept study. ACS Appl Mater Interfaces 13(11):12668–12678

Liu X, Li C, Lv J, Huang F, An Y, Shi L et al (2020) Glucose and H2O2 dual-responsive polymeric micelles for the self-regulated release of insulin. ACS Appl Bio Mater 3(3):1598–1606

Ma P, Dong X, Swadley CL, Gupte A, Leggas M, Ledebur HC et al (2009) Development of idarubicin and doxorubicin solid lipid nanoparticles to overcome PGP-mediated multiple drug resistance in leukemia. J Biomed Nanotechnol 5(2):151–161

Ma Y, Ji Y, Zhong T, Wan W, Yang Q, Li A et al (2017) Bioprinting-based PDLSC-ECM screening for in vivo repair of alveolar bone defect using cell-laden, injectable and photocrosslinkable hydrogels. ACS Biomater Sci Eng 3(12):3534–3545

Ma Y, Yao J, Liu Q, Han T, Zhao J, Ma X et al (2020) Liquid bandage harvests robust adhesive, hemostatic, and antibacterial performances as a first-aid tissue adhesive. Adv Func Mater 30(39):2001820

Ma W, Hua D, Xiong R, Huang C (2023) Bio-based stimuli-responsive materials for biomedical applications. Mater Adv 4(2):458–475

Martín-Saavedra F, Ruiz-Hernández E, Escudero-Duch C, Prieto M, Arruebo M, Sadeghi N et al (2017) Lipogels responsive to near-infrared light for the triggered release of therapeutic agents. Acta Biomater 61:54–65

Mehrali M, Thakur A, Pennisi CP, Talebian S, Arpanaei A, Nikkhah M et al (2017) Nanoreinforced hydrogels for tissue engineering: biomaterials that are compatible with load-bearing and electroactive tissues. Adv Mater 29(8):1603612

Menjoge AR, Kannan RM, Tomalia DA (2010) Dendrimer-based drug and imaging conjugates: design considerations for nanomedical applications. Drug Discov Today 15(5–6):171–185

Merino S, Martín C, Kostarelos K, Prato M, Vázquez E (2015) Nanocomposite hydrogels: 3D polymer–nanoparticle synergies for on-demand drug delivery. ACS Nano 9(5):4686–4697

Mitchell MJ, Billingsley MM, Haley RM, Wechsler ME, Peppas NA, Langer R (2021) Engineering precision nanoparticles for drug delivery. Nat Rev Drug Discovery 20(2):101–124

Mroz P, Tegos GP, Gali H, Wharton T, Sarna T, Hamblin MR (2007) Photodynamic therapy with fullerenes. Photochem Photobiol Sci 6(11):1139–1149

Ooi HW, Hafeez S, Van Blitterswijk C, Moroni L, Baker MB (2017) Hydrogels that listen to cells: a review of cell-responsive strategies in biomaterial design for tissue regeneration. Mater Horiz 4(6):1020–1040

Oroval M, Díez P, Aznar E, Coll C, Marcos MD, Sancenón F et al (2017) Self-regulated glucose-sensitive neoglycoenzyme-capped mesoporous silica nanoparticles for insulin delivery. Chemistry–A Eur J 23(6):1353–1360

Palmerston Mendes L, Pan J, Torchilin VP (2017) Dendrimers as nanocarriers for nucleic acid and drug delivery in cancer therapy. Molecules 22(9):1401

Parhi R (2020) Drug delivery applications of chitin and chitosan: a review. Environ Chem Lett 18(3):577–594

Qin Y, Chen J, Bi Y, Xu X, Zhou H, Gao J et al (2015) Near-infrared light remote-controlled intracellular anti-cancer drug delivery using thermo/pH sensitive nanovehicle. Acta Biomater 17:201–209

Rapp TL, DeForest CA (2021) Targeting drug delivery with light: A highly focused approach. Adv Drug Deliv Rev 171:94–107

Richards DJ, Tan Y, Coyle R, Li Y, Xu R, Yeung N et al (2016) Nanowires and electrical stimulation synergistically improve functions of hiPSC cardiac spheroids. Nano Lett 16(7):4670–4678

Samad A, Sultana Y, Aqil M (2007) Liposomal drug delivery systems: an update review. Curr Drug Deliv 4(4):297–305

Sawyer SW, Dong P, Venn S, Ramos A, Quinn D, Horton JA et al (2017) Conductive gelatin methacrylate-poly (aniline) hydrogel for cell encapsulation. Biomed Phys Eng Express 4(1):015005

Sedighi M, Mahmoudi Z, Ghasempour A, Shakibaie M, Ghasemi F, Akbari M et al (2023) Nanostructured multifunctional stimuli-responsive glycopolypeptide-based copolymers for biomedical applications. J Control Release 354:128–145

Tabata Y, Murakami Y, Ikada Y (1997) Photodynamic effect of polyethylene glycol–modified fullerene on tumor. Jpn J Cancer Res 88(11):1108–1116

Tan L, Li J, Liu Y, Zhou H, Zhang Z, Deng L (2015) Synthesis and characterization of β-cyclodextrin-conjugated alginate hydrogel for controlled release of hydrocortisone acetate in response to mechanical stimulation. J Bioact Compat Polym 30(6):584–599

Tao W, He Z (2018) ROS-responsive drug delivery systems for biomedical applications. Asian J Pharm Sci 13(2):101–112

Thoniyot P, Tan MJ, Karim AA, Young DJ, Loh XJ (2015) Nanoparticle–hydrogel composites: concept, design, and applications of these promising, multi-functional materials. Adv Sci 2(1–2):1400010

Üner M, Yener G (2007) Importance of solid lipid nanoparticles (SLN) in various administration routes and future perspectives. Int J Nanomed 2(3):289–300

Vasconcelos I, Santos T (2023) Nanotechnology applications in sepsis: essential knowledge for clinicians. Pharmaceutics 15(6):1682

Vlieghe P, Lisowski V, Martinez J, Khrestchatisky M (2010) Synthetic therapeutic peptides: science and market. Drug Discov Today 15(1–2):40–56

Wang X, Liu X, Li Y, Wang P, Feng X, Liu Q et al (2017) Sensitivity to antitubulin chemotherapeutics is potentiated by a photoactivable nanoliposome. Biomaterials 141:50–62

Wang J, Potocny AM, Rosenthal J, Day ES (2019) Gold nanoshell-linear tetrapyrrole conjugates for near infrared-activated dual photodynamic and photothermal therapies. ACS Omega 5(1):926–940

Wang R, Shi J, Song L, Zheng S, Liu X, Hong M et al (2021) Sustained antitumor immunity based on persistent luminescence nanoparticles for cancer immunotherapy. Adv Func Mater 31(52):2106884

Wang Y, Cui H, Esworthy T, Mei D, Wang Y, Zhang LG (2022) Emerging 4D printing strategies for next-generation tissue regeneration and medical devices. Adv Mater 34(20):2109198

Widakowich C, de Castro Jr G, De Azambuja E, Dinh P, Awada A (2007) Side effects of approved molecular targeted therapies in solid cancers. Oncologist 12(12):1443–1455

Wu M, Chen J, Huang W, Yan B, Peng Q, Liu J et al (2020) Injectable and self-healing nanocomposite hydrogels with ultrasensitive pH-responsiveness and tunable mechanical properties: implications for controlled drug delivery. Biomacromol 21(6):2409–2420

Wu C, Belenda C, Leroux JC, Gauthier MA (2011) Interplay of chemical microenvironment and redox environment on thiol–disulfide exchange kinetics. Chemistry–A Eur J 17(36):10064–10070

Xiao S, Guo L, Ai C, Shang M, Shi D, Meng D et al (2023b) Ph-/redox-responsive nanodroplet combined with ultrasound-targeted microbubble destruction for the targeted treatment of drug-resistant triple negative breast cancer. ACS Appl Mater Interfaces 15(7):8958–8973

Xiao R, Zhou G, Wen Y, Ye J, Li X, Wang X (2023) Recent advances on stimuli-responsive biopolymer-based nanocomposites for drug delivery. Compos Part B: Eng 111018

Xu L, Zhang H, Wu Y (2014) Dendrimer advances for the central nervous system delivery of therapeutics. ACS Chem Neurosci 5(1):2–13

Xu C, Lin T, Zhao X, Gan Y, Huang J, Zhang J et al (2023) Microenvironment responsive hypoxia-mimetic DFO composite hydrogel for on-demand neovascularization to promote tendon-to-bone healing. Compos B Eng 259:110726

Yang W, Liang H, Ma S, Wang D, Huang J (2019) Gold nanoparticle based photothermal therapy: development and application for effective cancer treatment. Sustain Mater Technol 22:e00109

Yi P, Wang Y, Zhang S, Zhan Y, Zhang Y, Sun Z et al (2017) Stimulative nanogels with enhanced thermosensitivity for therapeutic delivery via β-cyclodextrin-induced formation of inclusion complexes. Carbohyd Polym 166:219–227

Yin T, Liu J, Zhao Z, Zhao Y, Dong L, Yang M et al (2017) Redox sensitive hyaluronic acid-decorated graphene oxide for photothermally controlled tumor-cytoplasm-selective rapid drug delivery. Adv Func Mater 27(14):1604620

Yu J, Zhang Y, Ye Y, DiSanto R, Sun W, Ranson D et al (2015) Microneedle-array patches loaded with hypoxia-sensitive vesicles provide fast glucose-responsive insulin delivery. Proc Natl Acad Sci 112(27):8260–8265

Zhang L, Gu F, Chan J, Wang A, Langer R, Farokhzad O (2008) Nanoparticles in medicine: therapeutic applications and developments. Clin Pharmacol Ther 83(5):761–769

Zhang J, Liu M-Q, Zhang J-L, Li B-A, Wang X-Y, Han P (2015) Biotinylated epidermal growth factor surface modified lipid nanoparticles to enhance the targeting efficiency in liver cancer therapy. J Biomater Tissue Eng 5(2):135–141

Zhang M, Ma Y, Wang Z, Han Z, Gao W, Zhou Q et al (2019) A CD44-targeting programmable drug delivery system for enhancing and sensitizing chemotherapy to drug-resistant cancer. ACS Appl Mater Interfaces 11(6):5851–5861

Zhang L, Beatty A, Lu L, Abdalrahman A, Makris TM, Wang G et al (2020a) Microfluidic-assisted polymer-protein assembly to fabricate homogeneous functionalnanoparticles. Mater Sci Eng, C 111:110768

Zhang C, Hong S, Liu M-D, Yu W-Y, Zhang M-K, Zhang L et al (2020b) PH-sensitive MOF integrated with glucose oxidase for glucose-responsive insulin delivery. J Control Release 320:159–167

Zhao X, Hilliard LR, Mechery SJ, Wang Y, Bagwe RP, Jin S et al (2004) A rapid bioassay for single bacterial cell quantitation using bioconjugated nanoparticles. Proc Natl Acad Sci 101(42):15027–15032

Zhao X, Wei Z, Zhao Z, Miao Y, Qiu Y, Yang W et al (2018) Design and development of graphene oxide nanoparticle/chitosan hybrids showing pH-sensitive surface charge-reversible ability for efficient intracellular doxorubicin delivery. ACS Appl Mater Interfaces 10(7):6608–6617

Zhou W, Qiao Z, Nazarzadeh Zare E, Huang J, Zheng X, Sun X et al (2020) 4D-printed dynamic materials in biomedical applications: chemistry, challenges, and their future perspectives in the clinical sector. J Med Chem 63(15):8003–8024

Zhu J, Li F, Wang X, Yu J, Wu D (2018) Hyaluronic acid and polyethylene glycol hybrid hydrogel encapsulating nanogel with hemostasis and sustainable antibacterial property for wound healing. ACS Appl Mater Interfaces 10(16):13304–13316

Zhu Y, Yao Z, Liu Y, Zhang W, Geng L, Ni T (2020) Incorporation of ROS-responsive substance P-loaded zeolite imidazolate framework-8 nanoparticles into a Ca2+-cross-linked alginate/pectin hydrogel for wound dressing applications. Int J Nanomedicine 333–346

Chapter 7
Biocompatible Nanocomposites for Cancer Therapy

Souad A. Elfeky and Nayer Qenawi

Abstract One of the most exciting areas in cancer therapy has been the area of biocompatible nanocomposites, which is essentially the melding of nanotechnology and materials science to create better drug delivery mechanisms, and subsequently more effective treatments. These nanocomposites, usually consisting of polymers, ceramics, and metals, can be engineered to release chemotherapeutics in a controlled fashion and to target only tumour cells so as to lessen systemic side effects. They are biocompatible so they are less toxic and with the addition of targeting ligands through functionalisation the drug delivery becomes more specific. New developments in the creation of these nanocomposites have shown promise in overcoming several limitations of contemporary medicine, namely drug resistance and low bioavailability. This article reviews the current literature on several types of biocompatible nanocomposites, their proposed methods of action against cancer, and the various clinical trials being conducted to evaluate their effectiveness and toxicity. These novel materials could very well play a major role in the future of cancer treatment, leading to more efficient and individualized approaches in cancer therapy.

Keywords Biocompatible nanocomposites · Cancer therapy · Drug delivery systems · Controlled release · Tumor targeting

7.1 Introduction

7.1.1 Background

Cancer is a life-threatening disease that involves abnormal cell growth with the potential to invade or spread to other parts of the body. According to the World Health Organisation (WHO), cancer is the second leading cause of death globally, and was

S. A. Elfeky (✉) · N. Qenawi
Department of Laser Application in Metrology, Photochemistry, and Agriculture (LAMPA), National Institute of Laser Enhanced Sciences (NILES), Cairo University, Giza, Egypt
e-mail: dr_souad_elfeky@niles.cu.edu.eg

R. Balakrishnan et al. (eds.), *Biocompatible Nanocomposites*, Advanced Structured Materials 239, https://doi.org/10.1007/978-981-96-7450-3_7

responsible for an estimated 9.6 million deaths in 2018 (Cai and Liu 2021). The prevalence of cancer cases is increasing year by year, making it a prominent disease worldwide. With such devastating consequences, it is essential to explore advanced treatment options that can effectively combat cancer while minimising the harmful effects on the body (Debela et al. 2021). There are various forms of cancer therapy available, each tailored to the specific type of cancer a patient has. These therapies include radiation therapy, chemotherapy, hormone therapy, immunotherapy, targeted therapy, and stem cell transplant. The choice of treatment depends on several factors, such as the type and stage of cancer, as well as the individual patient's health and preferences (El-ghannam et al. 2024). Chemotherapy, for instance, involves the use of drugs that travel through the bloodstream to reach cancer cells throughout the body. While this method can be effective in killing cancer cells, it also poses risks to healthy cells, resulting in side effects like hair loss, mouth sores, and vomiting. On the other hand, radiation therapy utilises high doses of radiation to destroy cancer cells and shrink tumours. This approach has proven successful in treating various types of cancer, offering the possibility of cure, prevention of recurrence, or control of tumour growth (Anand et al. 2023). Despite the progress made in cancer therapy, achieving a complete cure remains challenging. Many treatment options are associated with potential harm to the body, emphasising the need for continuous advancements in the field. In recent years, nanomaterials and nanotechnology have experienced remarkable growth in the development of diagnostic and therapeutic tools. Biocompatible nanocomposites, in particular, have emerged as promising candidates for cancer therapy due to their versatility and multifunctionality in areas such as drug delivery, imaging and diagnostics, and therapy (Chehelgerdi et al. 2023). The use of biocompatible nanocomposites has led to superior treatment outcomes, improved patient management, and increased survival rates. These materials offer unique advantages, such as targeted drug delivery, enhanced imaging capabilities, and the ability to customise treatment based on individual patient needs. Furthermore, they exhibit properties that help overcome the challenges faced in traditional cancer therapies, such as reducing toxicity to healthy cells and increasing treatment efficacy (Wei et al. 2019). This chapter provides an overview of different types of biocompatible nanocomposites, highlighting their important properties and characterisations. It also explores the recent advancements and challenges in the field, offering insights into the potential of biocompatible nanocomposites for cancer therapy. Through continued research and innovation, the integration of nanomaterials and nanotechnology into cancer treatment holds great promise for improving patient outcomes and ultimately finding a cure for this devastating disease.

7.1.2 Importance of Biocompatible Nanocomposites in Cancer Therapy

Recently, researchers have focused on the development of "biocompatible nanocomposites" since conventional cancer therapy methods, such as chemotherapy and radiotherapy, have various limitations and side effects. Especially, to combat the highly prevalent and deadly disease, cancer, it is an emerging research area to improve the existing cancer therapy methods or to provide new alternative options (Dhilip Kumar and Abrahamse 2021). Nanocomposites refer to a material which is a combination of two or more constituent materials on a nanometre scale, while "biocompatible" means the ability of a material to perform with an appropriate host response in a specific application. Therefore, "biocompatible nanocomposites" can be defined as the technique to design and improve the material which is being used for the implant so that it can perform its desired functionality without having any side effects to the host and it can withstand in the body environment resisting degradation (Kargozar et al. 2023). Moreover, these advanced nanocomposites show advantages of enhancing physical, chemical and mechanical properties. This is because the nanosized particles or fillers provide a higher surface area to volume ratio and different spatial arrangements, which results in unique electrical, thermal, optical, magnetic, and mechanical properties compared to bulk materials (Khan et al. 2019). In terms of cancer therapy, the unique properties of biocompatible nanocomposites make them the ideal candidate for many applications such as drug delivery, imaging and diagnosis, photothermal therapy and theranostics (Lungu et al. 2019). For instance, these composites can be used as a drug carrier to improve the cancer treatment effect by increasing the drug bioavailability, reducing the degradation rate of drug and providing sustained drug release. Also, they can improve the diagnosis of cancer by providing better images and higher sensitivity of detection because of their high binding affinity, high specificity and loading capacity with signalling moieties (Lungu et al. 2019). Last but not least, the strong light absorption capability and precisely controllable release of the loaded drugs of nanocomposites make them excellent materials for the development of photothermal therapy which is the most commonly studied. So from the overall perspective of using the nanocomposites in cancer therapy, the combination of these new technology researches will lead to a multidisciplinary approach in the battle against cancer. The integration of nanotechnology into the ever-growing endeavour to eliminate cancer permits the development of new generations of tools for both the diagnosis and treatment of many cancer types.

7.1.3 Objectives of the Work

The main focus of this chapter is to give a thorough review of biocompatible nanocomposites and their uses in cancer treatment. Specifically, the chapter aims to:

- Explore the Fundamental Principles: To explain how biocompatible nanocomposites are designed and synthesised, with an emphasis on the materials that are used and the methodologies that are used to prepare these materials.
- Highlight Mechanisms of Action: To examine the potential of these nanocomposites to enhance drug delivery through superior pharmacokinetics, targeted release, and reduced systemic toxicity, ultimately striving for optimal therapeutic effectiveness.
- Review Recent Advances: To discuss some of the latest findings, as well as to highlight some of the uses, like targeted therapy, combination therapies and multifunctional platforms.
- Address Challenges and Limitations: To talk about the problems with the use of biocompatible nanocomposites today in the clinic, including biocompatibility issues, regulatory issues, and the ability to scale up production.
- Encourage Future Research: To identify the gaps in the existing literature and suggest possible directions for future research, hopefully as a catalyst for interdisciplinary investigation and innovation in the development of improved cancer therapies.

This chapter is completely focused on the applications of biocompatible nanocomposites in cancer therapy. Through the use of these novel materials, we can create new strategies in the fight against cancer. Hopefully, it will help researchers to understand how the advanced nanocomposites can be utilised as an innovative method for the current medication of cancer and from here new exciting breakthroughs will be discovered. It begins with an introduction to cancer therapy and the significance of using biocompatible nanocomposites. The next section explores the definition and characteristics of biocompatible nanocomposites, as well as the different types and synthesis methods.

7.2 Types of Biocompatible Nanocomposites

Comprising of nanosized fillers and matrixes, nanocomposites are heterogeneous materials. They are separated into metal matrix nanocomposites (MMNC), polymer matrix nanocomposites (PMNC), and ceramic matrix nanocomposites (CMNC) based on the inserted matrix material. Any shape, including nanofiber, nanoparticle, and nanosheet, as well as organic and inorganic sources, is possible for the integrated nanosized filler phase. Consequently, nanocomposites can be either organic-organic or organic–inorganic based on the phases that contribute (Omanović-Mikličanin et al. 2020). Classification of nanocomposites based on the constituent matrix (polymer, ceramic, or metal) and incorporated phases (Fig. 7.1) (Kargozar et al. 2024).

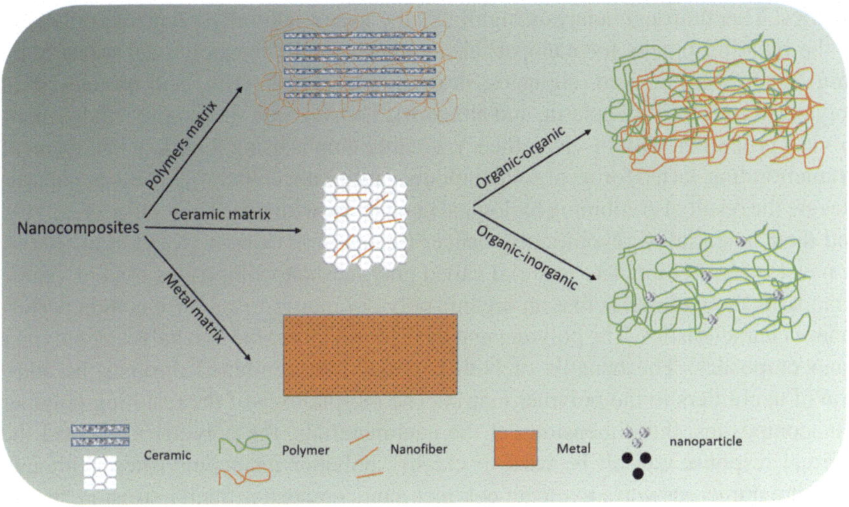

Fig. 7.1 Showing the classification of nanocomposites based on the constituent matrix (polymer, ceramic, or metal) and incorporated phases. Reproduced from reference (Kargozar et al. 2024) under the terms and conditions of the Creative Commons Attribution (CC BY) license (https://creativecommons.org/licenses/by/4.0/)

7.2.1 Polymer-Based Nanocomposites

Unchecked cell multiplication in the body causes tumours, which in turn cause the complex and widely distributed disease known as cancer. The challenging problem of developing effective cancer treatments has led scientists to look at cutting-edge approaches. One strategy that shows promise is the use of nanoparticles based on polymers, which are minuscule particles composed of chemicals that are safe for the body. These nanoparticles not only improve therapy efficacy but also lessen harmful effects on healthy cells by directly delivering potent anticancer drugs to cancer cells. (Al-Sahlawi et al. 2024) investigates the use of polymer-based nanoparticles for targeted cancer treatment. The benefits of polymer-based nanoparticles are underlined, including their great drug-loading ability, prolonged half-life, and active targeting of cancer cells with the least degree of damage to healthy tissues. More research and clinical studies are required to optimise the use of polymer-based nanoparticles in tailored cancer therapy and ultimately enhance patient outcomes in the field of oncology (Al-Sahlawi et al. 2024). Natural polymer-based nanotherapeutics have become a fascinating direction in this endeavour, garnering significant interest because of their intrinsic qualities. These characteristics include a broad spectrum of therapeutic payloads, longer circulation time, minimal toxicity, biodegradability, and biocompatibility. These devices' distinct morphology, size, and form allow for deep tissue penetration, which enhances both active and passive targeting techniques. Moreover, these nanotherapeutics leverage specific cellular and subcellular trafficking channels, giving precise control over drug release kinetics. This thorough analysis highlights the use of naturally occurring polymers as the building blocks for nanoparticle synthesis, including polysaccharides (e.g., chitosan, hyaluronic acid, alginates, dextran, and cyclodextrin) and protein-based polymers (e.g., ferritin, gelatin, and albumin). The research thoroughly studies their in-vitro properties with in-vivo efficacy, notably emphasising their key involvement in ameliorating varied forms of solid tumours within cancer therapy. These nanoparticles are the result of combining biological knowledge with material science creativity, and their potential to revolutionise cancer treatment is evident (Kuna et al. 2024). A novel type of composite material called polymer nanocomposites uses inorganic nanoparticles scattered across an organic polymer matrix to enhance certain performance characteristics. The polymer and the nanofillers are the two halves of polymer nanocomposites. The benefits of both materials are combined through the addition of nanofillers to the polymer matrix. The morphology of the resulting polymer nanocomposites, the dimensions of the nanomaterials, the polymer class, and the thermal response can all be used to classify polymer nanocomposites. Although there are numerous ways to create polymer nanocomposites, melt extrusion, in-situ polymerization, and solution dispersion are the most often used techniques. Excellent characteristics of polymer nanocomposites include electrical characteristics, barrier resistance, and magnetic efficiency. Polymer nanocomposites offer significant benefits. Due to their exceptional qualities and potential uses, polymer nanocomposites are used in medicinal applications (Khan et al. 2023). Studies are being conducted on synthetic polymers, biopolymers, and their nanocomposites; some of these materials are already in use in various medical fields. One can list a variety of synthetic materials, including polyesters, silicones, polyolefins, and fluorinated polymers. Biopolymers, including proteins like silk and fibroin and polysaccharides like chitosan, hyaluronic acid, starch, cellulose, and alginates, have also gained popularity and are being researched for use in medical applications. In addition to synthetic and biopolymers, their nanocomposites—hybrids made of a macromolecular matrix and an organic or mineral nanofiller—have garnered a lot of interest in the last several years in the domains of medicine and other sciences because of their exceptional qualities. (Feldman 2019) highlights the recent research conducted on polymers, biopolymers, nanocomposites, liposomes, hydrogels, nanogels, polymer micelles, and polymer hydrogels used as drugs or drug carriers for cancer therapy. It also highlights the polymers' ability to respond to both internal and external stimuli to become more active and efficient. With superior outcomes, they can take the place of traditional cancer medication carriers (Feldman 2019). Polymer-based nanocomposites represent a novel class of hybrid materials with numerous intriguing bioapplications, including drug delivery, tissue engineering, and a variety of biotechnological uses. For their design and development, a basic understanding of their structure, characteristics, and the interaction between polymers and matrixes is necessary. Additional adjustments should be made based on these fundamental requirements. Furthermore, their potential applications depend on the management of their chemical and physical properties as well as the evaluation of their biological interactions. The growing biomedical and biotechnological uses of polymer-based nanocomposites are covered (Tripathy 2017). The recent advancement in the battle against cancer that is particularly enticing is the combination of bioactive and nanotechnology. Natural bioactive compounds provide nutritional benefits as well as potential applications as chemotherapeutic drugs for the treatment of cancer. Medicinal compounds have been successfully delivered to specific areas using alginate, starch, xanthan gum, pectin, guar gum, hyaluronic acid, gelatin, albumin, collagen, cellulose, chitosan, and other biopolymers. Natural polymeric nanobiocomposites have attracted a lot of attention in the development of innovative anticancer medication delivery strategies because of their biodegradability. Systems of nanoparticles based on biopolymers can be made using a variety of methods. To improve therapeutic efficacy, these systems must be developed in a cost-effective and ecologically friendly manner that makes them more widely accessible, selective, and less dangerous. Therefore, it is crucial to have a thorough understanding of the many aspects and most recent advancements in natural polymeric nanobiocomposites that are used to deliver anticancer medications. (Mondal et al. 2023) gives a summary of the most recent findings and advancements in natural polymeric nanobiocomposites, with a focus on how they can be used to administer anticancer medications in a targeted and regulated manner (Mondal et al. 2023). The biomedical community has shown a special interest in carbohydrate polymer-based nanocomposites for breast cancer therapy applications because of their advantageous features, which include their natural abundance, biocompatibility, ease of fabrication, and good physicochemical properties. The physicochemical characteristics and biological activities of carbohydrate polymers and the nanocomposites derived from them were covered in a review. After that, a number of techniques for creating carbohydrate polymer-based nanocomposites were covered, along with their potential uses in the treatment of breast cancer and future directions (Moghaddam et al. 2023).

7.2.2 Metal-Based Nanocomposites

It has been discovered that anticancer drugs loaded onto metallic nanoparticles release more readily and sustainably, lowering the dosage needed for drug administration and avoiding drug accumulation in healthy cells. The integration of nanotechnology and biocompatible materials presents novel opportunities for the creation of cutting-edge treatments with increased patient outcomes, decreased side effects, and improved selectivity. Because of its biocompatibility and biodegradability, chitosan (CS), a polysaccharide with unique physicochemical properties, has favourable characteristics for regulated drug administration. For anticancer medications, chitosan nanocomposites show enhanced stability, enhanced biocompatibility, and prolonged release properties. Whereas the addition of silver (Ag) nanoparticles enhances the synthesised nanocomposite's antibacterial properties, the addition of gold (Au) nanoparticles causes the chitosan nanocomposite to exhibit photothermal properties. Investigating the latest developments in the application of Ag and Au nanoparticles, or their combination, in a chitosan matrix or its modified derivatives for the delivery of anticancer drugs is the aim of this review. The potential of a chitosan nanocomposite to transport different anticancer medicines, including paclitaxel, 6-mercaptopurine, 5-Fluroacil, curcumin, and doxorubicin, was studied. Additionally, different adjustments made to the surfaces of the nanocomposite and the chitosan matrix phase to improve pH sensitivity, loading efficiency, and targeted selectivity were emphasised (El-Meligy et al. 2024) Several targeting agents, including transferrin, folic acid, liposomes, antibodies, and carbohydrates, can be added to metal nanoparticles to modify them. For this reason, metal nanoparticles are very promising for a wide range of biomedical uses, including computed tomography (CT), photothermal therapy, photodynamic therapy, photoacoustic imaging, and magnetic resonance imaging (MRI). Notwithstanding their potential, safe clinical uses need addressing regulatory obstacles and safety concerns (Truong et al. 2024) Metal matrix nanocomposites, which consist of a ductile metal or alloy matrix into which nanoparticle reinforcement is inserted, are materials reinforced by nanoparticles. These composites have entirely different mechanical, chemical, and physical properties than the matrix material since they are composed of nanoparticles embedded in a metal/alloy matrix. Nanoparticles are used to enhance mechanical properties, wear resistance, and damping characteristics. Recently, scientists have started studying metal matrix nanocomposites, which are finding a variety of uses in structural components because of their enhanced properties brought about by nanoparticle embedding (Yu et al. 2019) A vast variety of matrix materials, such as aluminium, titanium, copper, nickel, iron, and so on, can be found in metal-based nanocomposites. Materials that can be used as reinforcement include carbides (like SiC, B_4C), nitrides (like Si_3N_4, AlN), oxides (like Al_2O_3, SiO_2), and elemental materials (like C, Si). Many studies have been conducted on the superior mechanical and physical features of metal-based nanocomposites, including increased strength, thermal stability, and specific modulus. Li recently reported that gas atomization was used to create an *in-situ* nano-TiB_2-decorated AlSi10Mg composite (NTD-Al) powder for selective laser melting (SLM). The metal-based nanocomposite demonstrated superior tensile strength (about 530 MPa), ductility (around 15.5%), and microhardness (approximately 191 $HV_{0.3}$) compared to the majority of wrought and tempered Al alloys that are typically manufactured (Li et al. 2017). MMNs combine the high strength and modulus capabilities of metal with the desirable attributes of ceramic, such as toughness and ductility. These come in two forms: metal matrix-CNT or nonlayered nanocomposite. Metal discontinuous reinforcement (nonlayered) nanocomposites and metal matrix-CNT composites are two types into which MMNs fall (Lateef and Nazir 2017). Cell membrane hybrid nanoparticles (NPs) are intended to enhance immunotherapy, thermal therapy, and drug delivery for many illnesses. Here, we describe the creation of unique biomimetic magnetic nanocarriers with IR780 near-infrared dyes integrated into the membranes and magnetic nanoparticles enclosed in vesicles. Three different cell membranes—glioblastoma (GL261), melanoma (B16F10), and red blood cell (RBC)—are examined. Hybrid nanocarriers containing synthetic lipids and a cell membrane are created. A comparison is made between multiple systems' biological applications. Different routes of delivery of nanoparticles were examined in *in-vivo* research. The effectiveness of nanoparticle-mediated PTT was demonstrated by intratumoural injection. Following intravenous or intraperitoneal delivery, the nanoparticles concentrate in the tumour, as demonstrated by MRI and near-infrared imaging. Both approaches show the multimodal theranostic possibilities for cancer therapy and benefit from MRI-guided PTT (Rocha et al. 2024).

7.2.3 Carbon-Based Nanocomposites

In the field of biomedicine, carbon nanomaterials have garnered a lot of interest for uses in tissue engineering, drug delivery, and biosensing. Carbon nanomaterials have the intrinsic ability to overcome many of the present limits in biomedical research, including poor biomolecule stability, limited sensitivity and selectivity of biosensors, and challenges with precise drug administration. These qualities include their unique structure and strong conductivity. Furthermore, a number of new nanomaterials have been combined with carbon nanomaterials lately to create carbon-based nanocomposites that can be used in biological research (Shin et al. 2024b). The integration of various nanomaterials with carbon-based structures is a crucial component in the composition of CNCs, which in turn influences their properties and functionalities. The selection and arrangement of these parts are essential in customising CNCs for particular uses. Among these are the carbon allotropes, which are notable for their ability to mix with other nanomaterials like metal oxides, polymers, or nanoparticles to produce desired properties. Examples of these include graphene, carbon nanotubes (CNTs), and nanodiamonds. Furthermore, because of its extraordinary mechanical strength and electrical conductivity, graphene a single sheet of carbon atoms organised in a hexagonal lattice is a common component. The mechanical strength and electrical conductivity of CNCs are enhanced by CNTs with cylindrical carbon atom configurations. The distinct surface chemistry and hardness of nanodiamonds improve the general characteristics of CNCs. Note that the synthesis techniques of chemical vapour deposition, sol–gel synthesis, and self-assembly, which allow for control over component distribution and arrangement, are closely related to the composition of CNCs. Customisation of CNCs for energy storage, nanomedicine, and environmental cleanup is made possible by the careful selection and exact blending of carbon-based structures with other nanomaterials (Sudharsana et al. 2024). Carbon nanomaterials have lately been combined with a number of new nanomaterials to create carbon-based nanocomposites that will be used in biological research. (Shin et al. 2024a) go over recent research on carbon-based nanocomposites and their uses in biomedicine in this review. They start out by talking about the typical carbon nanomaterials and nanocomposites made of both carbon and other cutting-edge nanomaterials. The uses of carbon nanomaterials and nanocomposites in the biomedical area are then examined in relation to various biomedical topics. We have talked about the most recent research on tissue engineering, drug delivery, and biosensors. In sum, it demonstrates the potential and applicability of carbon nanomaterials and their nanocomposites and points the way towards probable future avenues for the use of carbon-based nanocomposites in life sciences (Shin et al.

2024a). Herein, a unique targeted NIR-absorbing nanomaterial with covalent functionalisation was produced for cancer chemo-photothermal therapy. In order to do this, hyaluronic acid and dopamine-reduced graphene oxide nanoparticles were covalently linked before doxorubicin (DOX/HA-DOPA-rGO) was added. The generated nanomaterials have excellent photothermal capability, good encapsulation efficiency, and appropriate physicochemical characteristics. The *in-vitro* investigations demonstrated the cytocompatibility of the nanomaterials and their enhanced absorption by CD44-overexpressing breast cancer cells. Notably, DOX/HA-DOPA-rGO combined with NIR light demonstrated the promise of chemo-photothermal therapy by reducing the viability of breast cancer cells to only 23% (Lima-Sousa et al. 2024).

7.2.4 Ceramic-Based Nanocomposites

In order to overcome the brittle character of ceramics by nature and to maximise their advantageous features, new manufacturing approaches and design philosophies are required. Because of its unique mechanical properties, like strength and stiffness, researchers' focus has shifted to creating ceramic matrix composites, or CMCs. Traditionally, CMCs have been fibres scattered over an oxide or non-oxide ceramic matrix. Since the development of nanoparticles, traditional reinforcing media (microscale fillers) have given way to new opportunities at the nanoscale in the field of CMCs. As a result, a new class of CMCs known as ceramic matrix nanocomposites (CMNCs) has emerged at the nanometric level (Sharma et al. 2022). CMNC matrixes can be made of raw materials like alumina, SiC, SiN, etc. The majority of the nanocomposites' reinforcements are nanometric in size. For amorphous reinforcements, TiO2, clay, iron, and other metal powders are utilised. Because of their small particle sizes and well-researched chemical interactions, clays and silicates are the most often used reinforcements. Even minute additions of clays and silica layers alter the matrix architecture. There are numerous strategies intended for the integration of CMNCs. Recently developed methods are single-source precursors based on melting mixed raw materials and then pyrolyzing the resulting nanofibers. PM, the polymeric monomer approach, spray-pyrolysis, and vapour methods (CVD and PVD) are a few examples of recognised mechanisms (Kusdianto et al. 2021). A frequent and severe bone tumour that can afflict both young adults and children is called osteosarcoma (OS). Because of the high likelihood of lung metastases and radiation resistance, the prognosis for this kind of cancer is very dismal. Consequently, the discovery of creative therapies to prevent this form of cancer recurrence and metastasis is vital. This study investigated an OS treatment utilising a folic acid-functionalized modified version of curcumin (CMAP) with hydroxyapatite (CMAP-HA) (CMAP-HAF). The outcomes demonstrated that folic acid (FA) was successfully incorporated into the nanostructure, which had cytotoxic effects on adult human dermal fibroblasts (HDFa) and osteosarcoma cells (U2OS). The CMAP-HAF therapy reduced the expression of cyclin E by $16.8 \pm 2.3\%$. Moreover, 99.6% of U2OS cells internalised CMAP-HAF, which compromised the integrity of the plasma membrane, caused necrosis, affected cell motility, and inhibited the creation of clones. These results imply that CMAP-HAF is a therapeutic agent with promise for treating OS and presents a novel strategy for enhancing patient outcomes (Rodrigues et al. 2024).

7.3 Synthesis Methods of Biocompatible Nanocomposites

7.3.1 Sol–Gel Method

A sol–gel matrix (inorganic or hybrid organic–inorganic) and nanoparticles formed in the matrix by chemical reaction or thermally induced phase separation, or a polymer matrix with nanoparticles generated inside the organic network through sol–gel reactions, can comprise sol–gel nanocomposites that are obtained by "*in-situ*" methods. If "*ex-situ*" methods are used, nanocomposites could contain a matrix synthesised by other means and a filler obtained from sol–gel, a matrix synthesised by other means and sol–gel nanoparticles, and a matrix and filler from other sources. The preparation of a single batch containing the precursors for the matrix and the dispersed phase is the foundation of in-situ techniques for the synthesis of nanocomposites. It is consequently through chemical reactions and/or physical–chemical transformations that the various phases are realised. All the constituents are combined in a liquid state, and the precursors' hydrolysis and condensation to create a matrix—within which the dispersion phase is generated, either simultaneously or in a subsequent step—is what drives the creation of the nanocomposite. The dispersed phase can be produced inside the matrix by various chemical reactions, or it can be obtained by heat treatment, precipitation caused by solvent evaporation, or both. The in-situ technique has a distinct benefit in that there are fewer preparation stages. One significant disadvantage is the potential for interactions between the processes generating the two (or more) phases, which can make control over the formation of the micro/nanostructure extremely challenging, if not impossible. Conversely, *ex-situ* techniques necessitate a greater number of steps: typically, the matrix is treated in a single batch and the scattered phases in separate ones, after which the systems are combined. The goal is to achieve much finer control over the final microstructure by minimising interactions between various chemical processes. Various strategies can be used to achieve this, as will be demonstrated later, and the two (or more) systems may be mixed at different stages of their development (Guglielmi 2014). The following steps are included in a typical alkoxide precursor sol–gel process: Stable sol formation, casting or shape formation, gelation of the sols, ageing, drying, calcination, and, if required, sintering are the first seven steps. It should be noted that a wide range of precursors, in addition to alkoxides, can be used for sol–gel processing (Lin and Deng 1999). Diagram showing the several phases of the sol–gel process, from precursor to aerogel (Fig. 7.2) (Bokov et al. 2021).

Synthesis of Gelatin-BG-MWCNT Nanocomposites Derived from Sol–Gel Using Calcium Ethoxide and Chloride: To prepare gelatin–BG–MWCNT nanocomposites

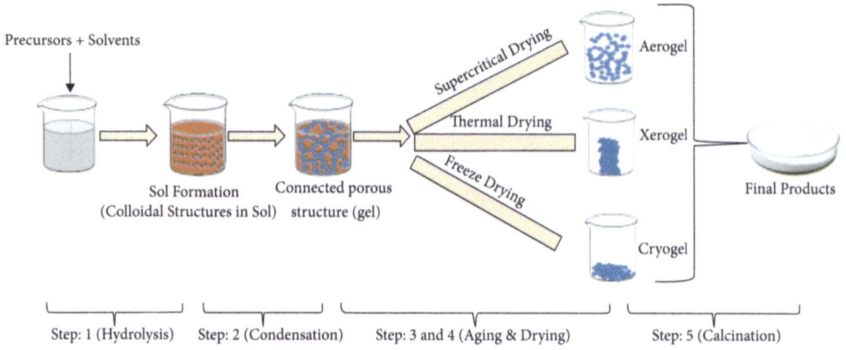

Fig. 7.2 Diagram showing the several phases of the sol–gel process, from precursor to aerogel. Reproduced from reference (Bokov et al. 2021) under the terms and conditions of the Creative Commons Attribution (CC BY) license (https://creativecommons.org/licenses/by/4.0/)

based on calcium chloride ($CaCl_2$), 20 mg/mL of MWCNT was dispersed in Pluronic F-127 (PF-127) solution, which was previously dissolved in water at a high temperature to a concentration of 20 mg/mL, followed by sonication for 2 h at 50 °C and stored at RT until further use. A 10% w/v concentration of gelatin type A (porcine skin) was dissolved in water. In the meantime, water and a catalytic quantity of 1 M HCl were added to TEOS, TEP, and $CaCl_2$ to create BG via the sol–gel method, which took 30 min at room temperature. 49 mg of $CaCl_2$, 248 μL TEOS, 22 μL TEP, and 100 mg of BG were prepared. To catalyse the hydrolysis of TEOS, 1 M HCl was added at a volume ratio (water/HCl) of 3, maintaining the mole ratio of water/TEOS at 4:1. Drop by drop, a viscous gelatin solution and MWCNT (1 weight percent) were added to the tertiary glass precursor's solution. The TEOS was subjected to in-situ hydrolysis and polycondensation while the sol-gelatin-MWCNT combination was aggressively agitated at 40 °C until it gelled. With a final molar composition of 70% SiO_2, 26% $CaCl_2$, and 4% P_2O_5, BG maintained a 50 weight percent organic–inorganic ratio. The gels were allowed to age for one day at 55 °C, and then they were vacuum-dried for one day at 60 °C. After that, the finished product was milled into a fine powder. In order to create gelatin–BG–MWCNT nanocomposites based on calcium ethoxide ($Ca(OEt)_2$), MWCNT stock dispersions (20 mg/mL MWCNT and PF-127) were made using PF-127. Dispersions were sonicated for 1 h at 50 °C and stored at RT until further use. A 10% w/v concentration of gelatin type A (porcine skin) was dissolved in water. After adding 1 weight percent of MWCNT to the gelatin solution, it was sonicated for 30 min at 50 degrees Celsius. Using a sol–gel method, BG (100 mg) was produced by hydrolysing TEOS (248 μL) and TEP (22 μL) with a catalytic quantity of 1 M HCl while vigorously swirling at room temperature. Separate calcium ethoxide solutions were made using 54 mg/mL of 2-ethoxyethanol. The hydrolysed BG precursors and calcium ethoxide were added to the gelatin–MWCNT mixture in a dropwise manner. With a final molar composition of 70% SiO2, 26% CaO, and 4% P_2O_5, BG maintained a 50 weight percent organic–inorganic ratio. The biomaterials were dried under vacuum at 40 °C after being aged for one day at room temperature. Using a pestle and mortar, the finished mixture was ground into a fine powder (Arambula-Maldonado and Mequanint 2024) Nanoparticles are one of the many components that can be added to a composite to improve its properties. Strong covalent bonds and weak physical interactions like hydrogen bonds, Van der Waals contacts, hydrophobic interactions, and electrostatic interactions can both be used by different moieties to connect with one another. Sol–gel was used to create chitosan/*in-situ* gelatin-magnesium silicate nanocomposites (Mansour et al. 2024).

7.3.2 Electrospinning Technique

Nanofibers made of polymers and polymeric nanocomposite materials have been produced primarily by spinning techniques. There are many different types of spinning procedures, such as electrostatic spinning, melt spinning, and wet spinning. Spinning processes typically feature controllable process parameters. In this instance, the polymers and nanocomposite nanofibers were spun via solution blow spinning. This technique blends the melt-blowing and electrospinning techniques. The nozzle, gas flow, collector, pumping mechanism, and syringe are all unique to the solution blow spinning configuration. Variations in the parameters, polymer type, and matrix concentration all affect the fibre diameter. Nano- or microfibers have also been created via centrifugal jet spinning. This configuration includes numerous fibre collectors, a DC motor, and a spinning chamber. Nanofibers are also formed on micro- or nanoscales by the electrohydrodynamic direct writing mechano-electrospinning process. Viscose ink is developed using both mechanical and electrical forces in this process, which produces the nanofibers. The associated arrangement features adjustable voltage sources and a configurable distance between the nozzle and substrate. The electrospinning method has received particular interest in the context of spinning approaches in order to create nanocomposite nanofibers. Superior physio-chemical characteristics of electrospun nanofibers have been achieved by the processing of polymers and polymer/nanocarbon nanocomposites. It has been stated that electrospinning is an easy, sensible, and adaptable way to create nanofibers (Kausar and Ahmad 2023). A conductive collector, a syringe pump, a high-voltage power source, and a spinneret—a hypodermic needle with a blunt tip—are the main parts. Direct current (DC) or alternating current (AC) are the two possible power source types. Surface tension causes the liquid to protrude from the spinneret during electrospinning, creating a pendant droplet. After the droplet is electrified, a charged jet is released from it due to the electrostatic repulsion between the surface charges with the same sign. Due to bending instabilities, the jet first extends in a straight line before undergoing violent whipping motions. A solid fibre or fibres are deposited on the grounded collector as a result of the jet's rapid solidification as it is stretched into finer dimensions. The electrospinning process can be broadly broken down into four steps: (i) charging the liquid droplet and creating a cone-shaped or Taylor cone-shaped jet; (ii) extending the charged jet in a straight line; (iii) thinning the jet in the presence of an electric field and creating electrical bending

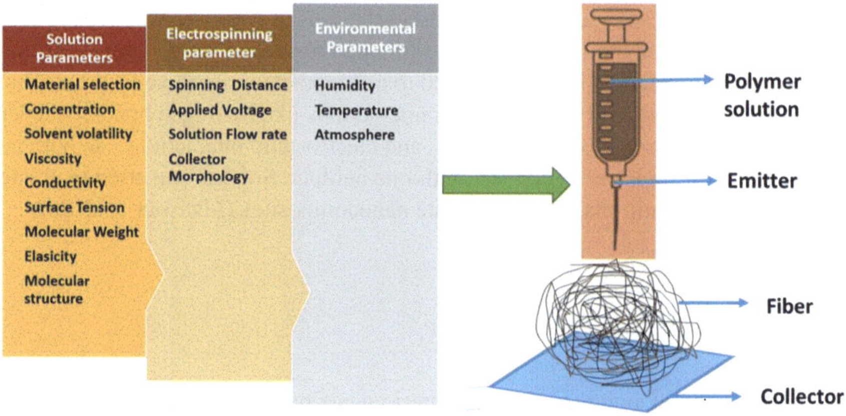

Fig. 7.3 Showing the fundamental spinning configuration, operating environment, and electrospinning process parameter. Reproduced from reference (Kailasa et al. 2021) under the terms and conditions of the Creative Commons Attribution (CC BY) license (https://creativecommons.org/licenses/by/4.0/)

instability, also called whipping instability; and (iv) solidifying and collecting the jet as solid fiber(s) on a grounded collector (Xue et al. 2019). The fundamental spinning configuration, operating environment, and electrospinning process parameter (Fig. 7.3) (Kailasa et al. 2021).

In-situ cancer treatment following surgical excision has garnered significant interest due to the simultaneous use of chemotherapy and heat. In this study, poly(ε-caprolactone) (PCL) is embedded with iron oxide nanoparticles (IONPs) and ammonium bicarbonate pH-responsive moieties that conjugate with the DOX anticancer drugs integrated into the core of nanofibers. Synthesis of electrospun fibres PCL@IONPs@DOX@Ammonium bicarbonate (PIAD) and PCL@IONPs@Ammonium bicarbonate (PIA). First, a 50:50 ratio was used to dissolve the blended solvent of DMF and chloroform ($CHCl_3$) at 10% (w/v) in order to prepare the PCL polymer solution. Subsequently, the solvent was allowed to come to room temperature and the various quantities of the NH_4HCO_3 saturated solution (0.005 and 0.01 mL) were added. The solvent was then vigorously agitated for 30 min using a magnetic stirrer. After adding DOX, stir for 30 min. IONPs were lastly inserted and dispersed by ultrasonication for 15 min in order to obtain the homogeneous solution. The mix solution was then introduced into a syringe pump that was attached to a high-voltage device. The same process was used to synthesise PIA without DOX. An infusion pump operating at a constant feed rate of 1 mL h^{-1}, a 12 mL syringe equipped with a 21-gauge needle with a distance of 15 cm between the needle tip and collector, and environmental conditions of 35–40% humidity and 25–30 °C were used to conduct electrospinning. The electrospun fibres were collected using a polyethylene sheet that was fastened to the collector at a distance of 15 cm from the needle tip. The collection plate was a negative electrode covered in aluminium foil (Vo et al. 2024).

7.3.3 Coprecipitation Method

Using a precipitating media, the coprecipitation method includes forming the oxo-hydroxide form from a solution of a salt precursor (metal salts, such as nitrates or chlorides) in a solvent (such as water). Following the occurrence of a critical concentration of species in solution, a brief nucleation burst is followed by a growth phase. Metal oxides have been synthesised using this technique. Alkaline coprecipitation of ferrous and ferric ions dissolved in stoichiometric volumes of water is a typical method for producing magnetic nanoparticles, such as magnetite. Typically, NaOH and NH_4OH are employed as bases for this process. The precipitating mediums of NaOH, KOH, and $(C_2H_5)_4NOH$ have demonstrated that the size, paramagnetic characteristics, and degree of agglomeration of synthesised magnetite nanoparticles are influenced by the alkali's nature, pH, and the rate at which alkaline solution is added. Furthermore, it has been observed that the use of surfactants can help further optimise the surface properties. This method's benefits include low cost, easy scaling up, flexibility in modulating core and surface features, mild reaction conditions (e.g., low synthesis temperature), and the ability to undertake direct synthesis in water. (Stankic et al. 2016). The typical procedure goes like this: choose the precipitant based on the materials to be precipitated; precipitate reactions; filtrate; wash; dry; and finally, obtain precipitates. This process's flowchart is displayed in (Fig. 7.4) (Z. Li et al. 2023a, b).

During the precipitation reaction, a homogeneous mixing of two or more cations in a solution is obtained using the coprecipitation method. This technique is frequently used to create composites with several metal components. Numerous processes, including nucleation, growth, coarsening, and agglomeration, make up the coprecipitation reaction. It is an important method for creating composites during precipitation that have several metal elements in a uniform solution. Because of its many benefits—including short diffusion routes, high product purity, reduced agglomeration, simplicity, uniform and controllable particle size, and cost-effectiveness—this approach is chosen over solid and vapour phase procedures. The creation of mesoporous iron-manganese bimetal oxide nanocomposites by the aeration coprecipitation approach is one recent example of this process. Manganese sulphate and ferric chloride are prepared separately in a molar ratio of 2:1. They are then mixed at 70 °C and stirred at 50 rpm. 10% NH_4OH is added to achieve a pH of 9, and the mixture is kept at 50 rpm for four hours to yield the precipitated nanocomposites. Similar to this, a $NiFeO_4$ nanocomposite was created using the coprecipitation approach, which involved utilising hydrazine hydrate as the reducing agent and a 1:1 molar ratio of $FeSO_4.7H_2O$ and $NiSO_4.7H_2O$ at a pH of 12 (Ghazzy et al. 2023). (Iffah et al. 2024) Fabricate a Zn-ferrite/HAp nanocomposite for use in drug delivery applications. The coprecipitation approach was utilised to synthesise the HAp/Zn-ferrite nanocomposite. The nanocomposite was able to load doxorubicin with a capacity of 27.61 mg/g, according to the results of the UV–Vis data analysis. It was also able to release doxorubicin with a relatively consistent release profile, reaching 79.15% in 300 min (Iffah et al. 2024).

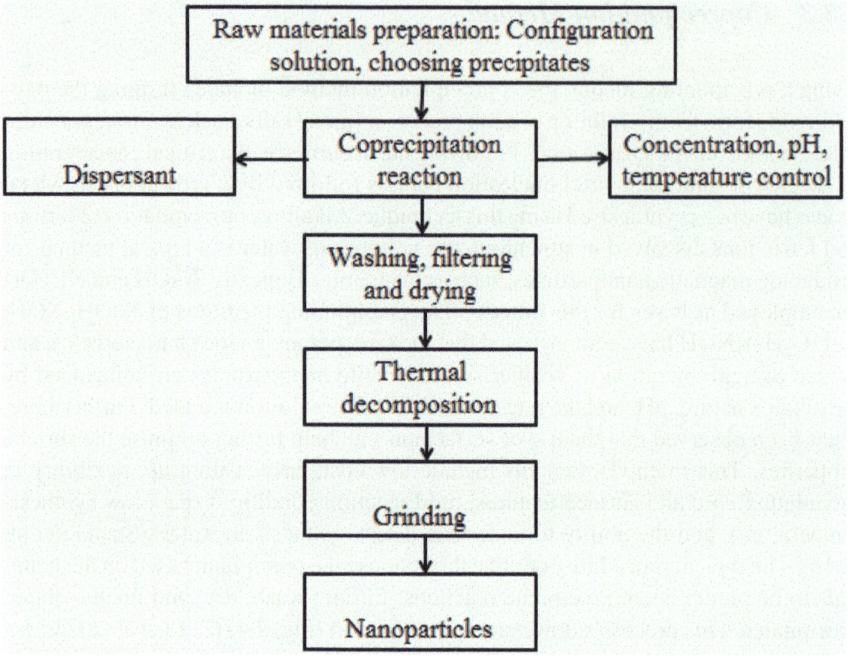

Fig. 7.4 Flowchart of coprecipitation process. Reproduced from reference (Z. Li et al. 2023a, b) under the terms and conditions of the Creative Commons Attribution (CC BY) license (https://creativecommons.org/licenses/by/4.0/)

7.3.4 *Chemical Vapour Deposition*

Chemical Vapour Deposition (CVD) is a potent technology that yields solid thin films and coatings of superior quality. Despite being widely employed in contemporary sectors, it is always evolving as new materials require adaptation. The accurate synthesis of high-purity polymeric thin films that can be conformally placed on a variety of substrates and inorganic thin films of 2D materials is pushing CVD synthesis to new heights today (Sun et al. 2021). The process of CVD involves using chemical processes to deposit material as a thin film from vapour species onto substrates. Numerous intricate variables, such as the system configuration, reactor layout, gas feedstock, gas ratios, reactor pressure, gas partial pressures, reaction temperature, growth time, temperature, etc., influence the process and kinds of different possible chemical reactions that take place in a CVD reactor. A popular bottom-up method for synthesising single- and few-layer graphene sheets is chemical vapour deposition (CVD). The first report of the growth of extremely crystalline graphite films on Ni substrates by thermal CVD on metals dates back to 1966. Subsequently, a single layer of graphite was deposited on the Pt surface via a CVD technique employing hydrocarbon breakdown at 530 °C. The CVD method of growing graphene is widely used because it is simple to set up in research labs, has shown effectiveness over an extended period of time in industrial settings, and has the potential to be scaled up in fabrication. Furthermore, the CVD approach is one of the best methods currently available for the synthesis of materials based on graphene, taking into account both ecological and financial concerns. These days, graphene-based compounds can be synthesised using a wide range of CVD techniques. These methods can be divided into seven main categories based on the characteristics of the processing parameters (pressure, temperature, wall/substrate, nature of precursor, depositing time, gas flow state, and activation/power source). There are seven main categories based on these characteristics (Saeed et al. 2020). For the first time, (Zhang et al. 2023) showed how to synthesise 2D MXene-MoS_2 nanocomposites using a chemical vapour deposition (CVD) technique, enabling fine design in terms of orientation and structure type. The computer outcomes demonstrate the metallic characteristics of nanocomposites. Due to their distinct 2D/2D structure, MXene-MoS_2 nanocomposites display a greater number of active catalytic sites, leading to enhanced electrochemical performance. These features are inherited from the parent materials, and a significantly reduced overpotential of around 69 mV is attained at a current density of 10 $mA \cdot cm^{-2}$. One common natural substance utilised in anticancer therapy is curcumin, which is derived from Curcuma longa. One important type of nanomaterials with distinctive physicochemical characteristics, single-walled carbon nanotubes (SWCNTs), is an efficient curcumin carrier. As a result, curcumin's traditional problems of limited water solubility and instability under physiological settings can be overcome while still using its anticancer properties. Using a Fe-Mo/Al_2O_3 catalyst, SWCNTs were grown in the current study without the use of hydrogen gas. They were mixed with curcumin after being optimally manufactured using the Chemical Vapour Deposition (CVD) technique (Mohammadi et al. 2024).

7.4 Characterisation Techniques for Biocompatible Nanocomposites

7.4.1 Scanning Electron Microscopy (SEM)

Scanning electron microscopy (SEM) is a highly adaptable, esteemed, and widely used method in both the industrial and research domains. This type of electron microscopy produces images with a high magnification and resolution by scanning the samples with a high-energy electron beam. Certain signals regarding the surface topography are produced by the interaction between the electron beam from the electron cannon and the sample's electrons. The signals from the secondary and backscattered electrons, which include information about the material, are analysed to create the SEM images. This method yields a wealth of information on the materials under study, including but not limited to elemental composition, surface morphology, and crystallinity. Theoretically, this method might yield resolutions smaller than an atom's radius when the energy and wavelength of the electrons are considered.

However, its working resolution is limited to the order of 1–2 nm because of restrictions deriving from the lenses and sample preparations (Spence 2013). SEM pictures in the thickness of the nanocomposites, when carefully analysed, show fine structures whose characteristics differ from those of the polymer matrix and whose dimensions line up with those of individual nanotubes (Fig. 7.5a). Furthermore noteworthy is the absence of nanotubes on the nanocomposites' surface. This indicates that the fiber-like components are "packed" within a matrix of macromolecules (Fig. 7.5b) (Rezvova et al. 2022).

To verify that CNTs were present in the pores and grain boundaries of the nanocomposites, a scanning electron microscopy (SEM) examination was conducted. The microstructure of the Ni/CNT nanocomposite made using Fibermax CNTs is depicted in SEM pictures in (Fig. 7.6) as an example (Carneiro and Simões 2020).

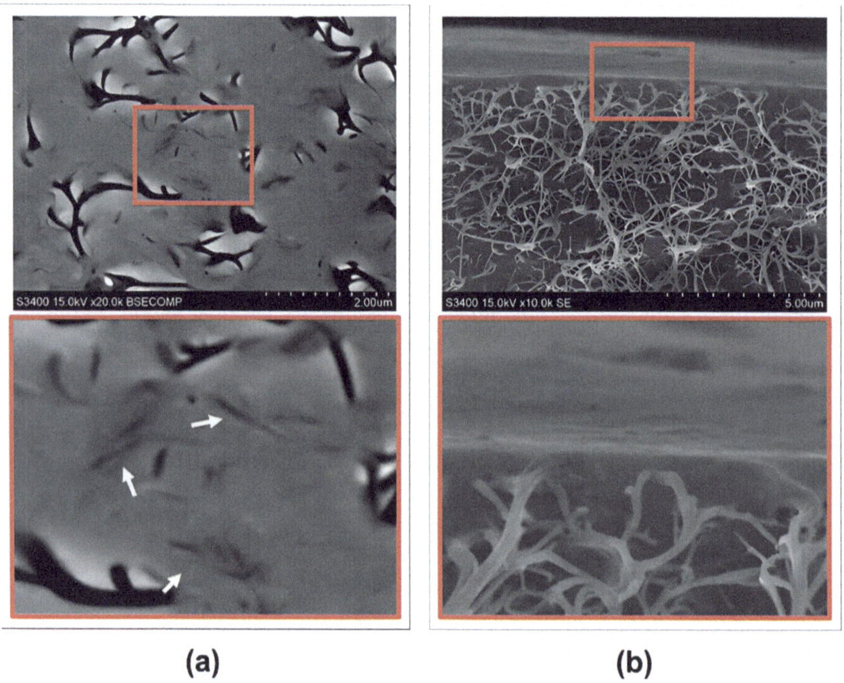

Fig. 7.5 Showing SEM pictures at 20,000 × and 10,000 × magnifications of the nanocomposite surface (**a**) and the sample fracture at the surface (**b**), respectively. Single nanotubes are indicated by arrows. Reproduced from reference (Rezvova et al. 2022) under the terms and conditions of the Creative Commons Attribution (CC BY) license (https://creativecommons.org/licenses/by/4.0/)

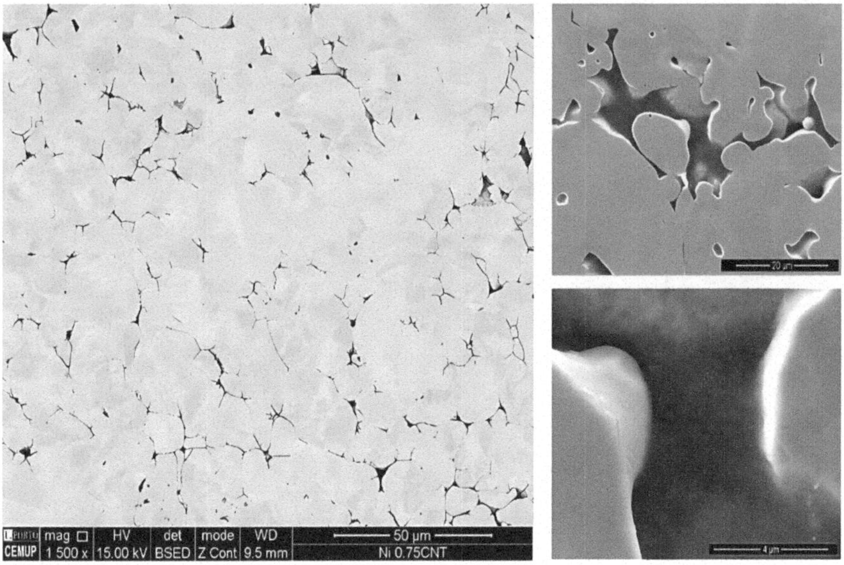

Fig. 7.6 Showing the presence of CNT agglomerates is confirmed by high-magnification pictures of the pores captured by scanning electron microscopy (SEM) of the nanocomposite microstructure. Reproduced from reference (Carneiro and Simões 2020) under the terms and conditions of the Creative Commons Attribution (CC BY) license (http://creativecommons.org/licenses/by/4.0/)

7.4.2 Transmission Electron Microscopy (TEM)

In the field of material science, transmission electron microscopy (TEM) has proven to be a very effective tool for investigating and learning about the structure of nanomaterials. Transmission electron micrographs can be used to provide quantitative measurements of parameters, morphological information, crystallographic features, grain size, size distribution, size homogeneity, lattice type, and chemical composition of phases. Thus, the most effective method for characterising nanomaterials, including nanoparticles and nanocomposites, is TEM (Asadi Asadabad and Eskandari 2015). In (Fig. 7.7a) shows the TEM images of GO that were created using the improved oxidation process. The image indicates that GO created large, smooth plates that were approximately 5 × 5 μm in size. The TiO_2-GO nanocomposite is shown in TEM pictures in (Fig. 7.7b), which indicates that the TiO_2 nanoparticles were evenly distributed over the graphene plane (Štengl et al. 2013).

Pd creates spherical and porous structures with an average size of 2–10 nm and does not form agglomerates, according to the TEM picture of PdNPs/PEI N-GQDs nanoparticles (Fig. 7.8). The TEM pictures showed that Pd exhibited spherical porous structures, while PEI N-doped GQDs were randomly distributed between the metals and kept their spherical porous shape (Gunes et al. 2023).

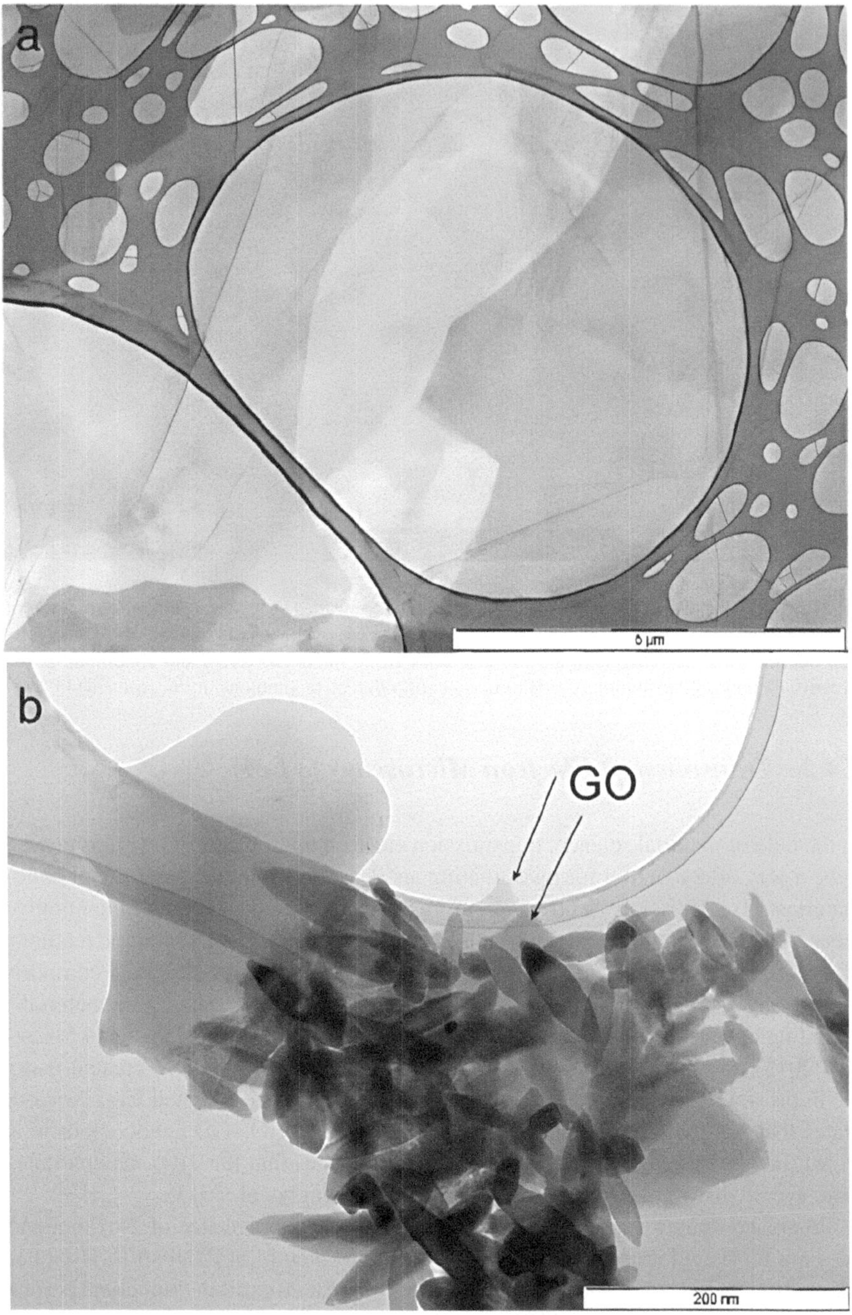

Fig. 7.7 TEM pictures of graphene oxide (GO) are shown in (**a**) and (**b**) of the TiO2-GO nanocomposite. Reproduced from reference (Štengl et al. 2013) under the terms of the Creative Commons Attribution 2.0 International License (https://creativecommons.org/licenses/by/2.0)

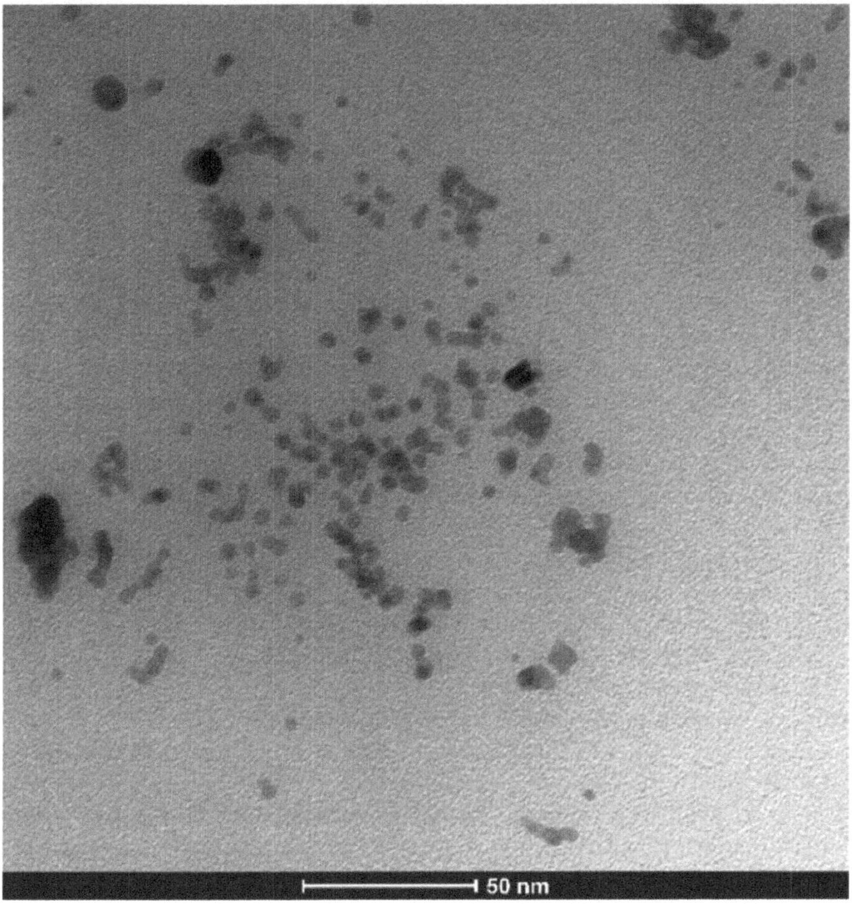

Fig. 7.8 TEM picture of PdNPs/PEI N-GQDs. Reproduced from reference (Gunes et al. 2023) under the terms and conditions of the Creative Commons Attribution (CC BY) license (https://creativecommons.org/licenses/by/4.0/)

7.4.3 X-Ray Diffraction (XRD)

A nondestructive method for obtaining comprehensive details about a material's chemical composition, physical characteristics, and crystalline structure is X-ray diffraction analysis or XRD. Its foundation is the constructive interference of a crystalline sample with monochromatic X-rays. When electrically charged particles with enough energy slow down, they produce X-rays, which are electromagnetic radiation with shorter wavelengths. In X-ray scattering (XRD), the produced X-rays are collimated and directed towards a sample of nanomaterial. The sample and incoming rays interact to form a diffracted beam, which is subsequently detected, processed, and tallied. A diffraction pattern is displayed by plotting the intensity of the diffracted rays scattered at various angles of the material.

Because of the material's particular chemistry and atomic organisation, each phase creates a different diffraction pattern. Each phase's diffraction pattern is simply added to create the diffraction pattern. The diffracted signal's pattern would be impacted by sample material imperfections. In this instance, the sample's imperfections are caused by the following factors: heterogeneity in composition, faults in crystal structure, microstains, and crystallite size (Raja et al. 2022). Even though XRD is most commonly used to analyse crystalline phases in materials qualitatively and quantitatively, much more information can be found by carefully examining the diffraction patterns or by using particular XRD settings, such as characterising solid solutions, determining the size and shape of crystallites, determining crystal orientation, determining internal elastic strains and stresses at different levels, determining the impact of temperature, characterising close surfaces, etc. (Gonon 2021). In the twenty-first century, the creation of polymer nanocomposites (PNCs) is a crucial part of material research. A crucial method for the investigation of the structure–property correlation of PNCs is X-ray diffraction spectroscopy (XRDS). These investigations are crucial since a wide range of fillers have been created as reinforcements in PNCs, such as layered materials like clay, graphite, graphene or its derivatives, carbon nanotubes (CNTs), and numerous other nanoparticles (NPs) and nanofibers. The use of XRDS to characterise PNCs depends on how the filler particles or the polymer's fundamental properties are impacted, which the technology can easily investigate. In a polymer matrix, NPs can function as impurities that prevent the creation of an ordered structure or as nucleating agents that boost the composite's crystallinity. XRD analysis can also be used to study the crystal phase transitions caused by the NPs. Based on the kind of nanofiller, the application of the XRD technique to the structure–property correlation investigation of PNCs is categorised in the following parts (Venkateshaiah et al. 2016).

7.4.4 *Fourier Transform Infrared Spectroscopy (FTIR)*

By identifying changes in functional groups, the Fourier transform infrared (FTIR) method of spectroscopy is able to identify modifications in the overall composition of biomolecules. FTIR is used to quantify how molecules rotate and vibrate in response to infrared radiation at a specific wavelength. Through the identification of structural variations in molecule binding, information regarding the existence of their interactions can be gleaned. The most often used FTIR-based techniques for characterisation are micro-spectroscopy FTIR, attenuated total reflectance (ATR–FTIR), and transmittance FTIR (Eid 2021). The qualitative and quantitative investigation of nanofillers and their nanocomposites has made considerable use of FTIR spectroscopy. FTIR spectroscopy will be used in qualitative investigations to characterise the nanofillers, which include metal oxide nanoparticles, carbonaceous nanostructure, clay, nanofiller synergy, and their polymer nanocomposites. FTIR will be used in quantitative analyses to demonstrate how a nanofiller affects the characteristics of nanocomposite films by observing variations in the FTIR spectra's band intensity (Jaleh and Fakhri 2016). FTIR analysis was used to characterise the nanofillers and nanocomposites. Figure 7.9 displays the FTIR spectra of nanofillers and nanocomposites. The nanocomposites containing 1.0% CNT, GNP, and C60 were chosen to symbolise the filler/epoxy composites. To determine the potential chemical reaction between polymers and nanofillers, the findings were compared to neat epoxy. FTIR spectra were utilised to look for the presence of hydroxyl groups on the surfaces of each nanofiller in order to assess the possibility of hydrogen bonding between the nanofiller and epoxy resin during the dispersion process. Figure 7.9 shows that neither raw CNTs nor GNPs had any major absorption peaks. Nevertheless, the broadband at 3682 cm^{-1} suggests the presence of an O–H bond, which was caused by moisture that was connected to the nanofiller surfaces and was either unbound or free. However, this bond was not seen in C60 particles; instead, a number of narrow peaks at 1180 and 1427 cm^{-1} were found to reflect C–C bonds. A number of peaks were found to aid in characterising the neat epoxy and nanocomposites. The polymer backbone's C–H stretch was responsible for the peak at 2920 cm^{-1}, whereas the DGEBA epoxy resin's aromatic ring stretching of $C = C$ was characterised by the peaks at 1607 and 1456 cm^{-1}. For each of the examined samples, the peak at 913 cm^{-1} vanishes, signifying the ring-opening polymerization that occurs during crosslinking and represents the curing process. Analysing the FTIR spectra closely, as shown in (Fig. 7.9), showed that all of the samples had the same distinctive peaks. No variations in peak value indicated the presence of new chemical interaction between the polymer and nanofillers (Wang et al. 2020).

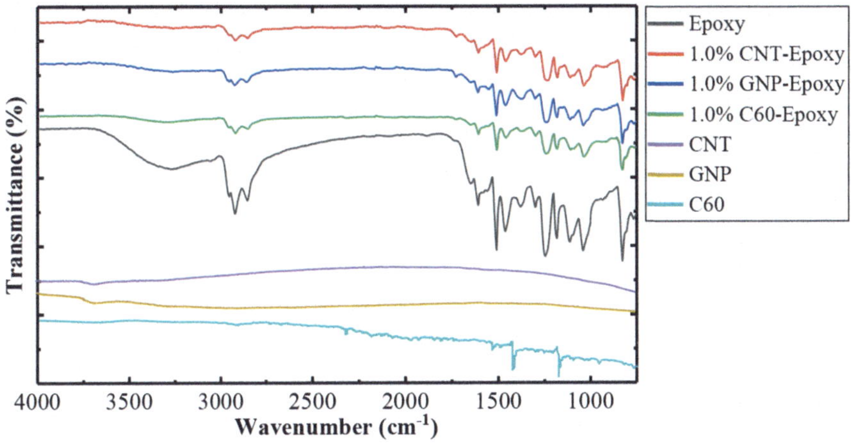

Fig. 7.9 The nanocomposites with different carbon nanofillers were analysed using FTIR. Reproduced from reference (Wang et al. 2020) under the terms and conditions of the Creative Commons Attribution (CC BY) license (http://creativecommons.org/licenses/by/4.0/)

7.4.5 Thermogravimetric Analysis (TGA)

TGA measures the amount and frequency of sample weight variations in relation to temperature and time in a controlled environment (such as nitrogen gas purging). Thermal stability (the material's strength at a specific temperature), oxidative stabilities (the material's rate of absorption of oxygen), and compositional features (such as fillers, polymer resin, and solvents) of the samples can all be examined using TGA. Additionally, the samples' weight growth or decrease correlates with several parameters. In general, adsorption or oxidation are blamed for weight growth, while decomposition, desorption, dehydration, desolvation, or volatilization are blamed for weight loss (Ng et al. 2018). The fundamentals and guiding ideas of thermogravimetric analysis (TGA) instrumentation are covered by (Loganathan et al. 2017). The many kinds of crucibles, furnaces, and balances used in TGA instruments are discussed. The fundamental data that may be obtained from a thermogravimetry thermogram is also thoroughly covered, along with the use of sophisticated TGA instruments for evolved gas analysis and how they can be applied to anticipate the mechanisms of thermal degradation of polymer nanocomposites. Furthermore, relevant case studies are provided for the application of TGA in areas including drug and functional moiety loading determination, gas adsorption investigations, catalyst relative strength, and thermal degradation kinetic study for polymer nanocomposites (Loganathan et al. 2017). Using thermogravimetric analysis, the thermal behaviour of the copolymers and their nanocomposite were investigated; the results are shown in (Fig. 7.10). The samples' estimated thermogram values show three stages of decomposition: in the first stage, a weight loss of 20% was computed for CPs at a temperature between 20 and 100 °C due to the loss of water, but the loss occurs at a temperature between 220 and 250 °C (40% loss) for NPs. Additionally, it was expected that the next stage would happen for CPs (30% loss) between 100 and 200 °C, but for NPs (30% loss) between 250 and 400 °C. At 400 and 600 °C, the final stage, CPs, were found to have a 50% loss, but NPs at this temperature yielded 0% loss. The heat stability of the copolymers and their nanocomposite could be the cause of these outcomes. Additionally, the Nps exhibit greater stability than the CPs; this may potentially be related to the nanoparticles existing within the polymer's epically matrix. High resilience to thermal degradation is demonstrated by NMMT (El-Bahnasawi et al. 2023).

7.4.6 X-Ray Photoelectron Spectroscopy Analysis (XPS)

For the examination of solid-state materials, one of the most popular and extensively utilised techniques is X-ray photoelectron spectroscopy or XPS. Based on Einstein's photoelectric effect, it calculates the kinetic energy of electrons released by photoelectric effects brought on by X-ray illumination. One of the distinctive properties of the elements to determine the quantity of elements and chemical bonds

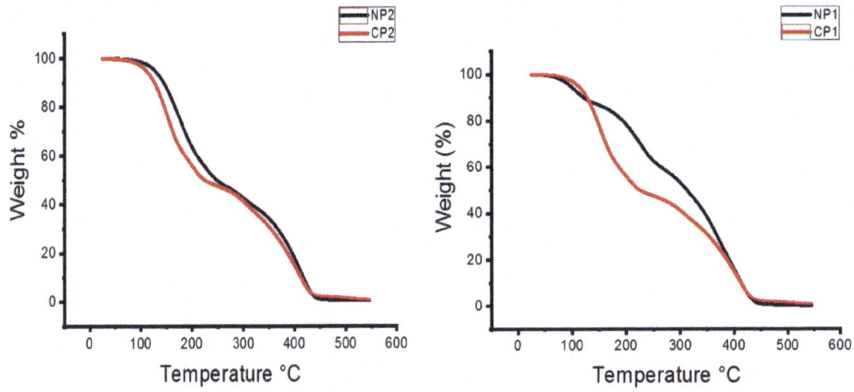

Fig. 7.10 Showing the polymer and its nanocomposites TGA analysis. Reproduced from reference (El-Bahnasawi et al. 2023). Under the terms and conditions of the Creative Commons Attribution (CC BY) license (http://creativecommons.org/licenses/by/4.0/)

in the system is the energy of released core electrons. XPS spectroscopy was used to examine the elemental compositions and bonding arrangements of PEI N-GQDs and PdNPs/PEI N-GQDs nanocomposites (Fig. 7.11a,b). The carbon, nitrogen, and oxygen-binding energies of PEI N-GQDs were determined to be 284.58, 400.08, and 531.58 eV, respectively (Fig. 7.11a). Moreover, in PEI N-GQDs, auger peaks for carbon, nitrogen, and oxygen (CKL1, NKL2, and OKL1) are detected at 1229.08, 1108, and 980 eV. The PdNPs/PEI N-GQDs nanocomposites' carbon, nitrogen, oxygen, and Pd bond configuration spectra are displayed in (Fig. 7.11b). The peaks at 283.19 (C 1 s), 398.64 (N 1 s), 530.13 (O 1 s), 558.76 (Pd 3p1), 336.55 (Pd 3d3), and 306.48 (Pd 4d5) eV are found when Fig. 7.11b is examined. In addition, the PdNPs/PEI N-GQDs nanocomposite exhibits the oxygen auger peak (O KL1) at 976.66 eV (Gunes et al. 2024).

7.4.7 UV–Visible Spectroscopy

The importance of ultraviolet–visible (UV–Vis) spectroscopy in understanding the optical characteristics of several significant nanofillers and the polymer nanocomposites they form is reviewed in this chapter. UV–Vis spectroscopy is used to comprehend the optical response of a polymer when it is subjected to the absorption of UV–Vis radiation of the electromagnetic spectrum, despite the fact that there are many characterisation techniques to examine the nanocomposites. The best way to assess the optical characteristics of polymer nanocomposites, such as their high or low refractive index, luminescence, UV absorption, and transparency, is to look at their UV–Vis spectra. The development of polymer nanocomposites with tunable optical characteristics for possible uses as sensors, colour filters, and optoelectronic devices is greatly

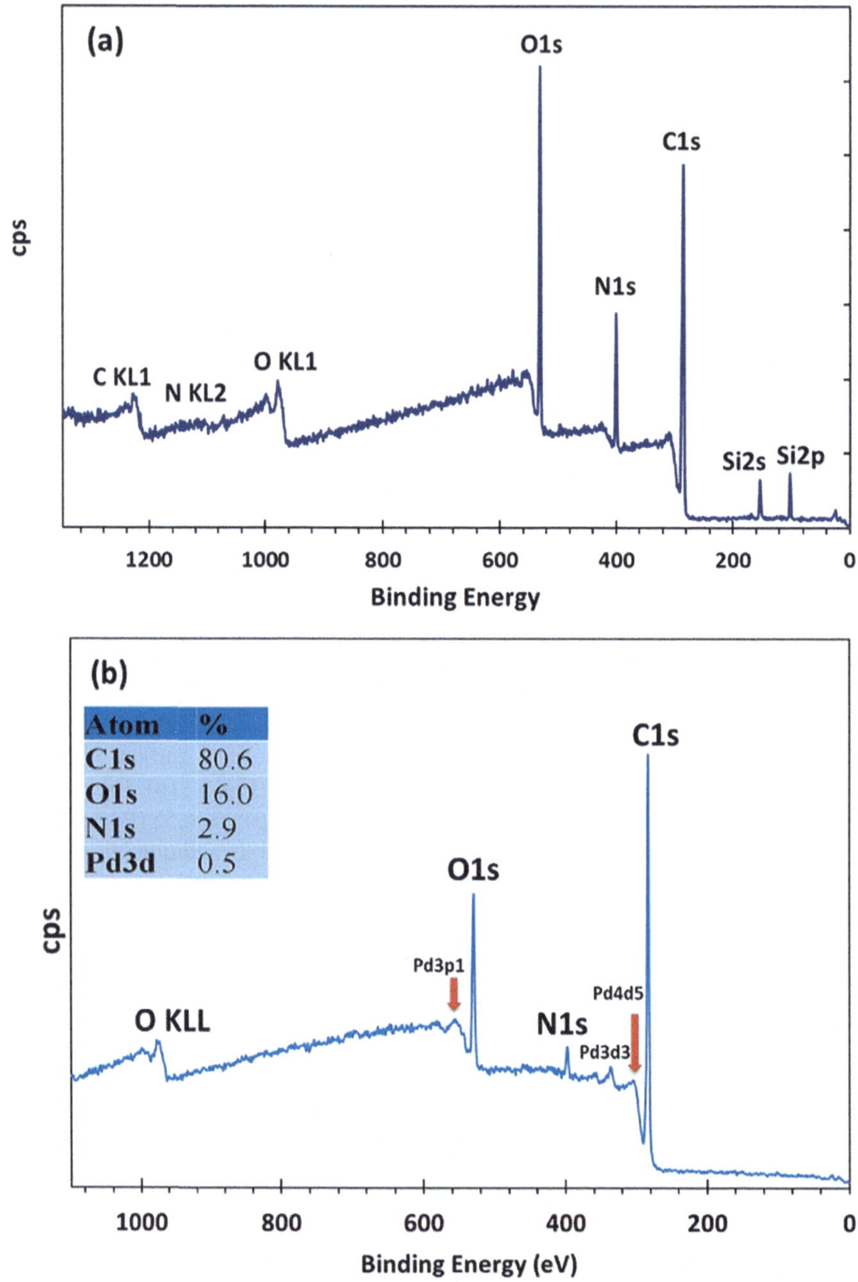

Fig. 7.11 X-ray photoelectron spectra of **a** PEI N-GQDs' carbon, nitrogen, and oxygen bond configurations and **b** the PdNPs/PEI N-GQDs nanocomposite's carbon, nitrogen, and oxygen bond configurations. Reproduced from reference (Gunes et al. 2024). Under the terms and conditions of the Creative Commons Attribution (CC BY) license (http://creativecommons.org/licenses/by/4.0/)

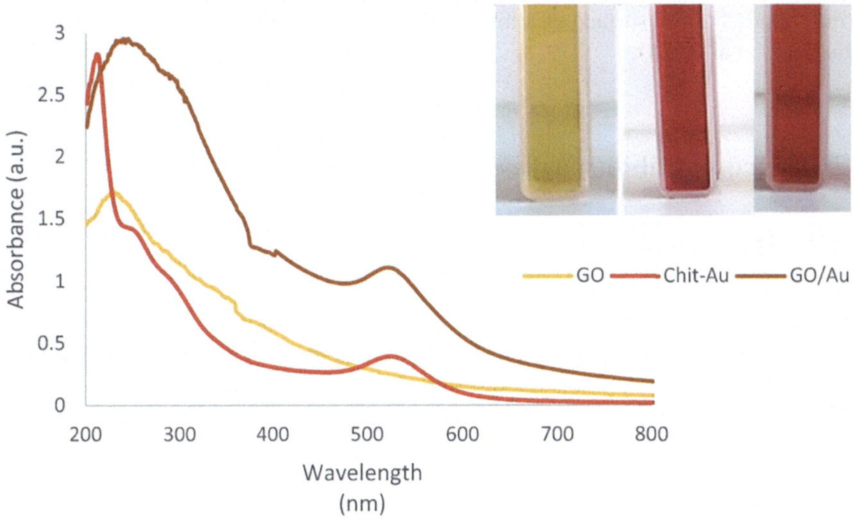

Fig. 7.12 UV Absorption of the GO, Cs-AuNPs and GO/Au nanocomposite. Reproduced from reference (Ramadan et al. 2024). Under the terms and conditions of the Creative Commons Attribution (CC BY) license (http://creativecommons.org/licenses/by/4.0/)

aided by these kinds of investigations. They go into great length about the UV–Vis absorption properties of nanofillers in appropriate polymer matrixes, including noble metal nanoparticles, semiconductor nanocrystals, nanooxides, nanoclays, and carbon nanotubes (Venkatachalam 2016). A modified Hummers process using potassium permanganate and sulphuric acid was used to oxidise graphite. This created manganese heptoxide (Mn_2O_7), an aggressive oxidant that oxidises the unsaturated double bond to make GO. Spectrophotometric analyses of GO solutions (Fig. 7.12) revealed absorption in the visible and near-infrared regions, but a prominent particular absorption peak in the ultraviolet region at a wavelength of 237 nm, which is indicative of a highly smoothed and exfoliated ultrathin structure of GO nanosheets. AuNPs exhibited a surface plasmon-specific resonance, peculiar to surface plasmon resonance, at approximately 530 nm (Fig. 7.12). Charge-transfer, physisorption, and electrostatic bonding were the ways in which Cs-AuNPs interacted with GO layers. UV–visible spectroscopy was used to characterise the GO/Au nanocomposites after GO was incorporated. This showed that the peak of the Cs-AuNPs showed a slight red-shift from 530 nm to roughly 540 nm, which strongly suggested that the mono-dispersion and surface modification remained persistent. The observation of a characteristic GO band at around 240 nm suggested the successful incorporation of Cs-AuNPs with GO nanosheets. In contrast to the Cs-AuNPs solution, the GO/Au nanocomposites solutions seem homogenous and significantly lighter in colour, which strongly suggests the integration of Cs-AuNPs with GO. Therefore, in order to change the surface property and photothermal performance of Cs-AuNPs, they are firmly bonded to the surface of a thin layer of GO through electrostatic interaction (Ramadan et al. 2024).

7.5 Biocompatibility Assessment of Nanocomposites

7.5.1 In-Vitro *Studies*

Since bone is the second most often transplanted tissue worldwide, there is a growing need for bone grafts, which has led to the development of synthetic scaffold grafting substitutes. Following injuries, fracture sites have elevated levels of oxidative stress, which impairs osteoblast function and impedes fracture remodelling and healing. In tissue engineering, free radical scavenging agents like cerium oxide nanoparticles have gained popularity as a means of combating oxidative stress. They characterised the biocompatibility of a porous, bioactive, free radical scavenging nanocomposite scaffold made of beta-tricalcium phosphate, cerium oxide nanoparticles, and poly (1,8 octanediol-co-citrate) in order to work towards the goal of creating a functional synthetic system for bone tissue engineering. Using both *in-vitro* and *in-vivo* models, (Dulany et al. 2020) investigated the compatibility of both human osteoblast cells and rat subcutaneous tissue with nanocomposite. It was discovered that the scaffolds facilitated cell adhesion, proliferation, mineralization, and infiltration and were biocompatible in both scenarios. They investigated the protective qualities of the nanocomposite scaffolds by simulating oxidative stress with hydrogen peroxide. This resulted in a decrease in cytotoxicity and a recovery of osteoblast cell mineralization *in-vitro* (Dulany et al. 2020). Biocompatibility assessments of the Chi-Au nanocomposites. (Fig. 7.13a) After a 24-h incubation period, the platelet activation culture on different nanocomposites was examined using an SEM. The non-activated platelet cells were seen to have a spherical shape (white arrow in the Chi-Au 50 ppm), whereas the flattened platelet cells represented the activated form (white arrow in the Control). The pseudopod form of the platelet cell was highlighted by the white arrow in Chi-Au 100 ppm. (Fig. 7.13b) IF staining was used to identify the CD68 macrophage marker, which is expressed in monocytes, after 96 h. Primary anti-CD68 antibodies were used to stain the monocytes, and secondary FITC antibodies were used to conjugate the staining (green fluorescence). Nuclei were identified using DAPI solution (blue fluorescence). 20 μm scale bars are used. (Fig. 7.13c) The quantity and level of activation of adherent platelet cells were measured. According to the findings, the combination of Chi and 50 ppm Au nanoparticles caused the least amount of platelet cell adhesion and activation. (Fig. 7.13d) The monocytes' CD68 expression was semi-quantified using fluorescence density as a basis. In the Chi–Au groups, there was a notable drop in CD68 expression. (Fig. 7.13e) The DCFH-dA probe was used to target the intracellular ROS generation of MSCs seeded on different nanocomposites, and flow cytometry was used to semi-quantify the results. One of

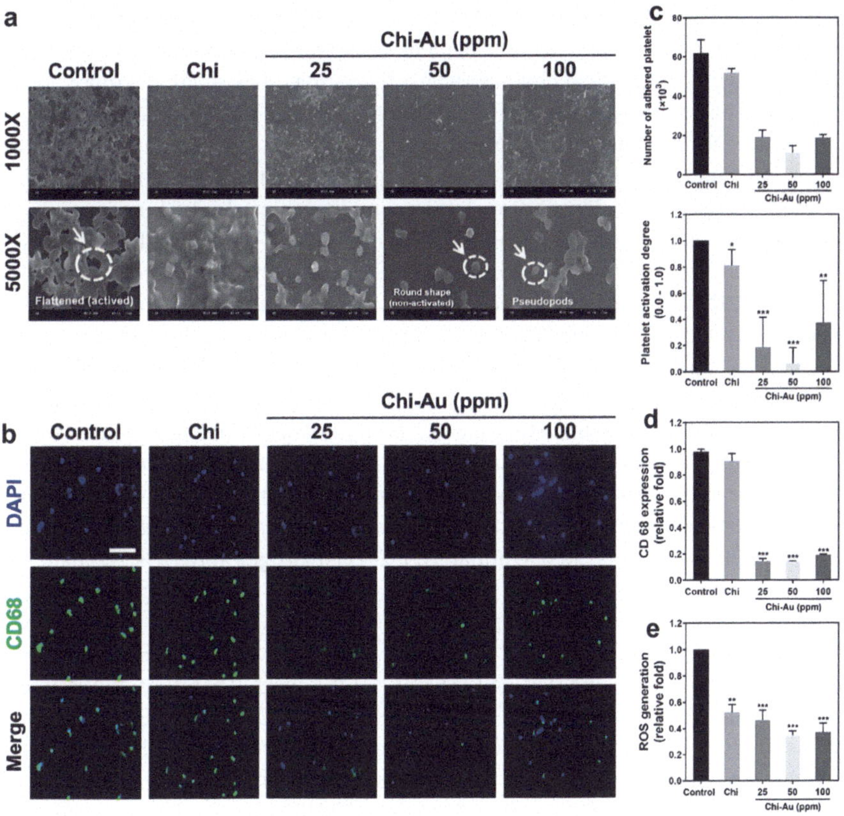

Fig. 7.13 Biocompatibility assessments of the Chi-Au nanocomposites. Reproduced from reference (Hung et al. 2022) under the terms and conditions of the Creative Commons Attribution (CC BY) license (http://creativecommons.org/licenses/by/4.0/)

three separate trials serves as a representation of the data. ** $p < 0.01$; *** $p < 0.001$; in comparison to the control group (TCPS) (Hung et al. 2022).

As a novel potential material for biomedical applications, the *in-vitro* biostability and biocompatibility of ethyl vinyl acetate (EVA) nanocomposites containing organically modified montmorillonite (organo-MMT) were studied. For four weeks, the materials were kept at 37 °C in oxidising and hydrolytic chemicals as part of an *in-vitro* treatment for neat EVA and EVA nanocomposites. Under these harsh environmental circumstances, the materials' *in-vitro* mechanical characteristics were evaluated. According to morphological studies, as the loading of the nanofiller increased, so did the degree of MMT dispersion and exfoliation. The greatest properties for nanofiller dispersion and exfoliation were shown by the EVA containing 1 weight percent organo-MMT. Compared to clean EVA and other EVA nanocomposites, this nanocomposite's surface degradation characteristics were observed to be smoother. In addition, the mechanical properties of the EVA nanocomposites are better than those of the clean EVA, and the *in-vitro* circumstances had less of an impact on these qualities. The addition of 1 weight percent of organo-MMT to the EVA produced the best *in-vitro* mechanical characteristics. It was hypothesised that a more convoluted diffusion pathway for oxidants and water molecules was introduced by a better distributed and exfoliated organo-MMT layered structure, which reduced their penetration into the EVA molecular chains. Consequently, the EVA polymer chains' degradation rates were slowed down, improving their biostability. Furthermore, the addition of 1 weight percent organo-MMT significantly increased the hardness of the hydrated EVA (exposed to PBS at 37 °C). According to the biocompatibility assessment, the EVA nanocomposites do not exhibit cytotoxicity, which means they meet the requirements needed to forward their development as a biomedical material (Osman et al. 2015). In order to evaluate a novel magnetic nanobiocomposite's potential for biological applications and hyperthermia, the current study set out to design and synthesise it. In order to do this, the PC-TG hydrogel (pectin and tragacanth gum) polymer was first created using $CaCl_2$ as a crosslinking agent. In order to improve the strength of the nanobiocomposite and because graphitic carbon nitride (CN) has pharmacological applications, natural silk fibroin (SF) protein and CN were added to the hydrogel in the second step. Ultimately, PC-TG hydrogel/SF/CN/Fe_3O_4 nanobiocomposite was synthesised and *in-situ* magnetised using Fe_3O_4 magnetic nanoparticles (MNPs) for an increased hyperthermia application. Following treatment with PC-TG hydrogel/SF/CN/Fe_3O_4 (1.75 mg/mL), the survival rate of HEK293T cells was 91.27% on the second day and 90.59% on the third. Therefore, this artificial magnetic nanocomposite is nearly biocompatible with HEK293T cells, with toxicity against these cells of less than 10.0%. Additionally, after 48 and 72 h, respectively, the viability percentage of BT549 cells at a concentration of 1.75 mg/mL decreased by roughly 19.0% and 20.0% in PC-TG hydrogel/SF/CN/Fe_3O_4. Consequently, this magnetic nanocomposite has the ability to impede BT549 cell development and lower the cell survival rate (Choopani et al. 2024) Two types of human cancer (HCT116) and normal (HUVCE) cells are studied in order to improve the anticancer and biocompatibility properties of TiO_2 NPs, ZnO NPs, ZnO-TiO_2 (NCs), and ZnO-TiO_2/reduced graphene oxide (RGO) NCs. ZnO-TiO_2/RGO NCs may now be made with a new process that makes use of Senna surattensis extract. The cytotoxicity and biocompatibility of synthesised NPs and NCs against HTC116 cancer and normal HUVEC cells were evaluated using MTT tests for exposure times of 24 and 48 h. These findings demonstrated that prolonged exposure or the production of reactive oxygen species caused the HCT116 cancer cells' cell viability to gradually decline. Moreover, the ZnO-TiO_2/RGO NCs showed the highest level of cytotoxicity, indicating promise for cancer treatment. Additionally, because of their low toxicity, all synthesised NPs and NCs have been shown to be safe for use in cancer therapy applications based on biocompatibility data on normal HUVEC cells. Furthermore, this work demonstrates that the plant extract's eco-friendliness, biocompatibility, and potential bioactivity give natural reducing and capping qualities for NPs and NCs (Alaizeri et al. 2024).

7.5.2 In-Vivo *Studies*

In the tissue engineering paradigm, scaffolds are vital because they help construct functional tissues and organs, guide cell growth, facilitate the formation of functional tissues and waste products, and provide temporary structural support. The usage of single-walled carbon nanotubes (SWNTs), particularly ultra-short SWNTs (US-tubes), has shown promise in the reinforcement of scaffold materials made of synthetic polymers. We describe here the *in-vivo* biocompatibility in a rabbit model of US-tube reinforced porous biodegradable scaffolds. Subcutaneous pockets and the femoral condyles of rabbits were implanted with US-tube nanocomposite scaffolds and control polymer scaffolds. At 4 and 12 weeks following implantation, the hard and soft tissue response was examined using histology, histomorphometry, and micro-computed tomography (micro CT). At both time points, the porous US-tube nanocomposite scaffolds showed positive responses from both hard and soft tissues. Defects using US-tube nanocomposite scaffolds showed three times more bone tissue ingrowth at 12 weeks than defects with polymer scaffold controls. The samples taken after 12 weeks also revealed improved connective tissue organisation and decreased inflammatory cell density. Qualitative variations between the two-time points were consistent with expected degradation owing to time progression; no significant quantitative difference in polymer degradation was identified among the various groups. The results of this work indicate that the presence of US-tubes may make nanocomposite scaffolds bioactively promoting osteogenesis, even though judgements on the osteoinductivity of US-tube nanocomposite scaffolds cannot be derived from it (Sitharaman et al. 2008). Following the guidelines of the UNE-EN ISO 10993–6:2017 Standard, "Biological evaluation of medical devices, Part 6: Tests related to local effects after implantation," a macroscopic observation of the intervention area was done prior to recovering the samples of the tissues implanted in the biomodels. The dorsal view of one of the biomodels 30 days after implantation is shown in (Fig. 7.14). During the three implantation periods, a comparable appearance was seen in every biomodel. In all three biomodels, hair regrowth was seen after 30 days of implantation (Fig. 7.14a). Following the trichotomy (Fig. 7.14b), it was noted that the surgical lesions had healed and that there was no longer any necrotic tissue or purulent or inflammatory discharge. Encasing the implanted materials facilitated the healing process. Upon tissue dissection, the interior surface (Fig. 7.14c) showed that the implanted materials were submerged in a more reddish-coloured tissue, devoid of pus and without a significant inflammatory reaction. The macroscopic picture demonstrated the material's biocompatibility with the surrounding tissue. Prior to the sample recovery, the macroscopic examination of the intervention biomodels revealed the emergence of new hair, a healthy external skin surface free of continuous solutions and exudates, and an interior surface devoid of any signs of purulent or inflammatory exudates. One barrier that keeps the rest of the body safe is the skin. In the face of trauma, it is capable of regeneration. It goes through the phases of remodelling, proliferation, and inflammatory response that enable it to revert to its original state before the shock. On the other hand, chronic inflammation or severe

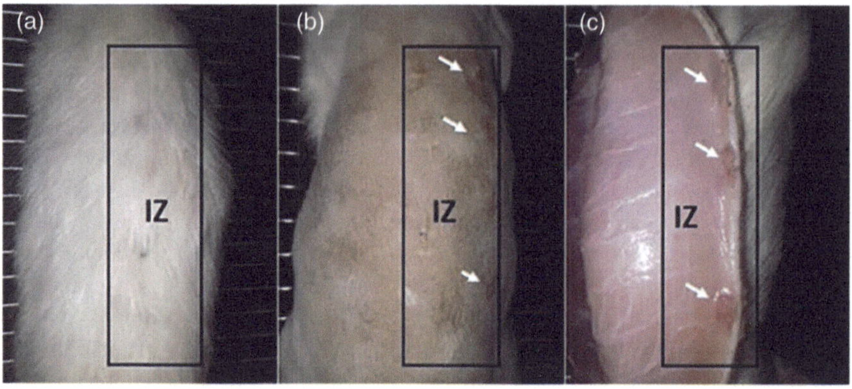

Fig. 7.14 Macroscopic picture of a 30-day-old biomodel implanted. **a** Dorsal picture. Dorsal picture displaying a trichotomy (**b**). **c** The skin's inner surface. Zone of Implantation: IZ. Implant locations are indicated by white arrows. Reproduced from reference (Castro et al. 2022) under the terms and conditions of the Creative Commons Attribution (CC BY) license (http://creativecommons.org/licenses/by/4.0/)

trauma can change the healing process, resulting in cuts and ulcers as well as pus production. Furthermore, colour changes and tiny bulges with induration—which signify a fibrous encapsulation of the material—made it easy to detect the insertion zones. The fact that the substance is encapsulated and encircled by fibrous tissue implies that the healing process is typical when a foreign body is present. According to earlier studies on soft tissue-embedded materials, this tissue will remain until it is removed (Castro et al. 2022).

The dorsal area, or implantation zone, in the biomodels is shown in (Fig. 7.15). For every biomodel, hair regrowth without purulent exudate showed a robust recovery. The wounds generated by the material implantation healed without a strong immune response, according to staining analyses performed later. Furthermore, the material's encapsulation in collagen fibre capsules suggests that the healing process is progressing since the collagen fibres enable the material to be fixed without inciting an aggressive immune response until it is removed for histological examination (Castro et al. 2023).

Low levels of hydrogen peroxide (H_2O_2) in cancer cells lead directly to limited efficacy of chemodynamic treatment (CDT). Despite tremendous efforts over the years, developing a CDT agent and ensuring the H_2O_2 supply within the tumour cell remain formidable challenges for researchers. In this work, a nanozyme-based bimetallic metal–organic frameworks (MOFs) tandem reactor is constructed to increase intracellular H_2O_2 levels, thereby improving CDT, by utilising the H_2O_2 level maintenance mechanism in cancer cells. Additionally, the up-conversion nanoparticles (UCNPs) loaded into the MOFs can perform photocatalysis and generate hydrogen under near-infrared excitation. This increases cellular susceptibility to radicals induced from H_2O_2, inhibits the energy of cancer cells, damages DNA, and induces tumour cell apoptosis, all of which work in concert to improve

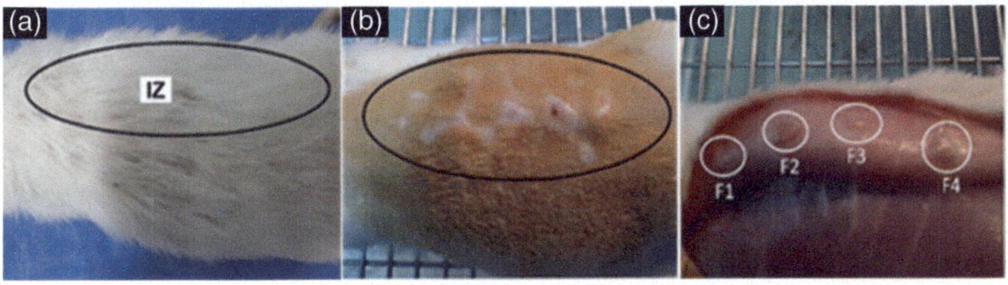

Fig. 7.15 Zone of subcutaneous dorsal implantation. **a** The dorsal region, which is hairy. **b** Dorsal region trichotomy. **c** Area of subdermal implantation. Zona implantation: black ovals. White circles indicate inserted blocks. Formulations 1, 2, 3, and 4 (F1–F4). Zone of Implantation: IZ. Reproduced from reference (Castro et al. 2023) under the terms and conditions of the Creative Commons Attribution (CC BY) license (http://creativecommons.org/licenses/by/4.0/)

the therapeutic efficacy of CDT. The evaluation of nanoparticles' *in-vivo* biocompatibility was conducted on female Kunming mice, with an average body weight of 20 g. After being suspended in saline, UCNPs, Zn–Co ZIF, and UCNPs@Zn–Co ZIF were injected into the tail vein. The mice were put to sleep after 14 days, and their organs were removed and ready for haematoxylin and eosin (HandE) staining so that they could be examined further. With great efficacy and minimal toxicity, the suggested bimetallic MOFs-mediated CDT and UCNPs-mediated hydrogen therapy function as a combination therapy (Zhong et al. 2024) An ultrasonic-responsive, *in-situ* gelling local delivery system (TPP ALG) was described, which combines tranilast poly(lactic-co-glycolic acid) (PLGA)/perfluoropentane (PFP) nanoparticles (TPP NPs) with alginate (ALG) solution to increase the efficacy of ICB after surgery. TPP NPs benefit from the *in-situ* gelling ALG by having a longer drug residence period, fewer repetitive deliveries, and increased safety. Before assessing TPP ALG's therapeutic efficacy, (Cao et al. 2024) performed a thorough *in-vivo* safety evaluation, looking at blood biochemical indicators, body weight changes, and the histology of important organs. The body weight of the mice did not significantly change after the treatment of TPP ALG. When TPP ALG-treated mice's blood biochemical profiles were compared to those of healthy mice, there were no appreciable differences in creatine kinase (CK), lactate dehydrogenase (LDH), aspartate aminotransferase (AST), alanine aminotransferase (ALT), albumin (ALB), alkaline phosphatase (ALP), creatinine (CREA), and UREA. Vital organs were taken at various times following TPP ALG injection in order to further assure safety, and haematoxylin and eosin were used for histological evaluation. There were no discernible pathogenic alterations found. The aforementioned results validated TPP ALG's good biosafety profile and bolstered its applicability for other therapeutic uses (Cao et al. 2024).

7.5.3 Cytotoxicity Evaluation

Bionanocomposites (BNCs) are nanocomposites based on biopolymers that have changed biological and physicochemical characteristics as well as a high surface-to-volume ratio and adaptability. BNCs find application in virology, cancer treatment, medicine delivery, cosmetics, and other fields. Biomaterials such as cellulose, chitin, starch, polylactic acid (PLA), sodium alginate, chitosan, and polyhydroxyalkanoates are utilised in the production of bionanocomposites. Bionanocomposite permeability, mechanical characteristics, and thermal stability have all improved with the addition of biomaterial; nonetheless, cytotoxicity and biocompatibility remain the main obstacles. As a result, numerous in-vitro investigations are carried out, such as flow cytometry studies to assess cell apoptosis, antimicrobial research using different agar medium assays, determination of minimum inhibitory doses, cell cultures, and assays for cell viability using the MTT assay. Determining lethal doses, examining organs, serum chemistry, and histology are examples of *in-vivo* research (Roy et al. 2021). Verifying the possible cytotoxicity of NiPB@AgNC is crucial for the effective use of these materials in biology and medicine. A fundamental test for assessing the cytotoxicity of any medications or nanomaterials is the *in-vitro* cell viability experiment with MTT reagents. They performed an *in-vitro* cell viability assay using MTT reagents in a variety of cancer (A549 and B16F10) and normal cell lines (EA.hy926) for 24 h in a dose-dependent manner in order to assess the cytotoxicity of nanocomposites (Fig. 7.16a–c). The results of cell viability demonstrate that NiPB@AgNC is biocompatible with both normal and cancer cells up to high levels of silver ([Ag] = 105 μM) (Fig. 7.16a–c). The synthesised nanocomposite may be beneficial for additional biological applications due to its biocompatibility. In this instance, the nanocomposite remains biocompatible even at elevated silver concentrations ([Ag] = 105 μM), suggesting the material's biocompatibility at elevated dosages (Mukherjee et al. 2017).

Superior mechanical and biological safety features are exhibited by Polyether Ether Ketone (PEEK), yet its biological inertness severely limits its use in biomedical applications. Active ceramic particles have been used in recent studies to improve biological activity and increase the bioactive composites' spectrum of applications in medical implants. In the current study, the PEEK matrix was reinforced using ceramic TiO_2 and SiO_2 nanoparticles (NP) functionalized with acrylic acid. Plastic injection moulding was used to create the PEEK/TiO_2/SiO_2 (PTS) nanocomposite. The average diameter of the PTS polymer composite's inhibitory zone is greater than that of the PEEK polymer composite, measuring 18.125 mm against E. coli and 16.375 mm against B. subtilis, respectively. MG-63 cell line was used for in-vitro direct and indirect cytotoxicity experiments, which revealed 5.70% cytotoxicity and 94.30% cell viability on PTS nanocomposite. As a result, the recently created composite is a viable option for biomedical uses (Thanigachalam and Subramanian 2022). Developing a new drug delivery system (DDS) is one of the hardest things to do in cancer treatment. Therefore, in order to enable the targeted delivery of doxorubicin (D), a common chemotherapeutic drug, into breast cancer cells, they developed

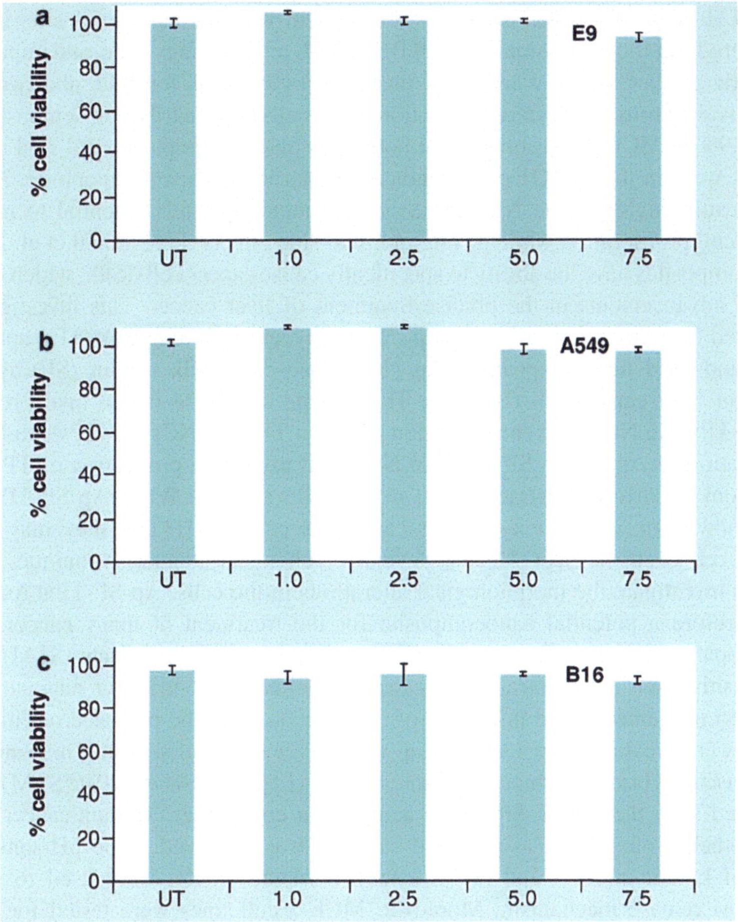

Fig. 7.16 The nickel-prussian blue@silver nanocomposite was tested for cell viability in **a** EA.hy926 (E9), **b** A549, and **c** B16F10 utilising varying doses over the course of 24 h using the 3-(4,5-Dimethylthiazol-2-yl)-2,5-diphenyltetrazolium bromide reagent. The volume of the nickel-prussian blue@silver nanocomposite is indicated numerically in µl/ml, where 1, 2.5, 5 and 7.5 µl/ml correspond to [Ag] = 14, 35, 70 and 105 µM, respectively. Reproduced from reference (Mukherjee et al. 2017) under the terms and conditions of the Creative Commons Attribution (CC BY) license (http://creativecommons.org/licenses/by/4.0/)

a new biocompatible and biodegradable zinc-based nanoscale metal–organic framework (Zn-NMOF) covered with folic acid (FA) functionalised chitosan (CS) in this study. Later *in-vitro* experiments were carried out on MCF-7 breast cancer cells and HFF-1 normal cells. These investigations included evaluations of cell survival, autophagy and apoptotic gene expression levels, cell cycle arrest, and apoptotic analyses. Results from *in-vitro* experiments showed that after 24 h of treatment with a 40 nM concentration of the NMOF-CS-FA-D nanocomposite, cell viability was

17.39% ± 6.34. Moreover, after 5 h of treatment with the nanocomposite, it was discovered that the expression levels of BECLIN1, an autophagy gene, and important apoptotic genes Caspase-9 and BAX increased by twofold, fourfold, and twofold, respectively. Furthermore, an examination of the cell cycle distribution demonstrated that 31.9% of MCF-7 cells underwent both early and late apoptosis, and 15.4 ± 2% of cells were in the sub-G1 phase, indicating that the cells were apoptotic. All of these results highlight the NMOF-CS-FA-D nanocomposite's potential to reduce cancer cell proliferation while posing minimal side effects (Fereydouni et al. 2024) Nanocomposites have the ability to specifically cause cancer cell death, which could lead to advancements in the precise treatment of liver cancer. This investigation evaluated the cytotoxicity of myogenic Talaromyces purpureogenus (TP), aptamer (Ap), and gold nanocomposites (Au NCs) encased in silk fibroin (SF) against a human liver cancer cell (HepG2). Three steps culminate in the final product, Ap-SF-TP@Au NCs. The encapsulation of SF on TP@Au NCs (SF-TP@Au NCs), conjugation of Ap within SF-TP@Au NCs, and myogenic production of TP@Au NCs from TP mycelial extract are all steps in this process. While Ap-SF-TP@Au NCs showed no cytotoxicity in non-cancerous cells (NIH3T3), they may have caused cell death in HepG2 cells. Several biochemical staining techniques were used to investigate the morphological alterations in the cells. Ap-SF-TP@Au NCs are therefore a potential nanocomposite for the treatment of many cancer cells (Elayappan et al. 2024) For the controlled release of sunitinib maleate (STM), new pH-sensitive and antibacterial drug carrier systems based on silver nanoparticles (AgNPs) implanted in the interlayer of laponite (Lap) in the presence of chitosan (CTS) were created. After loading Lap with silver ions and sunitinib maleate, the CTS-based hybrid bionanocomposite carrier (Lap@CTS@AgNPs@STM) was produced with the aid of citric acid acting as a crosslinker. The anticancer drug release behaviour of the as-prepared samples was observed to be pH-sensitive. Higuchi kinetic models and the Korsmeyer–Peppas model were used to study the STM release mechanism. Moreover, MCF-7 cell lines were tested for cytotoxicity *in-vitro* in order to demonstrate the efficacy of synthesised materials as multifunctional carrier systems for biomedical applications (Labib et al. 2024).

7.6 Drug Delivery Applications of Biocompatible Nanocomposites

7.6.1 Targeted Drug Delivery

The biomedical and pharmaceutical sciences have used nanocomposites more frequently in recent years due to the development of various forms of nanocomposites. Nanocomposites are an advantageous substitute for any one traditional material due to their distinct composition and nanoscale size range. An introduction to nanocomposites is given in this chapter, along with a discussion of the many varieties, including metal, ceramic, and polymer nanocomposites. The conversation also focuses on various controlled and targeted nanocomposite-based drug delivery systems, such as those for anti-inflammatory, antimicrobial, anticancer, anti-diabetic, and cardiovascular medications (Kaurav et al. 2018). Effective targeting of cancer cells with drug delivery via nanoparticles is crucial for improved therapeutic efficacy, decreased dose-limited toxicity, and safeguarding healthy cells from systemic toxicity. Anticancer drugs should be delivered to the tumour location using nanoparticles in a method that allows them to bypass all phagocytic barriers and enter the tumour tissues with the least amount of blood volume or activity loss. Moreover, the drug's active form should only harm malignant cells after being injected intracellularly, sparing healthy cells. Currently, there is a lot of study being done on the interaction between nanomaterials and tumour cells in an effort to better understand the problem of targeted medication administration using nanoparticles. Drug delivery mechanisms can be divided into two categories: passive targeting, which is powered by non-ligand mediated nanoparticles, and active targeting, which is powered by ligand-conjugated nanoformulations (Fig. 7.17) (Tiwari et al. 2023).

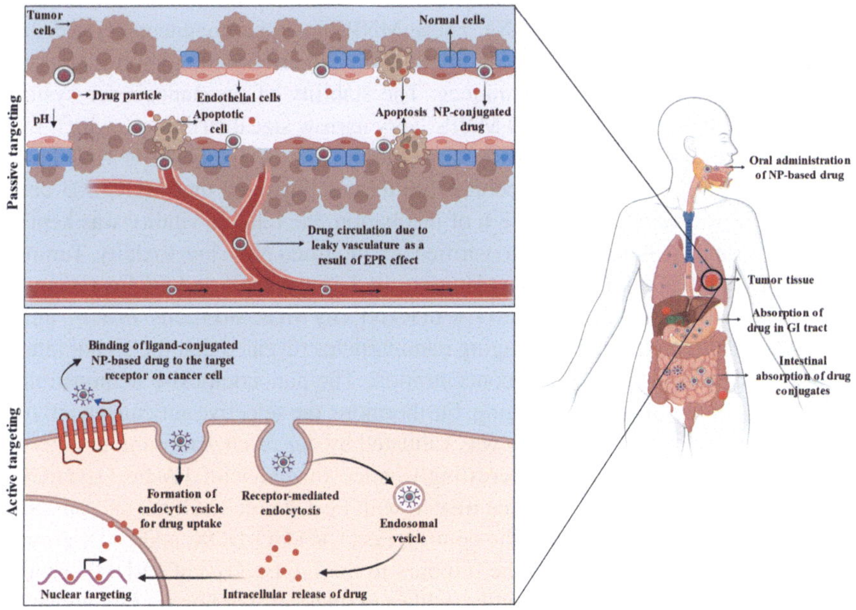

Fig. 7.17 NP-drug conjugate passive and active targeting mechanism. Because of inadequate lymphatic drainage during passive targeting, the drug-loaded nanoparticles are delivered through leaky blood vessels via the EPR effect, and the tumour's acidic microenvironment causes the drug to be released from the nanoparticles. Conversely, active targeting entails the binding of drug-laden nanoparticles loaded with ligands to the antigenic receptor on cancerous cells, which is then followed by receptor-mediated endocytic absorption. Reproduced from reference (Tiwari et al. 2023) under the terms and conditions of the Creative Commons Attribution (CC BY) license (https://creativecommons.org/licenses/by/4.0/)

It is a tried-and-true technique to employ biopolymers for sustained drug delivery. Because of their potential benefits in terms of biodegradation and biocompatibility, safety, environmental friendliness, and cost-effectiveness, the use of nanobiocomposites as sustained or targeted drug delivery devices has garnered a lot of attention. A number of drug delivery systems based on nanobiocomposite technology have entered the market recently. To prepare nanobiocomposites, the best materials at the moment are starch-based biopolymers, especially chitosan and alginate. The emulsion polymerization and in-situ solution polymerization techniques have proven to be very effective in the creation of nanobiocomposite materials. The next frontier for improving nanocomposites in targeted and controlled drug delivery processes is the development of novel techniques for the manufacture of nanobiocomposites with a particular arrangement of monomer units in three-dimensional space (Rahim et al. 2020). Through the use of nanomedicine, combined targeted drug delivery and prolonged drug release offer significant promise in cancer therapy and diagnostics. For such chemotherapeutic applications, systems based on folic acid coupled with graphene oxide-based magnetic nanoparticles (NPs) exhibit clear advantages. (D. Li et al. 2018) created FA-Fe_3O_4@nGO-DOX magnetic nanoparticles (MNPs) based on nanoscale graphene oxide (nGO) encapsulated Fe_3O_4, coupled with folic acid (FA), and loaded with doxorubicin (DOX). These MNPs had a homogeneous size distribution. The produced MNPs had a consistent size distribution and were assessed using a range of biophysical techniques. The stability of the nanocarrier system was enhanced by the Fe_3O_4@nGO MNPs' very narrow size distribution, which was a consequence of the nGO's uniform size. Studies on the FA-Fe_3O_4@nGO-DOX NPs' cell survival and *in-vitro* biocompatibility demonstrated that MGC-803 cells selectively absorbed them. After 48 h of incubation, the relative vitality was kept at almost 90%, and the hemolysis ratio verified our modified NPs' low toxicity. Tumour cells were specifically killed by the pH-controlled drug release and the FA receptor-mediated selective uptake of FA-Fe_3O_4@nGO NPs by MGC-803 cells. *In-vitro* analysis of the magnetic resonance imaging nanoparticles revealed that the signal intensity dropped with increasing NP concentration. The nanocomposite demonstrated remarkable efficacy in *in-vivo* imaging. Furthermore, the selective anticancer activity of the FA-Fe_3O_4@nGO-DOX NPs was validated by our *in-vivo* antitumour activity and histological examination. Interestingly, mice treated with FA-Fe_3O_4@nGO-DOX saw less weight loss than mice treated with Fe_3O4@nGO-DOX, and our NPs were very active. The tumours in the control, Fe_3O_4@nGO-DOX, and DOX groups showed less necrotic tissue than the tumours in the FA-Fe_3O_4@nGO-DOX group. FA-Fe_3O_4@nGO-DOX is a potentially stable and successful option for targeted drug delivery (D. Li et al. 2018). One major obstacle to more sophisticated drug delivery in the human body is the continued development of effective drug nanocarriers. The success of drug delivery may be increased by using cyanuric chloride (CC) as a linker with graphene-based nanomaterials. Here, a chemical precipitation process was utilised to create a superparamagnetic graphene oxide (SPMGO) nanocomposite using a straightforward technique. Cyanuric chloride worked well as a linker to functionalise the nanocomposite and load the drug. Methotrexate (MTX) immobilisation as a cancer treatment drug was implemented on the SPMGO and SPMGO/CC. The drug nanocarrier systems, SPMGO/MTX and SPMGO/CC/MTX, were shown to be hemocompatible and to boost the effectiveness of MTX against the Caov-4, HeLa, and MCF-7 cell lines in the *in-vitro* tests. The MTX nanocarriers exhibited regulated drug release and a notably high drug loading. Overall, the findings showed how promising SPMGO/CC/MTX nanocarriers are for targeted drug delivery, especially in MTX chemotherapy (Taheri-Kafrani et al. 2021). Conventional chemotherapeutic methods have drawbacks like poor drug stability, inaccurate cancer cell targeting, and unfavourable side effects. In this sense, hydrogels have demonstrated an appropriate remedy for the aforementioned shortcomings by providing an effective biocompatible drug delivery mechanism. This work used a graphene oxide/carboxymethylated chitosan (GO/CMCh) nanocomposite hydrogel to produce a biocompatible drug delivery method. The stability and therapeutic potential of quercetin (QR) and gallic acid (GA) were greatly increased by loading them onto the GO/CMCh. For QR and GA, respectively, high entrapment efficiency of 98% and 99.3% were attained. Moreover, the loading capacity of the GO/CMCh nanocomposite was impressive, attaining 200% for QR and 250% for GA. This novel strategy improved the stability of the loaded drugs and allowed for their targeted distribution, with a particular emphasis on the treatment of hepatocellular carcinoma at physiological pH levels. The GO/CMCh nanocomposites' prolonged drug release resulted in a controlled and extended therapeutic impact. Interestingly, the created nanocomposites showed low haemolysis ratios (less than 1%), guaranteeing their safety for possible therapeutic uses. These results offer a compelling justification for more research on the *in-vivo* uses of GA@GO/CMCh in drug delivery and cancer treatment (Salama et al. 2024). In contemporary nanomedicine, it is very desirable to produce multifunctional magnetic nanocomposites as a drug delivery mechanism for cancer therapy. Here, folic acid-bovine serum albumin conjugate (FA-BSA) was changed on nanocomposites to deliver an anticancer drug called 5-Fluorouracil (5-Fu) by combining quantum-sized Fe_3O_4 and layered double hydroxide (LDH). This resulted in a unique FA-BSA/Fe_3O_4@LDH. The synthesised nanocomposites exhibited strong magnetic properties, colloidal stability, dispersibility, and erythrocyte compatibility. FA-BSA/Fe_3O_4@LDH/5-Fu demonstrated pH responsiveness; in the pH 5.0 release media, FA-BSA/Fe_3O_4@LDH/5-Fu was released in considerably greater amounts and for a longer period of time than in the pH 7.4 release medium. According to the cellular studies, FA-BSA/Fe_3O_4@LDH did not appear to be significantly cytotoxic. This is likely because FA-BSA is present, which further improves the nanocomposite's biocompatibility. Additionally, the 2D HepG2 cell model and the 3D hepatoma cell microspheres model could be preferentially targeted by FA-BSA/Fe_3O_4@LDH/5-Fu, which would then internalise the cells efficiently by folate receptor-mediated endocytosis and exhibit good anticancer cell activity in a concentration-dependent manner. Consequently, the FA-BSA/Fe_3O_4@LDH nanocomposite that was created has the potential to be a unique multifunctional nanocomposite that can be used for pH-responsive drug release and magnetic-targeted drug delivery, both of which can improve the effectiveness of cancer therapy (Liu et al. 2024).

7.6.2 Passive Drug Delivery

According to the principle of passive targeting (Fig. 7.18), medications can be delivered by nanocarriers via passive diffusion or convection into the tumour mesenchyme or cells through the gaps in tumour capillary pores. The movement of molecules within a fluid is known as convection. Convection takes over as the main method of moving most macromolecules through the vascular pore spaces when the net filtration rate drops to zero. Low-molecular-weight substances, on the other hand, like oxygen, are mostly transferred by diffusion. Diffusion is the process by which molecules are moved across the cell membrane without using up cellular energy, according to a concentration gradient. Nonetheless, diffusion becomes the primary drug delivery mechanism due to the restricted convection across the tumour mesenchyme brought on by high interstitial pressure. A solid tumour's specific physiological phenomena, known as particle-mediated EPR effects, usually occur within 150 nm of the tumour and are characterised by a tumour's surrounding capillaries having a higher permeability than normal tissue and no functional lymphatic channels. Due to this physiological phenomenon, blood nanoparticles (about 250 nm in size) can more easily and efficiently travel through tumour capillary pores to the tumour site, where they can congregate and eventually penetrate into the tumour mesenchyme and intracellular compartments as time passes. The idea of passively targeting different tumours

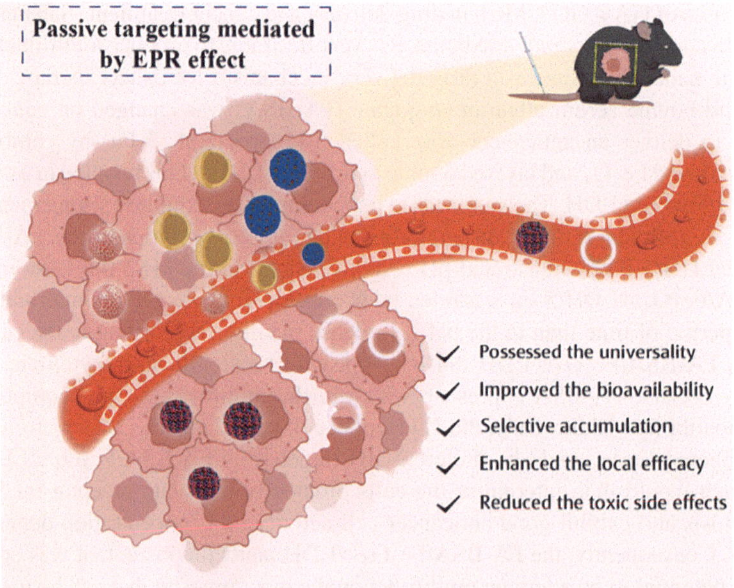

Fig. 7.18 Mechanism of passive Drug Delivery and its advantages. Reproduced from reference (Li et al. 2023a, b) under the terms and conditions of the Creative Commons Attribution (CC BY) license (https://creativecommons.org/licenses/by/4.0/)

with nanodrugs is based on this phenomenon. Nearly all fast-growing solid tumours are subject to the classical EPR effect, according to a significant number of reliable studies. Liposomes, polymeric nanoparticles, metal oxide nanoparticles, silicon dioxide nanoparticles, and other nanoparticles are the primary components of passive targeted drug delivery systems (Li et al. 2023a, b).

In order to collect NPs in the tumour tissue, passive targeting takes advantage of leaky vasculature and insufficient lymphatic drainage. This is based on the enhanced permeation and retention (EPR) effect. By quickly opsonizing extravasated particulate materials (drugs) and eliminating them from the bloodstream as foreign substances, the reticuloendothelial system (mononuclear phagocyte system, or MPS) stops the transport of medications to their intended locations. It is therefore possible to circumvent the opsonization process (which is accountable for RES absorption and MPS recognition of NPs) by using a well-designed nanocarrier as an efficient long-circulating delivery mechanism. For efficient drug delivery to the intended place, polyethylene glycol (PEG) coated nanocarriers are sometimes described as "stealthy" particles. Different PEGs with varying chain lengths and molecular weights are employed to regulate the thickness of the PEG coating and the efficacy of the grafting process. As PEGs get closer to the nanocarrier, their steric influence increases. A variety of PEG derivatives, including block copolymers, can also be used to modify the surface of nanocarriers. Additionally, the surfaces of nanocarriers can be altered using block copolymers of the poloxamer type. One product on the market is Genexol-PM, a micellar formulation made of block copolymers based on PEG (Pankhurst et al. 2003).

7.6.3 Active Drug Delivery

In order to provide more precise drug delivery, active targeting entails changing certain ligands, antibodies, or other molecules on the surface of nanoparticles to recognise and adhere to particular cells or tissues at the targeted region. For example, antibody-adorned nanoparticles can recognise and bind to particular antigens on the outside of tumour cells, allowing for more targeted drug delivery. The four primary categories of active targeting are aptamer-based targeting, small-molecule-based targeting, peptide-based targeting, and antibody-based targeting (Fig. 7.19). Through a bypass mechanism, the active targeting of nanoparticles *in-vivo* first targets the EPR effect before arriving at the targeted tumour site with surface changes.

Passive targeting strategies have a number of shortcomings. Consequently, a great deal of effort is being made to use different strategies to enhance the accumulation of NPs. One can target and penetrate a tumour cell specifically by using the phagocytosis/endocytosis mechanisms and the overexpressed receptors on the surface of the tumour cell. One example of this is the upregulation of the folic acid receptor and transferrin in a variety of cancer forms, which are both targets for the active DDS. The glycoprotein transferrin, which is connected to the cell membrane, controls cell development and aids in the intake of iron. Through enhancing the internalisation

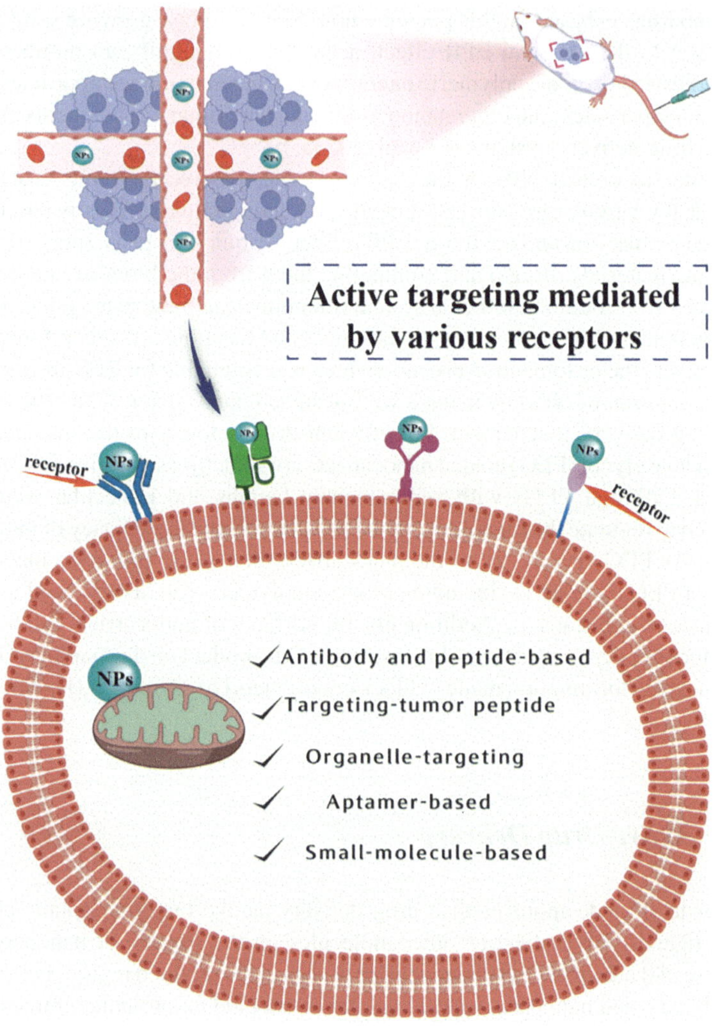

Fig. 7.19 Mechanism of active Drug Delivery. Reproduced from reference (Li et al. 2023a, b) under the terms and conditions of the Creative Commons Attribution (CC BY) license (https://creativecommons.org/licenses/by/4.0/)

of iron-loaded transferrin, the transferrin receptor promotes iron absorption. Drug distribution to tumours is also made easier by the folic acid receptor. According to the study, as cancer spreads, the folic acid receptor's density rises. Active targeting not only targets tumour-specific indicators but also can take advantage of processes like neoangiogenesis that promote the growth of tumours (Q A Pankhurst et al. 2003). Colon cancer is the leading cause of mortality worldwide, and the prevalence of the disease has not yet been reduced by conventional therapeutic approaches. In order to (active delivery) of metho-trexate (MTX) to colon cancer cells, a novel nanoscale metal–organic framework (NMOF) coated with folic acid (FA) and chitosan (CS) was created in this study. Drug-loading capacity (78%) and drug release way (10%˜ within first 5 h) was determined *in-vitro* following the qualification and quantification of the synthesised MOF-CS-FA-MTX nanocomposite using FT-IR, XRD, DLS, SEM, and TEM techniques, and after confirming the nanosize diameter of the nanocomposite. Within the first five hours, the release of MTX was approximately ten times more acidic than the typical pH, indicating that it was sensitive to pH. Less than 5% cell viability was shown by the MTT assay in HCT116 cells treated for 48 h with a 2.5 ng/mL concentration of MOF-CS-FA-MTX. Following treatment with 5 ng/mL MOF-CS-FA-MTX, the gene expression research revealed a six-fold increase in BAX and a two-fold decrease in Bcl2 gene expression. Additionally, the autophagy genes ATG5 and BECLIN1 showed a roughly five-fold increase in expression. The studies measuring cell cycle arrest (15.49% apoptosis) and apoptosis (20% apoptosis) during a 5-h period were perhaps approved for MOF-CS-FA-MTX's capacity to suppress cancer cells. These results demonstrate the NMOF-CS-FA-MTX's exceptional ability to suppress colon cancer cells (Khatibi et al. 2022). Over the past ten years, multifunctional nanosized metal–organic frameworks (NMOFs) have made significant progress in the development of drug delivery systems (DDSs). The utilisation of these material systems in drug delivery is hampered by their continued lack of accurate and selective cellular targeting and rapid release of the amount of pharmaceuticals that are merely adsorbed within and on the exterior surface of nanocarriers. Here, (Fytory et al. 2023) created a biocompatible Zr-based NMOF with a modified shell made of polyethyleneimine (PEI) and the hepatic tumour-targeting ligand glycyrrhetinic acid grafted on. Doxorubicin (DOX) is an anticancer drug that is effectively regulated and delivered to hepatic cancer cells (HepG2 cells) using an enhanced core–shell nanoplatform. The created nanostructure DOX@NMOF-PEI-GA exhibited a high loading capacity of 23%, an acidic pH-stimulated response, a longer drug release duration of 9 days, and enhanced selectivity towards the tumour cells. It's interesting to note that the hepatic cancer cell line (HepG2) and normal human skin fibroblast (HSF) showed very little toxic effect from the DOX-free nanostructures, but the hepatic tumour was more effectively killed by the DOX-loaded nanostructures, paving the way for effective cancer therapy applications and active drug delivery (Fytory et al. 2023).

7.7 Therapeutic Strategies Using Biocompatible Nanocomposites

7.7.1 Photothermal Therapy

The use of nanocomposites that are biocompatible in photothermal cancer therapy has been on the rise. Majority of the new approaches to cancer and other diseases are applied to this type of treatment. Light is directly employed in PTT through nanomaterials which convert it into heat, a process that causes hyperthermia at the affected site so as to bring about tissue specificity and targeted destruction. Presented below are some of the therapeutic strategies with regard to employing nanocomposite biocompatible materials in PTT:

(1) Engineering Optical Properties: A great interest has come around for biocompatible nanocomposites that have efficient absorbance within near-infrared light regions where both maximal tissue penetration and highest photothermal conversion efficiency are achieved. They nicely perform low-light absorption and generation as target therapies require but spare damaging healthy surrounding tissues (Duan et al. 2023).
(2) Surface Functionalisation for Directed Delivery: Nanocomposites functionalized with targeting ligands can be, for example, antibodies, peptides or aptamers, directed to specific organs and tumour sites, thus improving specificity and accumulation at the target site. Treatment precision is advanced since photothermal effect becomes specifically localized inside the tumour through targeted delivery while off-target effects on healthy tissues are minimized (Dehaini et al. 2016).
(3) Combination with Chemotherapy or Immunotherapy: Biocompatible nanocomposites can be incorporated with chemotherapeutic drugs or immunomodulatory agents to significantly boost synergistic treatment. Hyperthermia prompted by PTT can help improve drug release from nanocomposites or cancer cell sensitization to chemotherapy and hence achieve increased tumour cell killing as well as systemic toxicity reduction (He et al. 2015).
(4) Multimodal Imaging and Therapy: Nanocomposites offer the capability to combine imaging techniques like MRI, CT, or PAI with optical contrast imaging for directly observing how the treatment is progressing and also for precisely and accurately guiding PTT. Multimodal agents perform both imaging and therapy at the same time which makes it easier to determine the most optimal position of photothermal effect as well as thorough evaluation of treatment response (Lee et al. 2011).
(5) Stimulus-Responsive Release: Stimuli-responsive nanocomposites can be created that are able to release drugs at a specified location in response to a particular trigger (pH, temperature, enzyme), affording hierarchical spatiotemporal control over drug delivery in addition to sparing systemic exposure. PTT-induced hyperthermia then can catalyse drug release on demand from nanocomposites thus increasing efficacy of treatment (Mi 2020).

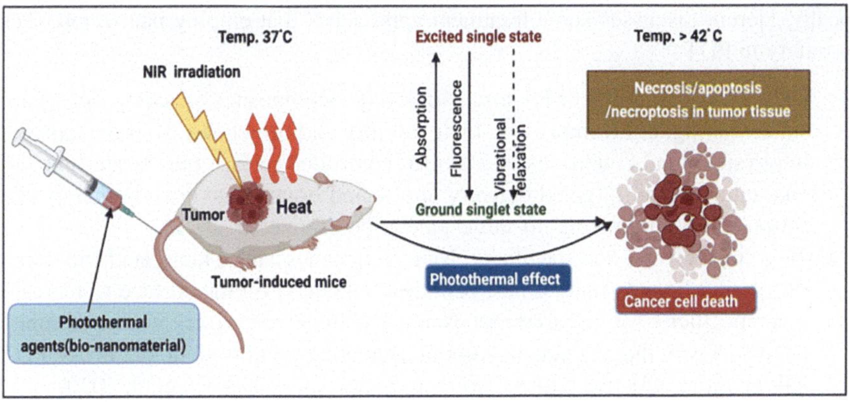

Fig. 7.20 Mechanism of PTT action. Reproduced from reference (Beniwal et al. 2024) under the terms and conditions of the Creative Commons Attribution (CC BY) license (https://creativecommons.org/licenses/by/4.0/)

(6) Biodegradability and Destination: Biocompatible nanocomposites are supposed to be biodegradable, clear the body after treatment and finally decrease the long-term toxicity and side effects. Desinging of nanocomposites by using biodegradable polymers or incorporating biodegradable components also assures safe degradation and clearance of carriers afterwards hyperthermia therapy(cataplasmosis) system (Karlsson et al. 2018).

(7) Preclinical and Clinical Translation: Implementing the transition of preclinical studies on biocompatible nanocomposites to clinical usage implies a solid assessment regarding safety, effectiveness, as well as biocompatibility. Running a complex preclinical program along with optimisation of formulation parameters is an indispensable stage in this translation process which makes it possible for very realistic PTT implementation to be developed eventually (Zhang et al. 2022).

Through making use of these healing strategies, biocompatible nanocomposites in the same line have a great chance to push forward the field of photothermal therapy and upgrade treatment outcomes for cancer and other diseases. In (Fig. 7.20) mechanism of PTT action (Beniwal et al. 2024).

7.7.2 *Photodynamic Therapy*

Nanocomposites that are biocompatible that are constructed in a variety of forms provide an all-inclusive means to upgrade the efficacy and safety of photodynamic therapy (PDT), a bright aspect to deal with various diseases, such as cancer or microbial infections. PDT is mainly about the photosensitisers activation (PS) using light so that ROS will be generated hence the destruction of tumour cells by cytotoxic effect locally. Herein discusses some treatment approaches that employ nanocomposites biosafety in PDT:

(1) Delivery and Stability of Photosensitiser: Nanocomposites biosafety can encapsulate/conjugate PS improving their stability and protection from degradation in physiological conditions. System for controlled release may be designed to guarantee a truly targeted delivery at affected organs maximising therapeutic action while minimising off-target side effect (Sun et al. 2022).
(2) Targeting delivery and specificity: Nanocomposites can be fabricated with functionalizing ligands (antibodies, peptides, aptamers) on the surface to increase their specificity for diseased tissues or cells. Targeted delivery will be administered in a way that PS gets maximum accumulation at tumour site hence PDT will be more effective while sparing normal tissues (Gierlich et al. 2020).
(3) Multifunctional imaging and therapy: Nanocomposites can also provide multi-imaging including fluorescence, MRI (magnetic resonance imaging), CT as well as optical coherence tomography and thus offer real-time monitoring and visualization of PDT. Multimodal platforms are helpful for accurate PS localization and monitoring, which eventually promotes precise treatment planning and therapeutic response assessment (Lee et al. 2011).
(4) Enhanced ROS Generation: ROS production under irradiation light in biocompatible nanocomposites may be improved through different engineering tricks transcending from increasing PS loading/distribution within nanocarrier matrix up to boosting development/efficiency of molecule light absorption/energy transfer. Presently this way results in better therapeutic outcomes as well as diminished resistance to treatment (Tian et al. 2022).
(5) Stimuli-Responsive Release of Soluble PS: Nanocomposites based on stimuli can be tailored to release PS in the condition of interest, e.g., pH, temperature or enzyme activity, thus improving spatial and temporal performance of PDT. Techniques for light-triggered drug release may also be included to let PS act just when required at a particular site (locale) and hence decrease systemic toxicity while treatment effectiveness is maximized (Blum et al. 2015).
(6) Combination Therapy: Additional therapeutic agents are better delivered using biocompatible nanocomposites which can be loaded into biodegradable nanofibers, for example, polymer meshes or hydrogels. These agents may include chemotherapy drugs, immunomodulators or more specificity (i.e., tumour-targeting) through nanoparticulate systems. Combination therapy approaches merge the different modality's mechanisms in cooperation that is quite often more cytotoxic effect and less resistance to the treatment (Songca 2023).
(7) Biodegradability and Clearance: Composure of biocompatible nanocomposites requires polymers with biodegradability facilitating safe destruction after treatment to prevent long-term toxicity as well as side effects. Just by copolymers containing degradable components will administer acceptable particles clearance after PDT cancer therapy system is employed (Wiwanitkit 2019).

(8) Translating nanocomposites from preclinical studies to clinical applications of biocompatible materials requires a thorough assessment of safety, efficacy, and biocompatibility. The first step towards clinical application is conducting adequate preclinical studies about the formulation and process parameters which are critical for successful translation. These instead clear the path for a possible clinical PDT approach (Metselaar and Lammers 2020).

Harnessing the therapeutic strategies presented in these nanoparticles makes them valuable agents in pushing forward the field of photodynamic therapy as well as improving treatment outcomes in cancer, microbial infections, and other disease system. Based on dual-mode conversion luminescence RENPs integrating with AIE photosensitiser and D-type neuropeptide, (He et al. 2024) created the phototheranostic nanoplatform RENPs@AIE-DNPY. When exposed to 808 nm light, the position of the glioma may be determined by the down-conversion NIR-II luminescence, while the up-conversion NIR-I luminescence may cause the AIE photosensitisers to produce ROS, which would prevent the growth of the orthotopic glioma tumour *in-situ*. The combination of D-type neuropeptide, AIE photosensitisers, and RENPs has been shown in both *in-vitro* and *in-vivo* experiments to be a potential option for *in-vivo* NIR-II fluorescence image-guided through-skull PDT treatments of brain tumours as described in (Fig. 7.21) (He et al. 2024).

7.7.3 Chemotherapy

(1) Site-Specific Drug Delivery: Biocompatible formulation for nanoparticles allows targeting only tumour cells and eliminates healthy tissues. Functionalisations with targeting ligands like antibodies, and or peptides increase their efficacy in cancer cells by ensuring the drug is localized at the tumour site while reducing off-target effects (Khalid et al. 2017).
(2) Controlled Drug Release: Nanocomposites are equipped with controlled release mechanisms that allow the chemotherapy drugs to be delivered more precisely. This could entail stimuli-responsive systems that would dock drugs in the presence of an environmental cue within the tumour microenvironment such as pH, enzyme level or redox potential to improve therapeutic efficiency and at the same time reduce systemic toxicity (Wu et al. 2024).
(3) Biocompatible nanocomposites for combination therapy whereby multiple therapeutic agents are co-delivered, allowing combination chemotherapy. Nanocomposites increase the efficacy of treatment and avoid drug resistance by incorporating synergistic drug combinations or targeting different pathways of tumourigenesis, along with decreased chance of recurrence (Yao et al. 2020).
(4) Nanocomposites encapsulating hydrophobic chemotherapy drugs are expected to increase the solubility and stability of these drugs in physiological conditions. Increased stability of the drug-micelle conjugate leads to longer retention in

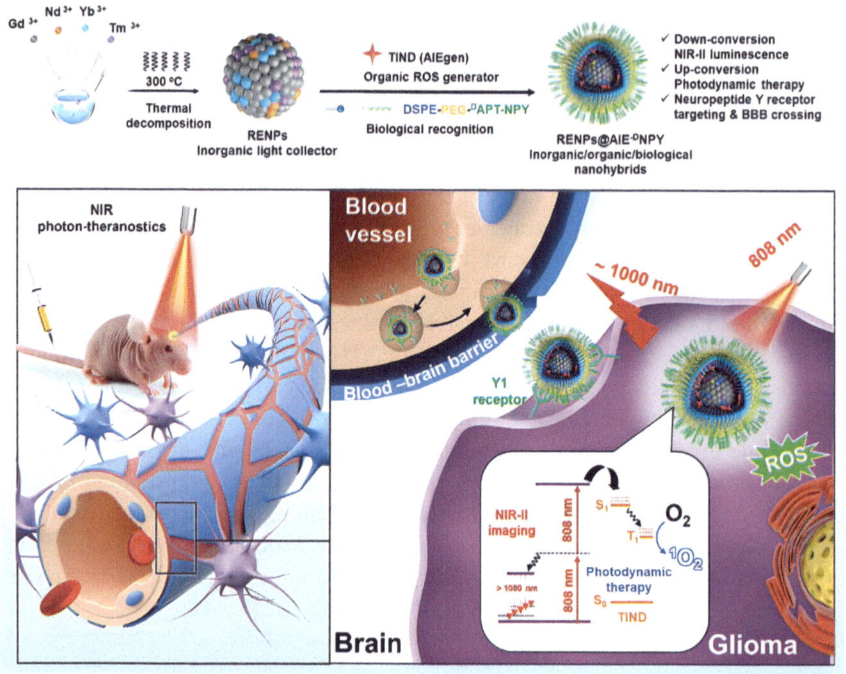

Fig. 7.21 Rare-earth doping nanoparticles (RENPs)@aggregation-induced emission (AIE)-^{D}NPY: production pathway and mechanism. AIEgen/RENP hybrid nanoprobes coated with ^{D}NPY, a D-type neuropeptide, have the ability to traverse the blood–brain barrier and target glioma sites precisely. The position of the glioma may be shown by the down-conversion near-infrared (NIR)-II luminescence under 808 nm light irradiation, whereas the up-conversion NIR-I luminescence may cause the AIE photosensitisers to produce reactive oxygen species, which would prevent the growth of the orthotopic glioma tumour in-situ. Reproduced from reference (He et al. 2024) under the terms and conditions of the Creative Commons Attribution (CC BY) license (https://creativecommons.org/licenses/by/4.0/)

circulation, thereby preventing premature drug degradation and maximising bioavailability with effective therapeutic benefits (El-Say and El-Sawy 2017).

Together, these tactics work to synergize the response of chemotherapy while reducing its systemic toxicity which allows for truly unique cancer-specific treatments. The phenomenon known as multidrug resistance (MDR) occurs when cancerous tumour cells become resistant to anticancer drugs following a single exposure. It is one of the main reasons for treatment failure and disease relapse, and it can happen with a wide variety of chemotherapy drugs with different chemical structures and modes of action. The use of several drug combinations to overcome multidrug resistance (MDR) has long been the subject of research; yet, this strategy is frequently linked to substantial adverse effects. Consequently, there is an urgent need for both the creation of new drugs to reverse MDR and boost the effectiveness of chemotherapy

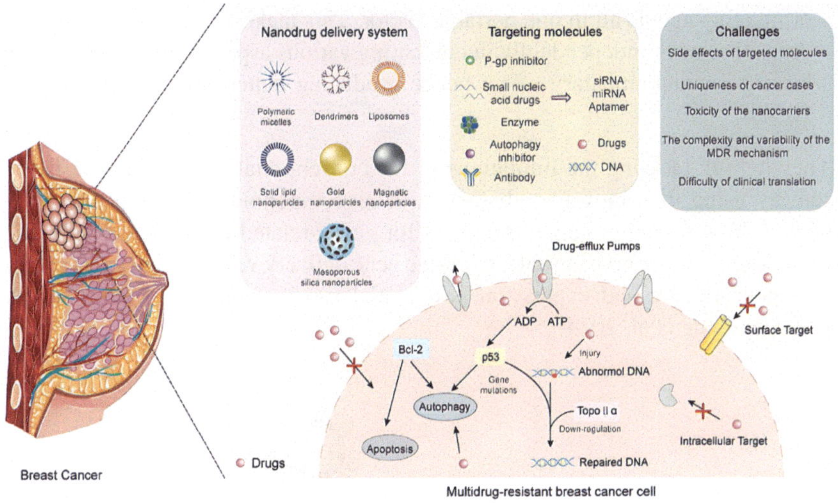

Fig. 7.22 Diagram of multidrug-resistant breast cancer cell. Reproduced from reference (Gao et al. 2024) under the terms and conditions of the Creative Commons Attribution (CC BY) license (https://creativecommons.org/licenses/by/4.0/)

for breast cancer as well as comprehensive study into the mechanisms underlying MDR as in (Fig. 7.22) (Gao et al. 2024).

7.7.4 Gene Therapy

(1) Targeted Delivery: One can apply nanotechnology in the delivery of genes or gene editing tools and make the nanocomposites biocompatible so that they target the affected cells or tissues only. One of the key advantages of surface modification is the ability to attach targeting ligands or antibodies to minimize side effects arising from off-target interactions (Chen et al. 2010).
(2) Enhanced Stability and Protection: Gene delivery vectors in nanocomposites protect the genetic molecules from degradation by enzymes as well as the host's immune system. The pen encapsulation has some advantages such as it provides protection of genetic material during circulation, enhances its stability and enables direct delivery to target progenitor cells (Hu et al. 2016).
(3) Controlled Release: Bio-responsive nanocarriers for gene delivery provide the targeted and stimuli-responsive release of therapeutic genes due to the presence of stimuli such as pH, temperature, and enzymes. This gives space–time regulation of gene delivery to dispense mishmash toxicity and to provide enhanced stature of treatment (Pan et al. 2021).
(4) Combination Therapies: Polymeric-based biocompatible nanocomposites allow for the co-encapsulation of multiple pharmaceuticals including genes, drugs and

or imaging agents all in one portable vector. This makes it possible to package drugs that work independently in combating various aspects of the disease thus having a complementary effect hence producing better results (Picchi et al. 2024).

They help in tackling the limitations of the conventional gene therapy delivery system by using the innovative characters of biocompatible nanocomposites thus providing the safe and effective treatment for genes-related diseases, cancers, and other diseases. Thus, choosing safe, effective gene delivery vectors that can precisely control gene release and expression is essential for the success of gene therapy (Fig. 7.23) (Gong et al. 2024).

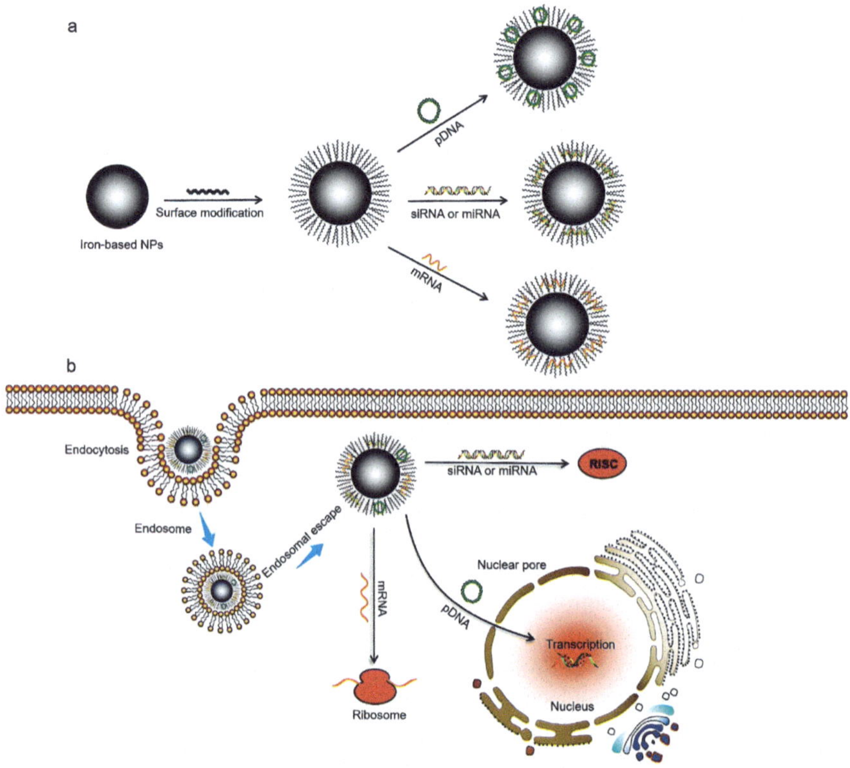

Fig. 7.23 An example of iron-based nanoparticles used in tumour treatment and gene transfer. Following surface modification, positively charged iron-based nanoparticles form an electrostatic bond with negatively charged DNA to form an iron-based nanoparticle/gene nanocomplex (**a**). Mice were given intravenously a specific dosage of the nanocomplex. Then, using active targeting or improved permeability and retention effect, the nanocomplex can precisely target tumour tissues. The loaded genes (pDNA, siRNA, mRNA, and miRNA) have antitumour effects when the nanocomplex gradually separates from the tumour cell (**b**). Reproduced from reference (Gong et al. 2024) under the terms and conditions of the Creative Commons Attribution (CC BY) license (https://creativecommons.org/licenses/by/4.0/)

7.8 Challenges and Future Perspectives

7.8.1 Biocompatibility Enhancement

Enhancing biocompatibility for nanocomposites poses several challenges and offers intriguing future perspectives: Enhancing biocompatibility for nanocomposites poses several challenges and offers intriguing future perspectives:

Challenges:

Immunogenicity and Toxicity: The properties of nanocomposites can also cause further immune responses or toxicity issues thus making biocompatibility to be delicately tested.

Long-term Stability: Another worthy challenge is to achieve adequate miscibility/coalescence and the stability of nanocomposites in physiological conditions over long periods of time.

Targeted Delivery: Thus, there is a need for developing precise versions of biomedical agents, which selectively target the intended site within the body without causing any kind of adverse reactions.

Regulatory Hurdles: Conformities with the requirements of the regulations involving biomedical uses require legal documentation of biocompatibility and safety parameters.

Future Perspectives:

Functional Diversity: Regarding the design of nanocomposites, the goal reached can imply multifunctionality where additional functions like drug delivery, imaging, and diagnostics can be incorporated into the same platform.

Personalized Medicine: This is likely to be achieved by developing nanocomposite formulations that are designed to fit the profile of an individual patient and therefore increase the chances of better results and reduced side effects.

Smart Nanocomposites: Technological progress may lead to the use of responsive materials and smart nanocomposites that change the scale of their therapeutic activity in response to the sensed physiological stimuli.

Biointerfacing Strategies: Creating novel plans aiming at increasing the selectivity of nanocomposites to interact with biological entities and systems may be useful in avoiding side effects.

Addressing these opportunities and forecasting future benefits can create significant value and bring about the revolutionary enhancement to the delivery of medical solutions based on biocompatible nanocomposites.

7.8.2 Controlled Drug Release

Challenges and Future Perspectives in Drug Release for Nanocomposites: Challenges and Future Perspectives in Drug Release for Nanocomposites:

Challenges:

Controlled Release: A more focused and fine control of the release characteristics of the drugs being released from nanocomposites to support the highest therapeutic value while at the same time minimising certain side effects.

Sustained Release: The need to develop nanocomposites that have long-term drug delivery duration capable of maintaining drug concentrations in the tissue at an effective rate.

Triggered Release: Designing stimuli-sensitive nanounits for controlled drug delivery where material's functions depend on specific biomolecular triggers –pH, temperature or enzyme functions.

Uniform Distribution: In order to minimize burst release and achieve consistent dosing throughout the nanocomposite material, it is necessary to control for the homogeneous distribution of the chosen drugs.

Future Perspectives:

Site-Specific Delivery: Nanotechnology may create the required opportunities that allow a drug, substance, or therapy to be delivered to targeted tissues or cells to enhance the effectiveness of a treatment or reduce the side effects of a treatment.

Real-time monitoring: The possibility of using sensors or imaging agents, incorporated into the nanocomposites for tracking both the drug release kinetics in the system and the therapeutic efficacy of the administered treatment.

Personalized Drug Delivery: The approaches to modify the position of nanocomposites through the molecular conjugate chemistry and develop optimal formulations for every single patient regarding his or her characteristics and disease markers to reach the maximum efficacy of the treatment.

Overcoming these challenges and building on the future trends may help to create effective nanocomposite-based drug delivery systems that can result in better therapeutic effectiveness and more improved quality of life for patients.

7.8.3 Clinical Translation

Challenges and Future Perspectives in Clinical Translation for Nanocomposites: Challenges and Future Perspectives in Clinical Translation for Nanocomposites:

Challenges:

Safety and Toxicity: Nanocomposites are predicted to be used in a lot of daily use products, and therefore, a safety concern for human use because of their toxicity has never been fully determined.

Biocompatibility: In discussing and or eliminating, fear over the immunogenic, inflammatory, or untoward reactions that may occur when a material comes in contact with biological systems.

Scalability and Manufacturing: Lit technological advancements in the manufacturing of nanocomposites, proficient and affordable methods to fabricate nanocomposite for clinical application on a large scale.

Regulatory Approval: Possible adverse effects and reactions of patient-population subjects, as well as pathways and procedures required to attain regulatory approval for clinical trials and ultimate commercial venture of effective, safe, and rightly-qualified trialled medications and treatments.

Future Perspectives:

Clinical Validation: Performing extensive studies to prove the effectiveness of the properties and features of nanocomposites in curing certain diseases or disorders and to also prove its safety.

Personalized Medicine: Tailoring of nanocomposite formulations for every patient and his condition, to derive the best results with having nil or lesser negative side effects.

Theranostic Integration: The design of nanocomposites with theranostics where diagnostic imaging or testing can be applied directly into the nanocomposite for monitoring treatment options in real-time.

Multimodal Therapies: Nanocomposites can be used in synergistic effects along with the conventional therapy or the other effective modality that is under development; this will help in overcoming multiple resistances and improving the efficacy of a particular therapy.

Overcoming these difficulties and adopting future visions may help advance clinical applications of nanocomposites, subsequently revealing their capacity to radically transform diagnostics, therapies, and individualized treatment strategies in healthcare.

7.8.4 Multifunctional Nanocomposites

Challenges and Future Perspectives in Multifunctional Nanocomposites:

Challenges:

Integration of Functionalities: Challenging the boundaries between science and engineering to enable fabrication of high-performance, multifunctional nanocomposites systems for multiple applications (e. g. therapeutic, diagnosis, monitoring).

Biocompatibility and Safety: Assuring that every functional component exhibits non-reactivity to the biological system as well as safety in isolation and in interaction with other components present in the nanocomposite. Synergistic Effects: An awareness of how there is potential for crossover between each of the functions to create cross-gains or cross-losses. Controlled Release: Designing and executing methods to ascertain that the release properties of each individual function become tailored to meet specific therapeutic or diagnosis end results.

Future Perspectives:

Tailored therapies will be directed to the specific body region, organ, or tissue. Loading of multifunctional nanocomposites with specific therapeutic or diagnostic payloads will be done keeping in view the profile of the individual patient and the type of disease.

Theranostic Advancements: Redefining the current capabilities of theranostics in the approach where treatment and bioassays of disease evolution or reaction to the administered treatment are performed concomitantly.

Smart Nanocomposites: Developing smart nanocomposite systems that can work in stimuli-sensitive manners to vary their inherent functions based on the nature of typical biological signals/signals or other triggers.

Cross-Disciplinary Collaboration: Encouraging multisector cooperation and team science across different fields (e.g., material science, biology, medicine) in order to enhance advancements in the nanocomposite field and analysis of multiple translated issues.

Overcoming these challenges and by taking into consideration future trends, multifunctional nanocomposites are poised to radically transform a wide spectrum of disciplines such as biomedical and diagnostic technologies as well as therapeutics.

7.9 Conclusion

Well, this is all about the biocompatible nanocomposites and how they are going to alter the nature of the fight against cancer. The use of nanotechnology combined with biocompatible materials has led to new drug delivery systems which may enhance the efficacy of a treatment and also reduce its side effects. With the ability to utilise the individual characteristics of a variety of nanocomposites, including targeted drug delivery, controlled release mechanisms, and the encapsulation of a wide range of therapeutic agents, researchers have opened up the door to newer, more efficient, and individualized treatments for cancer.

We have discussed a number of different biocompatible nanocomposites, such as those with polymers, silica, and various metal nanoparticles, which have all shown great potential in preclinical and clinical settings. The continued testing of biocompatibility shows the need to know that these materials can be used *in-vivo* without harming the surrounding healthy tissue too much.

This chapter has also touched on some of the modern dilemmas of the subject such as scalability, regulatory problems, and the requirement for serious clinical testing. However, with constant research, chemists, biologists, and clinicians will have to combine their efforts to conquer these hurdles, to allow the use of nanocomposite technologies in everyday cancer treatments.

All in all, biocompatible nanocomposites are the future of cancer treatment and can greatly improve the effectiveness of treatment and the quality of life in a patient. And as time marches on, it will take more and more creativity and research to use these materials to their fullest, and thus, a new era in cancer treatment will emerge.

References

Alaizeri ZAM, Alhadlaq HA, Aldawood S, Abduh NAY (2024) Green synthesis of ZnO-TiO2/RGO nanocomposites using Senna surattensis extract: a novel approach for enhanced anticancer efficacy and biocompatibility. RSC Adv 14(24):16685–16695. https://doi.org/10.1039/d4ra01634c

Al-Sahlawi F, Alabdali AYM, Chinnappan S, Al-Samydai A, Maki MAA (2024) Polymer-based nanoparticles in targeted cancer therapy: a review. J Appl Pharm Sci 14(9):57–68. https://doi.org/10.7324/JAPS.2024.172227

Anand U, Dey A, Chandel AKS, Sanyal R, Mishra A, Pandey DK, De Falco V, Upadhyay A, Kandimalla R, Chaudhary A, Dhanjal JK, Dewanjee S, Vallamkondu J, Pérez de la Lastra JM (2023) Cancer chemotherapy and beyond: current status, drug candidates, associated risks and progress in targeted therapeutics. Genes Dis 10(4):1367–1401. https://doi.org/10.1016/j.gendis.2022.02.007

Arambula-Maldonado R, Mequanint K (2024) Sol-Gel derived gelatin–bioactive glass nanocomposite biomaterials incorporating calcium chloride and calcium ethoxide. Polymers 16(6). https://doi.org/10.3390/polym16060747

Asadi Asadabad M, Eskandari M (2015) Transmission electron microscopy as best technique for characterization in nanotechnology. Synth React Inorg 45. https://doi.org/10.1080/15533174.2013.831901

Beniwal N, Verma A, Putta CL, Rengan AK (2024) Recent trends in bio-nanomaterials and non-invasive combinatorial approaches of photothermal therapy against cancer. Nanotheranostics 8(2):219–238. https://doi.org/10.7150/ntno.91356

Blum AP, Kammeyer JK, Rush AM, Callmann CE, Hahn ME, Gianneschi NC (2015) Stimuli-responsive nanomaterials for biomedical applications. J Am Chem Soc 137(6):2140–2154. https://doi.org/10.1021/ja510147n

Bokov D, Turki Jalil A, Chupradit S, Suksatan W, Javed Ansari M, Shewael IH, Valiev GH, Kianfar E (2021) Nanomaterial by Sol-Gel method: synthesis and application. Adv Mater Sci Eng 2021:5102014. https://doi.org/10.1155/2021/5102014

Cai Z, Liu Q (2021) Understanding the global cancer statistics 2018: implications for cancer control. Sci China Life Sci 64(6):1017–1020. https://doi.org/10.1007/s11427-019-9816-1

Cao Y, Zhong X, Wu N, Wan L, Tang R, He H, Wang C, Cheng H, Zhang Q, Zhong L, Wei X, Ren J, Sun Y, Li P (2024) An Ultrasound-responsive and in situ gelling hydrogel nanocomposite for boosting anti PD-L1 immunotherapy via remodeling aberrant ECM of post-surgical residual cancer. Adv Func Mater. https://doi.org/10.1002/adfm.202404941

Carneiro Í, Simões S (2020) Effect of morphology and structure of MWCNTs on metal matrix nanocomposites. Materials 13(23). https://doi.org/10.3390/ma13235557

Castro JI, Araujo-Rodríguez DG, Valencia-Llano CH, López Tenorio D, Saavedra M, Zapata PA, Grande-Tovar CD (2023) Biocompatibility assessment of polycaprolactone/polylactic acid/

zinc oxide nanoparticle composites under in vivo conditions for biomedical applications. Pharmaceutics 15(9). https://doi.org/10.3390/pharmaceutics15092196

Castro JI, Valencia Llano CH, Tenorio DL, Saavedra M, Zapata P, Navia-Porras DP, Delgado-Ospina J, Chaur MN, Hernández JHM, Grande-Tovar CD (2022) Biocompatibility assessment of polylactic acid (PLA) and nanobioglass (n-BG) nanocomposites for biomedical applications. Molecules 27(11). https://doi.org/10.3390/molecules27113640

Chehelgerdi M, Chehelgerdi M, Allela OQB, Pecho RDC, Jayasankar N, Rao DP, Thamaraikani T, Vasanthan M, Viktor P, Lakshmaiya N, Saadh MJ, Amajd A, Abo-Zaid MA, Castillo-Acobo RY, Ismail AH, Amin AH, Akhavan-Sigari R (2023) Progressing nanotechnology to improve targeted cancer treatment: overcoming hurdles in its clinical implementation. Mol Cancer 22(1):169. https://doi.org/10.1186/s12943-023-01865-0

Chen Y, Zhu X, Zhang X, Liu B, Huang L (2010) Nanoparticles modified with tumor-targeting scFv deliver siRNA and miRNA for cancer therapy. Mol Ther 18(9):1650–1656. https://doi.org/10.1038/mt.2010.136

Choopani L, Aliabadi HAM, Ganjali F, Kashtiaray A, Eivazzadeh-Keihan R, Maleki A, Salimibani M, Karimi AH, Salehpour N, Mahdavi M (2024) Fabrication of a magnetic nanocomposite based on natural hydrogel: pectin, tragacanth gum, silk fibroin, and integrated graphitic carbon nitride for hyperthermia and biological features. Carbohydr Polym Technol Appl 7:100495. https://doi.org/10.1016/j.carpta.2024.100495

Debela DT, Muzazu SG, Heraro KD, Ndalama MT, Mesele BW, Haile DC, Kitui SK, Manyazewal T (2021) New approaches and procedures for cancer treatment: current perspectives. SAGE Open Med 9:205031212110343. https://doi.org/10.1177/20503121211034366

Dehaini D, Fang RH, Zhang L (2016) Biomimetic strategies for targeted nanoparticle delivery. Bioeng Transl Med 1(1):30–46. https://doi.org/10.1002/btm2.10004

Dhilip Kumar SS, Abrahamse H (2021) Biocompatible nanocarriers for enhanced cancer photodynamic therapy applications. Pharmaceutics 13(11):1933. https://doi.org/10.3390/pharmaceutics13111933

Duan S, Hu Y, Zhao Y, Tang K, Zhang Z, Liu Z, Wang Y, Guo H, Miao Y, Du H, Yang D, Li S, Zhang J (2023) Nanomaterials for photothermal cancer therapy. RSC Adv 13(21):14443–14460. https://doi.org/10.1039/d3ra02620e

Dulany K, Hepburn K, Goins A, Allen JB (2020) In vitro and in vivo biocompatibility assessment of free radical scavenging nanocomposite scaffolds for bone tissue regeneration. J Biomed Mater Res—Part A 108(2):301–315. https://doi.org/10.1002/jbm.a.36816

Eid MM (2021) Characterization of nanoparticles by FTIR and FTIR-microscopy. In: Handbook of consumer nanoproducts. Springer Singapore, pp 1–30. https://doi.org/10.1007/978-981-15-6453-6_89-1

Elayappan PK, Kandasamy K, Sasikumar V, Bharathi M, Hirad AH, Alarfaj AA, Arulselvan P, Jaganathan R, Ravindran R, Suriyaprakash J, Thangavelu I (2024) Facile engineering of aptamer-coupled silk fibroin encapsulated myogenic gold nanocomposites: investigation of antiproliferative activity and apoptosis induction. Biotech Lett 46(5):871–885. https://doi.org/10.1007/s10529-024-03491-2

El-Bahnasawi AH, El-Segaey AA, Albohy SAH, El-Azabawy OE, Arafa EI, El-Koly NG, Al-Shafey HI (2023) Evaluation of newly copolymers and their montmorillonite nanocomposite as cold flow improver for petroleum lubricating oil. Sci Rep 13(1):14991. https://doi.org/10.1038/s41598-023-41802-1

El-ghannam G, Moawad M, Abo-Elfadl MT, Elfeky SA (2024) Beetroot extract@chitosan nanocomposite as a promising approach towards cancer therapy. Int J Biol Macromol 261:129700. https://doi.org/10.1016/j.ijbiomac.2024.129700

El-Meligy MA, Abd El-Monaem EM, Eltaweil AS, Mohy-Eldin MS, Ziora ZM, Heydari A, Omer AM (2024) Recent advancements in metallic Au- and Ag-based chitosan nanocomposite derivatives for enhanced anticancer drug delivery. Molecules 29(10). https://doi.org/10.3390/molecules29102393

El-Say KM, El-Sawy HS (2017) Polymeric nanoparticles: promising platform for drug delivery. Int J Pharm 528(1–2):675–691. https://doi.org/10.1016/j.ijpharm.2017.06.052

Feldman D (2019) Polymers and polymer nanocomposites for cancer therapy. Appl Sci (Switz) 9(18). https://doi.org/10.3390/app9183899

Fereydouni P, Al Mohaddesin A, Khaleghi S (2024) Targeted biocompatible Zn-metal–organic framework nanocomposites for intelligent chemotherapy of breast cancer cells. Sci Rep 14(1):18311. https://doi.org/10.1038/s41598-024-69457-6

Fytory M, Mansour A, El Rouby WMA, Farghali AA, Zhang X, Bier F, Abdel-Hafiez M, El-Sherbiny IM (2023) Core-shell nanostructured drug delivery platform based on biocompatible metal-organic framework-ligated polyethyleneimine for targeted hepatocellular carcinoma therapy. ACS Omega 8(23):20779–20791. https://doi.org/10.1021/acsomega.3c01385

Gao L, Meng F, Yang Z, Lafuente-Merchan M, Fernández LM, Cao Y, Kusamori K, Nishikawa M, Itakura S, Chen J, Huang X, Ouyang D, Riester O, Deigner H-P, Lai H, Pedraz JL, Ramalingam M, Cai Y (2024) Nano-drug delivery system for the treatment of multidrug-resistant breast cancer: current status and future perspectives. Biomed Pharmacother 179:117327. https://doi.org/10.1016/j.biopha.2024.117327

Ghazzy A, Naik RR, Shakya AK (2023) Metal–polymer nanocomposites: a promising approach to antibacterial materials. Polymers 15(9). https://doi.org/10.3390/polym15092167

Gierlich P, Mata AI, Donohoe C, Brito RMM, Senge MO, Gomes-da-Silva LC (2020) Ligand-targeted delivery of photosensitizers for cancer treatment. Molecules 25(22). https://doi.org/10.3390/molecules25225317

Gong Y, Hu X, Chen M, Wang J (2024) Recent progress of iron-based nanomaterials in gene delivery and tumor gene therapy. J Nanobiotechnology 22(1):309. https://doi.org/10.1186/s12951-024-02550-0

Gonon M (2021) Case studies in the X-ray diffraction of ceramics. In: Pomeroy M (ed) Encyclopedia of materials: technical ceramics and glasses. Elsevier, pp 560–577. https://doi.org/10.1016/B978-0-12-818542-1.00005-9

Guglielmi M (2014) Synthesis strategies for the preparation of sol-gel nanocomposites. In: Sol-Gel nanocomposites. Springer, New York, pp 51–82. https://doi.org/10.1007/978-1-4939-1209-4_3

Gunes BA, Kirlangic OF, Kilic M, Sunguroglu A, Ozgurtas T, Sezginer EK, Boyacioglu B, Unver H, Yildiz M (2023) Palladium metal nanocomposites based on PEI-functionalized nitrogen-doped graphene quantum dots: synthesis, characterization, density functional theory modeling, and cell cycle arrest effects on human ovarian cancer cells. ACS Omega. https://doi.org/10.1021/acsomega.3c10324

Gunes BA, Kirlangic OF, Kilic M, Sunguroglu A, Ozgurtas T, Sezginer EK, Boyacioglu B, Unver H, Yildiz M (2024) Palladium metal nanocomposites based on PEI-functionalized nitrogen-doped graphene quantum dots: synthesis, characterization, density functional theory modeling, and cell cycle arrest effects on human ovarian cancer cells. ACS Omega 9(11):13342–13358. https://doi.org/10.1021/acsomega.3c10324

He C, Lu J, Lin W (2015) Hybrid nanoparticles for combination therapy of cancer. J Control Release : Off J Control Release Soc 219. https://doi.org/10.1016/j.jconrel.2015.09.029

He X, Luo Y, Li Y, Pan Y, Kwok RTK, He L, Duan X, Zhang P, Wu A, Tang BZ, Li J (2024) D-type neuropeptide decorated AIEgen/RENP hybrid nanoprobes with light-driven ROS generation ability for NIR-II fluorescence imaging-guided through-skull photodynamic therapy of gliomas. Aggregate 5(1). https://doi.org/10.1002/agt2.396

Hu Q, Sun W, Wang C, Gu Z (2016) Recent advances of cocktail chemotherapy by combination drug delivery systems. Adv Drug Deliv Rev 98:19–34. https://doi.org/10.1016/j.addr.2015.10.022

Hung H-S, YangY-C, Chang C-H, Chang K-B, Shen CC, Tang C-L, Liu S-Y, Lee C-H, Yen C-M, Yang M-Y (2022) Neural differentiation potential of mesenchymal stem cells enhanced by biocompatible chitosan-gold nanocomposites. Cells 11:1861. https://doi.org/10.3390/cells11121861

Iffah F, Taufiq A, Pujiarti H, Nasikhudin Subadra STUI, Hidayat A, Diantoro M (2024) Fabrication of HAp/Zn-ferrite nanocomposite for drug delivery application. AIP Conf Proc 3074(1):070013. https://doi.org/10.1063/5.0211757

Jaleh B, Fakhri P (2016) Chapter 5—Infrared and Fourier transform infrared spectroscopy for nanofillers and their nanocomposites. In: Thomas S, Rouxel D, Ponnamma D (eds) Spectroscopy of polymer nanocomposites. William Andrew Publishing, pp 112–129. https://doi.org/10.1016/B978-0-323-40183-8.00005-7

Kailasa S, Reddy MSB, Maurya MR, Rani BG, Rao KV, Sadasivuni KK (2021) Electrospun nanofibers: materials, synthesis parameters, and their role in sensing applications. Macromol Mater Eng 306(11). https://doi.org/10.1002/mame.202100410

Kargozar S, Gorgani S, Nazarnezhad S, Wang AZ (2023) Biocompatible nanocomposites for post-operative adhesion: a state-of-the-art review. Nanomaterials 14(1):4. https://doi.org/10.3390/nano14010004

Kargozar S, Gorgani S, Nazarnezhad S, Wang AZ (2024) Biocompatible nanocomposites for post-operative adhesion: a state-of-the-art review. Nanomaterials 14(1). https://doi.org/10.3390/nano14010004

Karlsson J, Vaughan HJ, Green JJ (2018) Biodegradable polymeric nanoparticles for therapeutic cancer treatments HHS public access. Annu Rev Chem Biomol Eng 9:105–127. https://doi.org/10.1146/annurev

Kaurav H, Manchanda S, Dua K, Kapoor D (2018) Nanocomposites in controlled and targeted drug delivery systems p 19

Kausar A, Ahmad I (2023) Electrospinning processing of polymer/nanocarbon nanocomposite nanofibers—design, features, and technical compliances. In: J Compos Sci 7(7). https://doi.org/10.3390/jcs7070290

Khalid A, Persano S, Shen H, Zhao Y, Blanco E, Ferrari M, Wolfram J (2017) Strategies for improving drug delivery: nanocarriers and microenvironmental priming. Expert Opin Drug Deliv 14(7):865–877. https://doi.org/10.1080/17425247.2017.1243527

Khan I, Khan I, Saeed K, Ali N, Zada N, Khan A, Ali F, Bilal M, Akhter MS (2023) 7—Polymer nanocomposites: an overview. In: Ali N, Bilal M, Khan A, Nguyen TA, Gupta RK (eds) Smart polymer nanocomposites. Elsevier, pp 167–184. https://doi.org/10.1016/B978-0-323-91611-0.00017-7

Khan I, Saeed K, Khan I (2019) Nanoparticles: properties, applications and toxicities. Arab J Chem 12(7):908–931. https://doi.org/10.1016/j.arabjc.2017.05.011

Khatibi Z, Kazemi NM, Khaleghi S (2022) Targeted and biocompatible NMOF as efficient nanocomposite for delivery of methotrexate to colon cancer cells. J Drug Deliv Sci Technol 73:103441. https://doi.org/10.1016/j.jddst.2022.103441

Kuna K, Baddam SR, Kalagara S, Akkiraju PC, Tade RS, Enaganti S (2024) Emerging natural polymer-based architectured nanotherapeutics for the treatment of cancer. Int J Biol Macromol 262:129434. https://doi.org/10.1016/j.ijbiomac.2024.129434

Kusdianto K, Sari TD, Laksono MA, Madhania S, Winardi S (2021) Fabrication and application of ZnO-Ag nanocomposite materials prepared by gas-phase methods. IOP Conf Ser: Mater Sci Eng 1053(1):012023. https://doi.org/10.1088/1757-899x/1053/1/012023

Labib P, Mahdavinia GR, Dehghani A, Nobarzad RS, Sayyar Z, Jafari H, Najafzadeh K, Akinay Y, Akbari A (2024) Chitosan-based pH-sensitive antibacterial bionanocomposites with laponite immobilized silver nanoparticles for delivery of sunitinib maleate to breast cancers. Appl Organomet Chem 38(9):e7585. https://doi.org/10.1002/aoc.7585

Lateef A, Nazir R (2017) Metal nanocomposites: synthesis, characterization and their applications, pp 239–256

Lee D-E, Koo H, Sun I-C, Ryu J, Kim K, Kwon I (2011) ChemInform abstract: multifunctional nanoparticles for multimodal imaging and theragnosis. Chem Soc Rev 41:2656–2672. https://doi.org/10.1039/c2cs15261d

Li D, Deng M, Yu Z, Liu W, Zhou G, Li W, Wang X, Yang D-P, Zhang W (2018) Biocompatible and stable GO-coated Fe3O4 nanocomposite: a robust drug delivery carrier for simultaneous tumor MR imaging and targeted therapy. ACS Biomater Sci Eng 4(6):2143–2154. https://doi.org/10.1021/acsbiomaterials.8b00029

Li J, Wang Q, Xia G, Adilijiang N, Li Y, Hou Z, Fan Z, Li J (2023) Recent advances in targeted drug delivery strategy for enhancing oncotherapy. Pharmaceutics 15(9). https://doi.org/10.3390/pharmaceutics15092233

Li XP, Ji G, Chen Z, Addad A, Wu Y, Wang HW, Vleugels J, Van Humbeeck J, Kruth JP (2017) Selective laser melting of nano-TiB2 decorated AlSi10Mg alloy with high fracture strength and ductility. Acta Mater 129:183–193. https://doi.org/10.1016/j.actamat.2017.02.062

Li Z, Sun Y, Ge S, Zhu F, Yin F, Gu L, Yang F, Hu P, Chen G, Wang K, Volinsky AA (2023) An overview of synthesis and structural regulation of magnetic nanomaterials prepared by chemical coprecipitation. Metals 13(1). https://doi.org/10.3390/met13010152

Lima-Sousa R, Melo BL, Mendonça AG, Correia IJ, de Melo-Diogo D (2024) Hyaluronic acid-functionalized graphene-based nanohybrids for targeted breast cancer chemo-photothermal therapy. Int J Pharm 651:123763. https://doi.org/10.1016/j.ijpharm.2023.123763

Lin YS, Deng SG (1999) Sol-gel preparation of nanostructured adsorbents. In: Dąbrowski A (ed) Adsorption and its applications in industry and environmental protection, vol 120, pp 653–686. Elsevier. https://doi.org/10.1016/S0167-2991(99)80568-0

Liu Z, Wang X, Zhang C, Lin K, Yang J, Zhang Y, Hao J, Tian F (2024) Folic acid-coupled bovine serum albumin-modified magnetic nanocomposites from quantum-sized Fe3O4 and layered double hydroxide for actively targeted delivery of 5-fluorouracil. Int J Biol Macromol 256:128385. https://doi.org/10.1016/j.ijbiomac.2023.128385

Loganathan S, Valapa RB, Mishra RK, Pugazhenthi G, Thomas S (2017) Chapter 4—Thermogravimetric analysis for characterization of nanomaterials. In: Thomas S, Thomas R, Zachariah AK, Mishra RK (eds) Thermal and rheological measurement techniques for nanomaterials characterization. Elsevier, pp 67–108. https://doi.org/10.1016/B978-0-323-46139-9.00004-9

Lungu II, Grumezescu AM, Volceanov A, Andronescu E (2019) Nanobiomaterials used in cancer therapy: an up-to-date overview. Molecules 24(19):3547. https://doi.org/10.3390/molecules24193547

Mansour AM, Abou Hammad AB, Balkhtb AO, Elhelali TM, Abouelnaga AM, El Nahrawy AM (2024) Chitosan/In situ Gelatin-Mg3Si2O5(OH)4 nanocomposites via Sol-Gel method: preparation, characterization and antimicrobial properties. Arab J Sci Eng 49(1):1015–1024. https://doi.org/10.1007/s13369-023-07935-1

Metselaar JM, Lammers T (2020) Challenges in nanomedicine clinical translation. Drug Deliv Transl Res 10(3):721–725. https://doi.org/10.1007/s13346-020-00740-5

Mi P (2020) Stimuli-responsive nanocarriers for drug delivery, tumor imaging, therapy and theranostics. Theranostics 10(10):4557–4588. https://doi.org/10.7150/thno.38069

Moghaddam FD, Heidari G, Zare EN, Djatoubai E, Paiva-Santos AC, Bertani FR, Wu A (2023) Carbohydrate polymer-based nanocomposites for breast cancer treatment. Carbohydr Polym 304:120510. https://doi.org/10.1016/j.carbpol.2022.120510

Mohammadi A, Hosseini MS, Bagheri F, Safari H, Shadfar Y, Sharafi A, Rezaeejam H, Aghaei A, Danafar H (2024) Synthesis of curcumin loaded single walled carbon nanotubes: characterization and anticancer effects in vitro. Results Chem 7:101370. https://doi.org/10.1016/j.rechem.2024.101370

Mondal A, Nayak AK, Chakraborty P, Banerjee S, Nandy BC (2023) Natural polymeric nanobiocomposites for anti-cancer drug delivery therapeutics: a recent update. Pharmaceutics 15(8). https://doi.org/10.3390/pharmaceutics15082064

Mukherjee S, Das S, Nuthi S, Patra CR (2017) Biocompatible nickel-prussian blue@silver nanocomposites show potent antibacterial activities. Future Sci OA 3(4). https://doi.org/10.4155/fsoa-2017-0048

Ng HM, Mohamad saidi N, Omar FS, Kasi R, Subramaniam R, Bashir S (2018) Thermogravimetric analysis of polymers, pp 1–29. https://doi.org/10.1002/0471440264.pst667

Omanović-Mikličanin E, Badnjević A, Kazlagić A, Hajlovac M (2020) Nanocomposites: a brief review. Heal Technol 10(1):51–59. https://doi.org/10.1007/s12553-019-00380-x

Osman AF, Alakrach AM, Kalo H, Azmi WNW, Hashim F (2015) In vitro biostability and biocompatibility of ethyl vinyl acetate (EVA) nanocomposites for biomedical applications. RSC Adv 5(40):31485–31495. https://doi.org/10.1039/C4RA15116J

Pan X, Veroniaina H, Su N, Sha K, Jiang F, Wu Z, Qi X (2021) Applications and developments of gene therapy drug delivery systems for genetic diseases. Asian J Pharm Sci 16(6):687–703. https://doi.org/10.1016/j.ajps.2021.05.003

Picchi DF, Biglione C, Horcajada P (2024) Nanocomposites based on magnetic nanoparticles and metal-organic frameworks for therapy, diagnosis, and theragnostics. ACS Nanosci Au 4(2):85–114. https://doi.org/10.1021/acsnanoscienceau.3c00041

Pankhurst QA, Connolly J, Jones SK, Dobson J (2003) Applications of magnetic nanoparticles in biomedicine. J Phys D Appl Phys 36(13):R167. https://doi.org/10.1088/0022-3727/36/13/201

Rahim M, Haris M, Saqib N (2020) An overview of polymeric nano-biocomposites as targeted and controlled-release devices. Biophys Rev 12. https://doi.org/10.1007/s12551-020-00750-0

Raja PB, Munusamy KR, Perumal V, Ibrahim MNM (2022) 5—Characterization of nanomaterial used in nanobioremediation. In: Iqbal HMN, Bilal M, Nguyen TA (eds) Nano-bioremediation : fundamentals and applications. Elsevier, pp 57–83. https://doi.org/10.1016/B978-0-12-823962-9.00037-4

Ramadan MA, Gad S, Sharaky M, Faid AH (2024) Laser photostability of chitosan coated gold-GO nanocomposite and its role as a nano-therapeutic agent for control breast cancer growth. Discov Appl Sci 6(4):170. https://doi.org/10.1007/s42452-024-05808-2

Rezvova MA, Nikishau PA, Makarevich MI, Glushkova TV, Klyshnikov KY, Akentieva TN, Efimova OS, Nikitin AP, Malysheva VY, Matveeva VG, Senokosova EA, Khanova MY, Danilov VV, Russakov DM, Ismagilov ZR, Kostjuk SV, Ovcharenko EA (2022) Biomaterials based on carbon nanotube nanocomposites of Poly(styrene-b-isobutylene-b-styrene): the effect of nanotube content on the mechanical properties, biocompatibility and hemocompatibility. Nanomaterials 12(5). https://doi.org/10.3390/nano12050733

Rocha JVR, Krause RF, Ribeiro CE, Oliveira NC, Ribeiro de Sousa L, Leandro Santos J, Castro S, Valadares MC, Cunha Xavier Pinto M, Pavam MV, Lima EM, Antônio Mendanha S, Bakuzis AF (2024) Near infrared biomimetic hybrid magnetic nanocarrier for MRI-guided thermal therapy. ACS Appl Mater Interfaces. https://doi.org/10.1021/acsami.4c03434

Rodrigues MA, Marinho JPN, Arantes Fernandes Vieira L, Gomes DA, de Sousa EMB (2024) Internalized hydroxyapatite nanoparticles conjugated with a modified form of curcumin functionalized with folic acid promote bone tumor cell necrosis. ACS Appl Nano Mater 7(18):21304–21318. https://doi.org/10.1021/acsanm.4c02467

Roy S, Chakraborty P, Das S, Chakraborty S (2021) Cytotoxicity and biocompatibility of bionanocomposites, pp 565–586. https://doi.org/10.1016/B978-0-12-821280-6.00013-1

Saeed M, Alshammari Y, Majeed SA, Al-Nasrallah E (2020) Chemical vapour deposition of graphene—synthesis, characterisation, and applications: a review. Molecules 25(17). https://doi.org/10.3390/molecules25173856

Salama AM, Alqarni Z, Hamed YS, Yang K, Nour HF, Lu J (2024) Biocompatible graphene oxide/carboxymethylated chitosan nanocomposite hydrogel for controlled release of polyphenolic antioxidants. J Drug Deliv Sci Technol 95:105630. https://doi.org/10.1016/j.jddst.2024.105630

Sharma N, Saxena T, Alam SN, Ray BC, Biswas K, Jha SK (2022) Ceramic-based nanocomposites: a perspective from carbonaceous nanofillers. Mater Today Commun 31:103764. https://doi.org/10.1016/j.mtcomm.2022.103764

Shin M, Lim J, Park Y, Lee JY, Yoon J, Choi JW (2024a) Carbon-based nanocomposites for biomedical applications. RSC Adv 14(10):7142–7156. https://doi.org/10.1039/d3ra08946k

Shin M, Lim J, Park Y, Lee J-Y, Yoon J, Choi J-W (2024b) Carbon-based nanocomposites for biomedical applications. RSC Adv 14(10):7142–7156. https://doi.org/10.1039/D3RA08946K

Sitharaman B, Shi X, Walboomers XF, Liao H, Cuijpers V, Wilson LJ, Mikos AG, Jansen JA (2008) In vivo biocompatibility of ultra-short single-walled carbon nanotube/biodegradable polymer nanocomposites for bone tissue engineering. Bone 43(2):362–370. https://doi.org/10.1016/j.bone.2008.04.013

Songca SP (2023) Combinations of photodynamic therapy with other minimally invasive therapeutic technologies against cancer and microbial infections. Int J Mol Sci 24(13). https://doi.org/10.3390/ijms241310875

Spence JCH (2013) High-resolution electron microscopy. OUP Oxford

Stankic S, Suman S, Haque F, Vidic J (2016) Pure and multi metal oxide nanoparticles: synthesis, antibacterial and cytotoxic properties. J Nanobiotechnology 14(1):73. https://doi.org/10.1186/s12951-016-0225-6

Štengl V, Bakardjieva S, Grygar TM, Bludská J, Kormunda M (2013) TiO2-graphene oxide nanocomposite as advanced photocatalytic materials. Chem Cent J 7(1):41. https://doi.org/10.1186/1752-153X-7-41

Sudharsana C, Anvarsha N, Kalyani P (2024) Carbon-based nanocomposites: a comprehensive review of their multifunctional applications. In: Parvulescu V, Anghel EMM (eds) Nanocomposites—properties, preparations and applications. IntechOpen. https://doi.org/10.5772/intechopen.114402

Sun L, Yuan G, Gao L, Yang J, Chhowalla M, Gharahcheshmeh MH, Gleason KK, Choi YS, Hong BH, Liu Z (2021) Chemical vapour deposition. Nat Rev Methods Prim 1(1):5. https://doi.org/10.1038/s43586-020-00005-y

Sun N, Wen X, Zhang S (2022) Strategies to improve photodynamic therapy efficacy of metal-free semiconducting conjugated polymers. Int J Nanomedicine 17:247–271. https://doi.org/10.2147/IJN.S337599

Taheri-Kafrani A, Shirzadfar H, Abbasi Kajani A, Kudhair BK, Jasim Mohammed L, Mohammadi S, Lotfi F (2021) Functionalized graphene oxide/Fe3O4 nanocomposite: a biocompatible and robust nanocarrier for targeted delivery and release of anticancer agents. J Biotechnol 331:26–36. https://doi.org/10.1016/j.jbiotec.2021.03.005

Thanigachalam M, Subramanian AVM (2022) Evaluation of PEEK-TiO2- SiO2 nanocomposite as biomedical implants with regard to in-vitro biocompatibility and material characterization. J Biomater Sci Polym Ed 33(6):727–746. https://doi.org/10.1080/09205063.2021.2014028

Tian H, Zhang T, Qin S, Huang Z, Zhou L, Shi J, Nice EC, Xie N, Huang C, Shen Z (2022) Enhancing the therapeutic efficacy of nanoparticles for cancer treatment using versatile targeted strategies. J Hematol Oncol 15(1). https://doi.org/10.1186/s13045-022-01320-5

Tiwari H, Rai N, Singh S, Gupta P, Verma A, Singh AK, Kajal Salvi P, Singh SK, Gautam V (2023) Recent advances in nanomaterials-based targeted drug delivery for preclinical cancer diagnosis and therapeutics. Bioengineering 10(7). https://doi.org/10.3390/bioengineering10070760

Tripathy J (2017) Polymer nanocomposites for biomedical and biotechnology applications. In: Tripathy K, Sahoo B (ed) Properties and applications of polymer nanocomposites: clay and carbon based polymer nanocomposites. Springer Berlin Heidelberg, pp 57–76. https://doi.org/10.1007/978-3-662-53517-2_4

Truong TT, Mondal S, Doan VHM, Tak S, Choi J, Oh H, Nguyen TD, Misra M, Lee B, Oh J (2024) Precision-engineered metal and metal-oxide nanoparticles for biomedical imaging and healthcare applications. Adv Colloid Interface Sci 332:103263. https://doi.org/10.1016/j.cis.2024.103263

Venkatachalam S (2016) Chapter 6—Ultraviolet and visible spectroscopy studies of nanofillers and their polymer nanocomposites. In: Thomas S, Rouxel D, Ponnamma D (eds) Spectroscopy of polymer nanocomposites. William Andrew Publishing, pp 130–157. https://doi.org/10.1016/B978-0-323-40183-8.00006-9

Venkateshaiah A, Nutenki R, Kattimuttathu S (2016) X-ray diffraction spectroscopy of polymer nanocomposites, pp 410–451. https://doi.org/10.1016/B978-0-323-40183-8.00014-8

Vo QNQ, Rezk AI, Chun S, Park CH, Kim CS (2024) PH-sensitive composite nanofibers of poly(ϵ-caprolactone) loaded with iron oxide nanoparticles/ammonium bicarbonate as a nanocarrier toward efficient doxorubicin release for postsurgical cancer treatment. Mater Adv 5(5):2128–2139. https://doi.org/10.1039/d3ma01026k

Wang X, Tang F, Cao Q, Qi X, Pearson M, Li M, Pan H, Zhang Z, Lin Z (2020) Comparative study of three carbon additives: carbon nanotubes, graphene, and fullerene-C60, for synthesizing enhanced polymer nanocomposites. Nanomaterials 10(5). https://doi.org/10.3390/nano10050838

Wei W, Zhang X, Zhang S, Wei G, Su Z (2019) Biomedical and bioactive engineered nanomaterials for targeted tumor photothermal therapy: a review. Mater Sci Eng, C 104:109891. https://doi.org/10.1016/j.msec.2019.109891

Wiwanitkit V (2019) Biodegradable nanoparticles for drug delivery and targeting, pp 167–181. https://doi.org/10.1007/978-3-030-06115-9_9

Wu J, Xue W, Yun Z, Liu Q, Sun X (2024) Biomedical applications of stimuli-responsive "smart" interpenetrating polymer network hydrogels. Mater Today Bio 25:100998. https://doi.org/10.1016/j.mtbio.2024.100998

Xue J, Wu T, Dai Y, Xia Y (2019) Electrospinning and electrospun nanofibers: methods, materials, and applications. In: Chemical reviews, vol 119, no 8. American Chemical Society, pp 5298–5415. https://doi.org/10.1021/acs.chemrev.8b00593

Yao Y, Zhou Y, Liu L, Xu Y, Chen Q, Wang Y, Wu S, Deng Y, Zhang J, Shao A (2020) Nanoparticle-based drug delivery in cancer therapy and its role in overcoming drug resistance. In: Frontiers in molecular biosciences, vol 7. Frontiers Media S.A. https://doi.org/10.3389/fmolb.2020.00193

Yu WH, Sing SL, Chua CK, Kuo CN, Tian XL (2019) Particle-reinforced metal matrix nanocomposites fabricated by selective laser melting: a state of the art review. Prog Mater Sci 104:330–379. https://doi.org/10.1016/j.pmatsci.2019.04.006

Zhang P, Xiao Y, Sun X, Lin X, Koo S, Yaremenko A, Qin D, Kong N, Farokhzad O, Tao W (2022) Cancer nanomedicine toward clinical translation: obstacles, opportunities, and future prospects. Med 4. https://doi.org/10.1016/j.medj.2022.12.001

Zhang R, Sun Y, Jiao F, Li L, Geng D, Hu W (2023) MXene-MoS2 nanocomposites via chemical vapor deposition with enhanced electrocatalytic activity for hydrogen evolution. Nano Res 16(7):8937–8944. https://doi.org/10.1007/s12274-023-5602-5

Zhong J, Zheng X, Wen Y, Li Y, Zhang J, Kankala RK, Wang S, Chen A (2024) NIR-switchable local hydrogen generation by tandem bimetallic MOFs nanocomposites for enhanced chemodynamic therapy. Regen Biomater 11. https://doi.org/10.1093/rb/rbad097

Chapter 8
Biocompatible Nanocomposites for Biosensing Applications

Ankitha Menon, Anson T. Mathew, M. Haritha, V. G. Gayathri, Sony Devassy, and Saju Joseph

Abstract In recent years, there has been a growing exploration of biocompatible nanocomposites, encompassing inorganic and organic nanomaterials as well as nanohybrids, for their potential applications in various fields. These applications include biosensors, electromagnetic shielding in aerospace, sensor devices, drug delivery systems, and food packaging. This interest is largely attributed to their superior properties, which include high mechanical strength, thermal stability, barrier effectiveness, antimicrobial activity, and antioxidant capabilities. A biosensor represents a groundbreaking diagnostic instrument that integrates the principles of biology and technology to identify and measure specific analytes. This capability facilitates real-time monitoring, early disease detection, and the development of novel applications. Recent advancements in biocompatible materials have brought a substantial transformation in the evolution of biosensors. Nonetheless, the implementation of biocompatible nanocomposites is hindered by challenges related to scalability,

A. Menon
Department of Chemistry, Indian Institute of Technology Palakkad, Palakkad, Kerala, India
e-mail: 202204003@smail.iitpkd.ac.in

A. T. Mathew · M. Haritha · V. G. Gayathri
Department of Biological Sciences and Engineering, Indian Institute of Technology Palakkad, Palakkad, Kerala, India
e-mail: 172503001@smail.iitpkd.ac.in

M. Haritha
e-mail: 172504001@smail.iitpkd.ac.in

V. G. Gayathri
e-mail: 172314001@smail.iitpkd.ac.in

S. Devassy
St. Peter's College, Kolencherry, Ernakulam, Kerala, India
e-mail: sony.devassy@stpetercollege.ac.in

S. Joseph (✉)
School of Nanoscience and Nanotechnology, Mahatma Gandhi University, Kottayam, Kerala, India
e-mail: sajujoseph@mgu.ac.in

R. Balakrishnan et al. (eds.), *Biocompatible Nanocomposites*, Advanced Structured Materials 239, https://doi.org/10.1007/978-981-96-7450-3_8

manufacturing methodologies, and long-term stability. This book chapter emphasizes the transformative potential of biocompatible materials in enhancing biosensing and other technological applications across diverse domains within science and technology.

Keywords Nanocomposites · Biosensors · Biopolymers · Tissue Engineering · Chitosan

8.1 Introduction

Biocompatible nanocomposites are advanced materials characterized by a mixture of components with at least one component phase within the nanoscale range (up to 100 nm), ensuring spacing among comparable or distinct phases to optimize their functional properties for biological applications. Nanocomposites exhibit diverse beneficial characteristics viz. a larger surface-to-volume ratio, plasmonic properties, outstanding electrical conductivity, resistance to flames and chemicals, redox activity, enhanced optical properties, and catalytic efficiency (Malaki et al. 2019; Said et al. 2020; Khane et al. 2024; Armentano et al. 2018). The core concept of nanocomposites lies in engineering interfaces between the nanoscale components and the matrix. The well-known biocompatible matrices are polymers, hydrogels, or ceramics. Biocompatible nanocomposites hold great promise for various biomedical applications due to their capacity for biological modification and unique nanoscale properties. Biocompatible materials can come into direct contact with skin or tissues without inducing negative effects in imaging (Soufi and Iravani 2020), targeted drug delivery (Chen et al. 2014; Yuan et al. 2016), artificial implants (Qiu et al. 2013), photothermal therapy (Zhang et al. 2015), tissue engineering (Yang et al. 2021; Shinoka et al. 1995) and biosensing applications (Soares et al. 2017). These diverse applications necessitate the careful design of novel materials tailored to meet specific requirements while ensuring they are eco-friendly, biodegradable, and biocompatible with all forms of life.

Biopolymer nanocomposites have recently garnered significant interest from researchers due to their broad applicability in addressing challenges in the healthcare industry and remediation efforts (Yi et al. 2022). Derived from eco-friendly sources or produced by living organisms, biopolymers primarily consist of polysaccharides and polypeptides. These natural polymers possess exceptional physical, chemical, and biological attributes, including renewability, biocompatibility, biodegradability into non-toxic byproducts, and antimicrobial properties. Biopolymeric nanocomposites outperform nanoparticles and polymers in thermophysical properties, mechanical strength, and physicochemical attributes. These composites are formed by incorporating inorganic nanofillers, such as nanoclays, carbon-based materials, metal, and metal oxide nanoparticles, into a biopolymer matrix. Beyond their inherent biodegradability and biocompatibility, they offer flexibility, mechanical resilience, electrical conductivity, and barrier properties against gases and vapors (Khane et al.

2024). However, many polymers alone cannot meet these performance demands. As a result, recent research on biopolymers has focused on integrating high-performance nanofillers to enhance their functionality. For instance, Fu et al. (2017) utilized lactic acid-based nanofibers containing hydroxyapatite (HA) to support bone mineralization, promoting osteoblast growth and differentiation. Similarly, Duarah et al. (2016) developed a self-expanding stent from bio-based polyurethane polymer and CD/silver nanocomposite. This stent, demonstrating both hematocompatibility and antimicrobial activity, shows promise for use in endoscopic procedures based on laboratory findings.

Wang et al. (2016) developed a microneedle which is a self-degrading patch for delivering anti-PD-1 (programmed death-1 pathway) therapy. This patch consisted of hyaluronic acid (HA), a biocompatible component combined with dextran nanoparticles (pH-sensitive) encapsulating anti-PD-1 and a glucose oxidase-catalase (GOx-CAT) enzyme system. pH is lowered when gluconic acid is released due to the oxidation of glucose. This reduction of pH enhances the dissociation and the release of anti-PD-1, while CAT regenerated O_2 and neutralized unwanted hydrogen peroxide (H_2O_2). Designed for painless epidermal penetration, the patch delivered its therapeutic payload efficiently to the tumor microenvironment. Su et al. (2014) described an enzyme-responsive hydrogel formed via GOx-triggered gelation of N-hydroxyimide–heparin conjugates for cell-specific drug delivery. The conjugate, N-hydroxy-5-norbornene-2,3-dicarboximide (HONB), underwent reduction by D-glucose in the presence of GOx to produce a carbon-centered radical, leading to the polymerization of a heparin-HONB conjugate bound to doxorubicin (DOX). The resulting hydrogel was stable at room temperature and released DOX in response to heparinase levels associated with tumor angiogenesis and metastasis. Cell toxicity experiments on HeLa, HepG2, and NIH-3T3 cell lines demonstrated promising outcomes, with the drug-loaded gel (Hep(DOX)SN) enabling targeted delivery to cancer cells overexpressing heparinase, while reducing premature drug release and minimizing damage to healthy cells.

Recently carbon-based nanocomposites have gained immense attention for the fabrication and development of biosensors (Shin et al. 2024a). In one of the works, Cardoso et al. (2022) developed screen-printed electrodes that were modified with surface-functionalized single-walled carbon nanotubes (SWCNTs) for the development of an immunosensor of antibodies that act against SARS-CoV-2. Carboxyl groups were introduced on SWCNTs so as to conjugate with the amine group of o-phenylenediamine. This platform was capable of sensing SARS-CoV-2 antibodies with a detection limit of 0.7 $pg.mL^{-1}$ in a linear range of 1.0 $pg.mL^{-1}$ to 10 $pg.mL^{-1}$. Yang et al. (2021) sulphonate multiwalled carbon nanotubes (MWCNTs) were added to a hydrogel composed of polypyrrole for sensitive detection of a neurotransmitter dopamine. The inclusion of CNTs into the hydrogel enhanced the biocompatibility which aided in the detection of the analyte from PC12 cells.

These advancements in nanocomposite-based therapeutic platforms and enzyme-responsive delivery systems underscore their immense potential as smart, multifunctional tools for targeted drug delivery, cancer treatment, and regenerative medicine, paving the way for innovative solutions to address complex biomedical challenges.

8.2 Synthesis

Biodegradable polymers are termed "green materials" due to their shorter life span when compared to synthetic polymers (Mukherjee et al. 2023). Synthetic biodegradable polymers have been developed to achieve specific physical and chemical properties, allowing their use in a variety of fields, including tissue engineering scaffolds, coatings, wound care, enzyme immobilization, gene delivery, and cosmetics, even though naturally occurring biopolymers are among the most abundant resources (Ramesh Kumar et al. 2020; Zhuang et al. 2024; Satchanska et al. 2024), in this chapter we have discussed well established and latest methodology for synthesis of biodegradable polymers (Fig. 8.1).

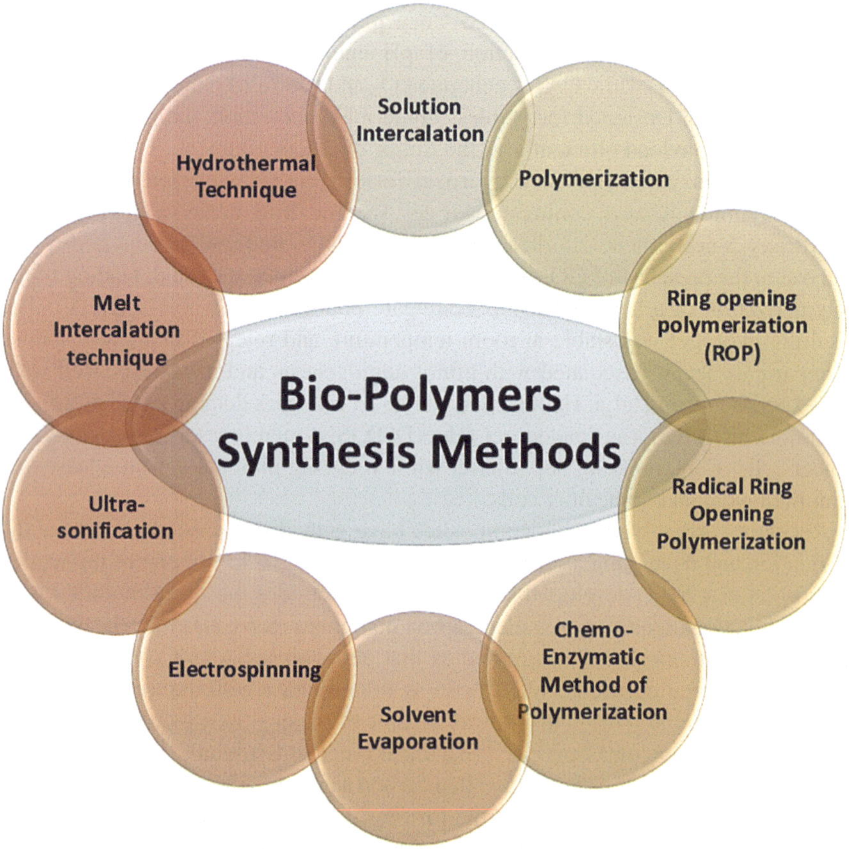

Fig. 8.1 Different methodologies for the preparation of biodegradable polymers (Kim and Chung 2024; Cakmak 2024)

8.2.1 *Solution Intercalation*

Solution intercalation is initiated by selecting such polymers which have well-known biocompatibility and inherent nontoxicity to achieve this we need nanoparticles or fillers into a polymer matrix which also enhances properties such as mechanical strength and thermal stability (Jamróz et al. 2019; Barra et al. 2023). Given the fact that of their abundance, renewability, and biodegradability, proteins and starches are well-known examples of biopolymers that are frequently used in the making of bio-nanocomposites (Ahsan et al. 2023; Flórez et al. 2023). These biopolymers are usually dissolved or distributed in an appropriate solvent, like water, chloroform, or toluene, these solvents facilitate the swelling of inorganic micro-filler silicate platelets by penetrating the interlayer spaces of the silicate structure. The polymer chains intercalate and remove the solvent in the silicate interlayer when the biopolymer and swelling nanoparticle solution mix. A layered silicate bio-nanocomposite is created when the solvent is removed but the intercalated structure remains intact resulting in the formation of a layered silicate bio-nanocomposite (Mittal 2009; Nandhini et al. 2024). An example of solution intercalation work in which biopolymer Hydroxy-propyl methylcellulose (HPMC) was intercalated with Montmorillonite (Cloisite®- Na +) clay mineral used as filler material and final application of product used for wound dressing (Darder et al. 2021).

8.2.2 *Polymerization*

Biodegradable polymers are synthesized through various polymerization methods, each offering distinct advantages and applications.

8.2.2.1 Ring-Opening Polymerization (ROP)

Cyclic monomers are polymerized into an acyclic monomeric unit is called ROP (Penczek et al. 2021), at this point the ring system of cyclic monomers is opened to form a long polymer chain, with the polymeric chain's terminal end acting as a reactive center. Three propagation centers are available radical, anionic, and cationic is present. This approach has been adopted as one of the most flexible techniques for producing biopolymers in large numbers. The biodegradable polymer polylactic acid is made using ROP in the procedure described below.

PLA (polylactic acid) is produced by the chain growth ROP of lactide. The ROP of lactide is carried out using a variety of catalysts and initiators. One popular and reliable catalyst is stannous octoate [$Sn(Oct)_2$]. Lactide and glycolide ROP lead to the formation of PLGA (poly lactic-co-glycolic acid). In this case, $Sn(Oct)_2$ acts as the catalyst and the formed PLGA has a high molecular weight. The ROP of PDS (poly-dioxanone), a moderately weak and quickly biodegradable polymer, can result in PDS. However, compared to PLA and PGA, it produces less acidic products during breakdown (Meghana et al. 2023; Panchal and Vasava 2020).

8.2.2.2 Radical Ring-Opening Polymerization

Free radical ring-opening polymerization (FROP) is the process of adding free radicals as building blocks produced through different processes that use different initiator molecules to create polymers. The free radicals that start which are produced newly add monomer units to the polymer chain, causing the chain to expand. One of the most adaptive types of polymerization, FROP enables simple interactions between polymeric free radical chain ends and other molecules or substrates. This process is used to create materials based on ε-caprolactone. For example, the FROP synthesis process is used to create poly(HEMA-g-CLk) nanoparticles. Here, poly(HEMAg-CLk) NPs are created by combining ε-caprolactone with 2-hydroxyethyl methacrylate (HEMA) (Meghana et al. 2023; Ferrari et al. 2011).

8.2.2.3 Chemoenzymatic Method of Polymerization

Chemoenzymatic is a popular process for the manufacturing of active pharmaceutical ingredients. There are researchers who have developed optically active polymeric prodrugs by employing this method. These are nonsteroidal anti-inflammatory prodrugs having high molecular weight substrates (Panchal and Vasava 2020). This technique combines an efficient enzymatic approach with traditional polymerization, PLA monomers can also be synthesized with the same methodology (Meghana et al. 2023; Ahmed et al. 2019).

8.2.2.4 Photo-Initiated Radical Polymerization

Photo-initiated ring-opening polymerization is a fascinating approach because it offers fast polymerization rates under physiological conditions with spatiotemporal control. By doing this, the reaction medium no longer needs a hazardous cross-linker. This process is a more environmentally friendly method which is frequently used to synthesize the majority of the polymers that are often used in biomedical applications (Meghana et al. 2023; Acik and Karatavuk 2020).

8.2.3 Solvent Evaporation

For preparing microparticles or nanoparticles with biodegradable polymers, the solvent evaporation process is commonly used. Initially, polymer solution in a volatile solvent (e.g., chloroform, ethyl acetate, or dichloromethane) is set up and then the drug is introduced into that polymer solution. An oil in water (o/w) emulsion is formed when the resulting polymeric solution or dispersion, which is then added to an aqueous phase that contains an emulsifying agent, and the entire system is then stirred. To create a suspension with particles, the organic solvent of the resulting emulsion is evaporated (George et al. 2019; Bhatia 2016). Temperature, stirring speed, polymer concentration, stabilizer type, and the viscosity of the organic and aqueous phases all have a significant effect on the size of the resultant particles (Tice and Gilley 1985; Zohri et al. 2009). However, scaling up this procedure is challenging as it is only applicable to lipid-soluble drugs (Soppimath et al. 2001). Polymers such as PLGA (Kizilbey 2019; Ospina-Villa et al. 2019), PLA (Rachmawati et al. 2016; Zhang et al. 2018), Poly (β-caprolactone) (PCL) (Lemarchand et al. 2006), ethyl cellulose (Abbaspoor et al. 2018), and others are used in the solvent evaporation procedure to prepare particle systems.

8.2.4 Electrospinning

Electrospinning is a commonly used approach for synthesizing nanofibers from biodegradable polymers, which allows the development of materials for applications in healthcare engineering, biological filtration, and drug delivery (Kai et al. 2014). The technique turns melted polymer or solution into continuous fibers by using electrostatic forces. Because of its wide range of applications, the process is perfect for creating fibers with adjustable surface morphologies, porosity, and mechanical characteristics of biodegradable polymers (Ashammakhi et al. 2006; Dong et al. 2009). The majority of studies confirmed that the primary factors influencing fiber diameter and shape are polymer viscosity. Larger fiber diameters may be the consequence of increased viscosity brought on by high polymer molecular weight or concentration (Doshi and Reneker 1995; Fong et al. 1999). When the viscosity becomes too high, the movement of the polymer solution may be obstructed and the droplet dry out at the tip. However, if the viscosity is too low, the lack of chain entanglement may cause the fiber jet to shatter into droplets. For instance, beads and droplets rather than nanofibers were produced when the PLGA concentration in (Tetrahydrofuran + Dimethylformamide) was 0.10 g/mL or less (Katti et al. 2004). The gelatin nanofibers were examined by Ki et al. Fiber diameter increases rapidly with increasing polymer concentration in the 8–12% weight percentage range. To put it another way, there is a nonlinear relationship between fiber diameter and polymer content. Gelatin fibers (76–169 nm) were produced that were uniform and bead-free (Ki et al. 2005).

8.2.5 Ultra-Sonification

Ultrasonication, the application of ultrasound energy to chemical systems, has emerged as a potent technique in the synthesis and processing of biodegradable polymers which enhances their properties and increases their performance in different applications (Soman et al. 2023). Ultrasonic baths and ultrasonic tools are the main pieces of equipment used for the synthesis of polymer nanocomposites where the ultrasound generators convert electrical energy into ultrasonic energy, which is basically mechanical energy (Raso et al. 1999). Ultrasound-assisted methods were implemented by Boro et al. (2022) to develop PLA/ZnO nanocomposites with flowerlike morphological characteristics. These nanocomposites exhibited outstanding mechanical, optical, and thermal properties at very low ZnO loadings. The enhanced properties were brought about by homogeneous nanofiller dispersion and strong interfacial contacts between ZnO and the PLA matrix. Additionally, the nanocomposites demonstrated effective antibacterial activity against Escherichia coli and Listeria monocytogenes, making them a powerful substance for use in food packaging. Orasugh et al. created a nanocomposite using hydroxypropylmethylcellulose and jute cellulose nano-fibrils using an ultrasound-assisted preparation process. Due to their low water vapor permeability and high mechanical strength, the nanocomposites were suitable for use in food packaging and medicine delivery (Orasugh et al. 2018).

Sharififard et al. proposed a nanocomposite consisting of chitosan, activated carbon, and iron nanoparticles made via sonic route for the effective removal of $Cd2+$ from water, with promising future prospects in heavy removal of metal (Sharififard et al. 2018).

8.2.6 Melt Intercalation Technique

The melt intercalation method involves the annealing of polymeric materials, the incorporation of fillers, and the compression of the composite to attain a homogeneous distribution. The synthesis of nanocomposites processed via melt intercalation exhibits temperature ranges between 190 and 220 °C (Vaia et al. 1993). Consequently, achieving efficient dispersion and exfoliation is highly contingent upon the optimization of processing parameters. Degradation may be mitigated by operating at reduced temperatures or implementing more thermally stable modifications. To produce intercalated or exfoliated nanocomposites, the polymer may penetrate the filler interlayers as a result of weak electrostatic forces and its compatibility with the polymer matrix (Alexandre and Dubois 2000; Alsayed and Ashraf 2020).

Melt extrusion has effectively facilitated the intercalation of C18-mmt modified silicate layers within PLA matrix materials (Sinha Ray et al. 2002). The hydroxylated interaction of silicate layers, despite their minimal loading in the polymer matrix, led to increased flocculation and improved parallel layer stacking. The thermomechanical properties of organofunctionalized Ni–Al layered double hydroxide/PS nanocomposites have been investigated (Suresh et al. 2017; Shaikh et al. 2024).

8.2.7 *Hydrothermal Technique*

This method works best when there is a large amount of vapor pressure (more than 1 atm) and high heat (about 180 °C). Under these operating conditions, the solvent properties of various compounds, such as solubility and dielectric constant, can exhibit significant variation. This variation contributes to the in-situ formation of nanocomposites with separate and controlled nanostructured morphology (Sedira and Mendaci 2019). Researchers synthesized hybrid nanocomposites of PANI intercalated V_2O_5 xerogel using hydrothermal methods. Hydrothermal methods have produced carbon dot-reinforced molecularly imprinted polymer (MIP) nanocomposites (Liu 2016). Applying an organic layer of MIPs onto the surface of nanomaterials can result in the formation of polymer nanocomposites. This method can significantly degrade organic layers under strong UV light exposure (Lai et al. 2016). The hydrothermal method requires additional research due to its energy intensity and potential to damage imprinted spaces at elevated temperatures.

8.3 Properties

8.3.1 *Biosensing Applications*

Biosensors are analytical devices that combine a biological recognition element with a transducer to detect and quantify specific analytes with high sensitivity and selectivity. In the field of biosensing, biomaterial-based nanomaterials have emerged as a focal point of interest, as they open avenues for enhanced impact studies while boosting sensitivity, specificity, and cost-effectiveness. A biopolymer is composed of a bioreceptor and a transducer or electrode. Signal detection takes place at the bioreceptor, where the binding of the analyte triggers a response in the output. Biosensors have garnered significant attention in numerous studies as a cost-effective, precise, and highly sensitive platform for biochemical analysis, integrating specific biorecognition mechanisms with signal transduction processes. Nanocomposites based on conductive polymers have become integral to biosensing applications due to their remarkable characteristics, including superior conductivity, high adsorption potential, electrocatalytic efficiency, and enhanced bioreagent loading capabilities. Examples of such polymers include polyindole, polyaniline, polyfuran, and polycarbazole. Moreover, metal nanoparticles such as gold and silver often prepared using polysaccharides as scaffolding materials, have shown immense promise in advancing biosensor technologies.

A report highlights that nanocomposites synthesized using xylan biopolymer exhibit exceptional sensitivity and selectivity toward Hg^{2+} ions. This is attributed to the hydroxyl and carboxyl functional groups in xylan, which significantly enhance its binding affinity for mercury ions (Luo et al. 2015). The detection of H_2O_2 poses a critical challenge due to its association with various conditions such as brain injuries, asthma, and cancer. Lee et al. have introduced an efficient H_2O_2 vapor sensor (Fig. 2a), utilizing single-walled carbon nanotubes (SWCNTs) combined with porphyrin nanofibers, capable of operating effectively at room temperature (Lee et al. 2020). The COVID-19 pandemic has prompted significant research into the development of point-of-care testing (POCT) biosensors, with electrochemical biosensors receiving considerable attention. These biosensors function by detecting alterations in electrochemical signals caused by interactions between the target and the probe, enabling both quantitative and qualitative analysis. The conductive properties of carbon nanomaterials are particularly beneficial in the development of highly conductive electrodes, which enhance the biosensor's ability to detect even minute changes in electrochemical signals. In one study, carboxylated single-walled carbon nanotubes (SWCNTs) were applied to carbon-based screen-printed electrodes (SPEs) to design (Fig. 2b) an electrochemical biosensor for the detection of anti-SARS-CoV-2 antibodies (Anti-S protein) (Cardoso et al. 2022). A flexible enzymatic electrochemical glucose biosensor was developed by combining graphene sponge (GS) with glucose oxidase (GOx) and chitosan (CTS) (Fig. 3c). The sponge-like structure of graphene is particularly valued in the biomedical field due to its large surface area, high loading capacity, excellent conductivity, biocompatibility, and mechanical strength. GS was synthesized from pure graphene using a hydrothermal method, and GOx and CTS were loaded onto the porous GS structure. This structure effectively absorbed, dispersed, and protected the GOx. The GS-based complex was then applied to a flexible polyimide electrode for glucose detection. The resulting flexible biosensor demonstrated excellent bending properties and was capable of detecting glucose in human sweat with a low detection sensitivity of 2.45 mM through amperometric sensing. This sensitivity was superior to that of biosensors developed using carbon nanotubes (CNTs) (Li et al. 2023). To enhance the conductivity of reduced graphene oxide (rGO), a nanocomposite was created by combining rGO with molybdenum disulfide (MoS_2) nanoparticles, a transition metal dichalcogenide (TMD) nanomaterial, for use in an electrochemical biosensor. In this study, the rGO-MoS_2 carbon-based nanocomposite was further integrated with a conductive polymer, poly(3,4-ethylenedioxythiophene) (PEDOT), to develop an electrochemical biosensor for detecting parathyroid hormone (Juang et al. 2023). A carbon-based nanocomposite consisting of gold and reduced graphene oxide (rGO) was utilized as an electrode for the electrochemical detection of miRNA-122 (Kasturi et al. 2021). In another study, a probe DNA was functionalized onto a gold-decorated laser-induced graphene electrode for the detection of miRNA-141 (Torati et al. 2023). Recent studies have highlighted the combination of newly reported nanomaterials with carbon-based materials to enhance conductivity and facilitate electron transfer reactions. For instance,

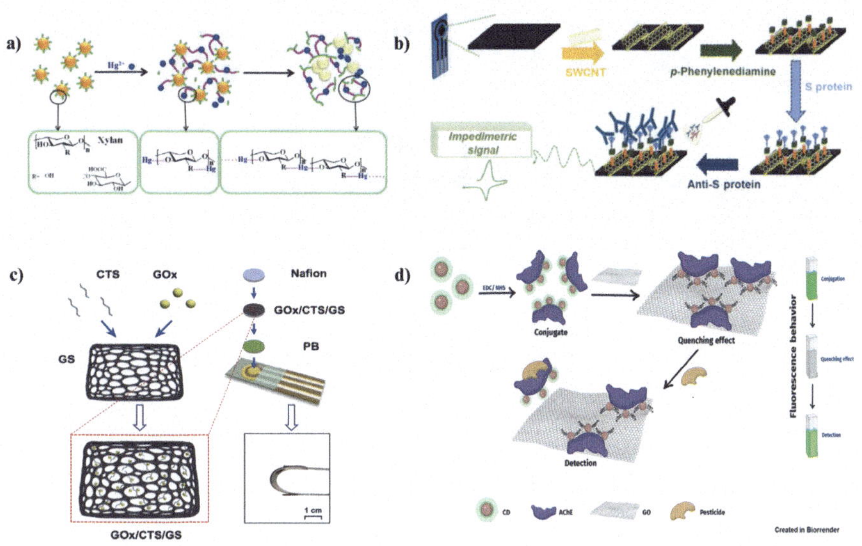

Fig. 8.2 **a** Proposed mechanism showing the interaction between Hg^{2+} with the functional groups of xylan, Reproduced with permission from Luo et al. (2015). **b** A schematic diagram of an electrochemical SARS-CoV-2 antibody biosensor utilizing screen-printed electrodes (SPE) integrated with single-walled carbon nanotubes (SWCNTs), Reproduced with permission from Cardoso et al. (2022). **c** The scheme of a flexible electrochemical glucose biosensor composed of glucose oxidase (GOx), chitosan (CTS), and graphene sponge (GS), Reproduced with permission from Li et al. (2023). **d** schematic diagram of the fluorescent biosensor composed of AChE, CD, and GO for detecting organophosphate pesticides, reproduced with permission from Gaviria et al. (2022)

MXene and nickel oxide were integrated with reduced graphene oxide (rGO) to create an electrode for an electrochemical biosensor designed to detect influenza viruses (H1N1 and H5N2) and the viral surface protein hemagglutinin (Reddy et al. 2022).

Alongside electrochemical biosensors, the fluorescent technique has emerged as a promising approach for biosensor development, utilizing the distinct fluorescence properties of certain materials. Recombinase polymerase amplification (RPA)-mediated fluorescent biosensors have been developed utilizing the quenching properties of graphene oxide (GO) and FAM-labeled primers for the detection of RNA interference (RNAi) transgenic plants. In this biosensor, GO served two primary purposes. Firstly, it enhanced RPA efficiency by adsorbing molecules onto its blanket-like structure, promoting local aggregation. Secondly, it suppressed fluorescent signals from FAM-labeled primers in the absence of target sequences. The biosensor remained non-fluorescent when target DNA sequences were absent. However, in the presence of target sequences, fluorescence was restored due to the binding of targets and their subsequent release from GO. This increased RPA efficiency enabled highly sensitive detection, achieving a detection limit of 1.5 ng of genomic DNA. These findings suggest that the biosensor is capable of rapid, on-site detection of RNAi soybeans

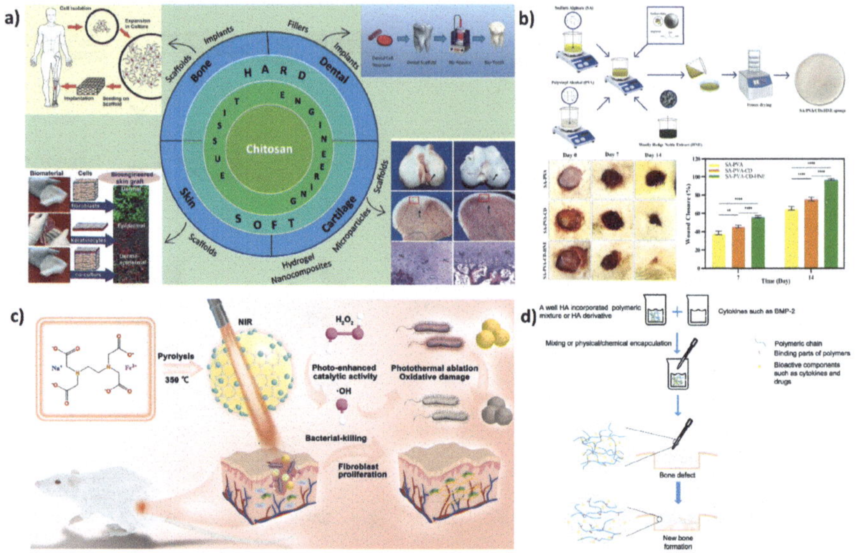

Fig. 8.3 **a** A schematic representation of chitosan-based nanocomposites used for various tissue engineering, Reproduced with permission from Aguilar et al. (2019). **b** Schematic representation of alginate-based nanocomposites used for wound healing, Reproduced with permission from Abbasinia et al. (2024). **c** Schematic representation of carbon dots nanozymes for photoenhanced antibacterial therapy and wound healing, Reproduced with permission from Liu et al. (2022). **d** A schematic representation of Hyaluronic acid-based nanocomposites used for bone repair, Reproduced with permission from Zhai et al. (2020)

with high sensitivity (Li et al. 2022). A fluorescent biosensor was developed to detect organophosphate pesticides by combining two fluorescence properties of carbon nanomaterials: the fluorescence emission of carbon dots (CDs) and the quenching ability of graphene oxide (GO) (Fig. 2d). In this biosensor, acetylcholinesterase (AChE) and CDs were conjugated through EDC-NHS coupling between the amine groups of AChE and the carboxyl groups of the CDs. The resulting conjugates were then attached to GO. In the absence of pesticides, the fluorescence emitted by CDs was quenched by GO. However, in the presence of pesticides, the CDs were released from GO, restoring their fluorescence. By monitoring this fluorescence recovery, quantitative detection of pesticides was achieved, with a detection limit as low as 0.14 ppb for chlorpyrifos, an organophosphate pesticide. These advancements in biosensor development highlight the innovative integration of nanomaterials and fluorescence-based mechanisms, enabling highly sensitive, selective, and efficient detection of a wide range of analytes, including pesticides, biomolecules, and viruses, with potential applications in environmental monitoring, healthcare, and food safety.

8.3.2 Tissue Engineering

Biocompatible nanocomposites with their exceptional physicochemical properties hold immense promise in enhancing various aspects of tissue engineering, starting from designing scaffold and mechanical support to cell interaction and regeneration. Mainly, the bioactivity and biocompatibility of various nanocomposites helps in cell adhesion, proliferation, and differentiation for more advanced applications in tissue interaction and regeneration (Shin et al. 2024b).

A significant development in regenerative medicine in tissue engineering, which combines engineering and biological science concepts to provide novel approaches for replacing or repairing damaged tissues and organs. Tissue engineering includes various branches such as bone tissue engineering, cardiac tissue engineering, neural tissue engineering, wound healing, and skin regeneration (Howard et al. 2008; Bakhshandeh et al. 2017; Kwon et al. 2018).

8.3.2.1 Alginate-Based Nanocomposites

Alginate gels have recently emerged as the most widely utilized form of alginate in tissue engineering. Alginate-based hydrogels have also become well-known among the various biomaterials that have been researched because of their unique qualities and remarkable adaptability in a wide range of tissue engineering applications (Sahoo and Biswal 2021). These gels are formed through ionic cross-linking, primarily using calcium ions, which interact with the carboxyl groups of alginate chains to create a three-dimensional network under physiological conditions. This process enables the encapsulation of cells and bioactive compounds while preserving their viability. Furthermore, the mild gelation technique not only protects the integrity of biological components but also facilitates easy customization of the alginate matrix to meet specific tissue engineering needs (Sharma et al. 2023; Reakasame and Boccaccini 2018).

8.3.2.2 Chitosan-Based Nanocomposites

Chitosan is known to be a key material for creating advanced bio-scaffolds and hydrogels for tissue engineering. Its unique structure allows for modifications to enhance mechanical strength, biodegradability, and controlled release of bioactive molecules. Blending chitosan with other polymers or incorporating nanoparticles further expands its utility in tissue engineering. Recent studies over the past decade have focus on its use in bone tissue engineering, hard tissue engineering, dental implants, dental tissue engineering, dental fillers, and cartilage tissue engineering (Aguilar et al. 2019; Kim 2023; Ressler 2022).

8.3.2.3 Carbon-Based Nanocomposites

In tissue regeneration, carbon nanomaterials are efficiently utilized for their stiffness and mechanical strength that enhances the performance of scaffolds for load-bearing tissues such as tendons, cartilages and bones. Moreover, iron-doped carbon dots (Fe-CDs) have been widely explored for efficient wound healing (Liu et al. 2022).

In the context of neural tissue regeneration, carbon-based materials, specifically reduced graphene oxide (rGO) incorporated into gelatin-methacrylate (GelMA) hydrogels, have been utilized to engineer nerve guidance conduits (NGCs) that facilitate neuronal growth and improve functional recovery in animal models (Zhang et al. 2023). Likewise, for muscle tissue regeneration, scaffolds composed of carbon nanomaterials, such as carbon nanotube (CNT)-based cryogels and biohybrid artificial muscles, have been developed to promote the alignment and contraction of muscle cells, thereby simulating the functional characteristics of muscle tissue (Velasco-Mallorquí et al. 2020).

8.3.2.4 Hyaluronic Acid-Based Nanocomposites

Apart from biocompatibility, biodegradability, and cellular interaction, hyaluronic acid (HA) poses some exceptional properties, such as providing a hydrophilic environment and ease of modification (Gupta et al. 2019). The hydrophilic environment increases the water absorption capacity and helps in creating a hydrated environment that is conducive to cell attachment, differentiation, and proliferation, making HA-based hydrogels ideal for tissue engineering. HA can be chemically modified to introduce functional groups for cross-linking and to incorporate bioactive molecules, further enhancing its applicability in bone tissue engineering (Zhai et al. 2020; Lin et al. 2020).

HA-based hydrogels are highly advantageous for bone tissue engineering due to their ability to promote bone formation by adsorbing growth factors like bone morphogenetic proteins (BMPs), which enhance mesenchymal stem cell (MSC) differentiation and bone regeneration. Their mechanical properties are tunable, enabling use in both load-bearing and flexible applications. These hydrogels can incorporate bioactive molecules to improve osteogenic differentiation and integration with surrounding tissues. Additionally, they are biocompatible and biodegradable, reducing the risk of adverse effects and inflammation, making them a versatile and effective choice for regenerative medicine (Ding et al. 2022).

8.3.3 Drug Delivery Applications

The emergence of nanotechnology has transformed the area of drug delivery, providing novel methods to address the challenges faced by traditional drug delivery techniques. Biocompatible nanocomposites have emerged as a viable method due to their distinctive features and potential to improve therapeutic efficacy. These nanocomposites are synthetic materials made of biocompatible polymers and nanoscale particles, intended to engage safely and efficiently with biological systems (Kalayil et al. 2024).

Biocompatible nanocomposites have numerous benefits, such as enhanced drug solubility, targeted distribution, controlled release, and diminished side effects. Utilizing the elevated surface area-to-volume ratio of nanoparticles enables these materials to attain enhanced drug loading and release kinetics. The integration of biodegradable polymers guarantees that the nanocomposites are ecologically sustainable and may be safely assimilated by the body (Ozcelikay et al. 2021; Shariatinia et al. 2020).

Most common biocompatible nanocomposites used for the formation of drug delivery systems are chitosan-based nanocomposites, alginate-based nanocomposites, collagen-based nanocomposites, hyaluronic acid-based nanocomposites and graphene oxide-based nanocomposites (Shariatinia et al. 2020; Bellier et al. 2022).

This chapter examines the essential principles of biocompatible nanocomposites, focusing on their composition, production techniques, and applications in drug administration systems. We will examine the latest accomplishments and problems in this domain, emphasizing the potential of these nanocomposites to revolutionize pharmaceuticals and personalized treatment.

8.3.3.1 Chitosan-Based Nanocomposites

Chitosan (CS) is a versatile biopolymer produced from chitin via the partial deacetylation of acetyl groups in an alkaline environment. This copolymer consists of two monomeric units: β-(1 → 4)-2-acetamido-D-glucose (N-acetyl-D-glucosamine or GlcNAc) and β-(1 → 4)-2-amino-D-glucose (D-glucosamine or GlcN). Various functional groups like amino groups and carboxyl groups provide ample opportunities for desirable modifications. Moreover, its biocompatibility, biodegradability, and hypoallergenic properties render it an exemplary choice for diverse biomedical applications (Yassue-Cordeiro et al. 2018).

Many studies have utilized various combinations of materials with chitosan for example, chitosan combined superparamagnetic iron oxide(Fe_3O_4) was used for the pH-responsive delivery of curcumin for cancer, it showed an encapsulation efficiency of 86% and the system showed maximum release at pH 4.5 (Prabha and Raj 2016). Chitosan-based nanocomposites provide numerous substantial advantages for drug delivery. They possess inherent biocompatibility and biodegradability, facilitating safe metabolism within the body and minimizing unwanted effects. Chitosan improves drug stability by safeguarding encapsulated medications from degradation and prolonging their shelf life. These nanocomposites can be designed for precise drug delivery, enhancing efficacy and reducing adverse effects, while facilitating controlled and sustained release for a stable therapeutic impact. The mucoadhesive characteristics of chitosan facilitate its attachment to mucosal surfaces, which is especially advantageous for gastrointestinal medication administration. The existence of functional groups facilitates diverse chemical changes, permitting tailoring for particular requirements. Moreover, chitosan is hypoallergenic and non-toxic, rendering it a secure vehicle for medicines, particularly in sensitive applications. These characteristics render chitosan-based nanocomposites a compelling and efficient option for sophisticated drug delivery systems (Mikušová and Mikuš 2021).

8.3.3.2 Alginate-Based Nanocomposites

Alginate-based nanocomposites are extensively utilized in drug delivery systems owing to their distinctive characteristics. Alginate is a natural biopolymer extracted from seaweed, recognized for its biocompatibility, biodegradability, low toxicity, and gentle gelation properties. These characteristics render alginate an exemplary candidate for the development of micro- and nanosystems for regulated and targeted drug delivery (Iravani and Varma 2022).

Alginate-based nanocomposites can be synthesized into multiple forms, including microparticles, nanoparticles, microgels, and composite gels. These formulations can be customized for particular drug delivery requirements, offering increased stability, prolonged release kinetics, and greater cellular absorption. Alginate's capacity to gel with multivalent cations facilitates the encapsulation and oral administration of physiologically active compounds, such as pharmaceuticals and proteins (Lai et al. 2024).

Recent developments in alginate-based nanocomposites have demonstrated their efficacy in cancer treatment, wound healing, and colon-targeted drug delivery. The integration of other materials such as cellulose, chitosan, or inorganic nanoparticles can further improve their mechanical characteristics, mucoadhesion, and adjustable release kinetics (Thomas et al. 2023).

8.3.3.3 Hyaluronic Acid-Based Nanocomposites

Hyaluronic acid (HA)-based nanocomposites have gained interest in drug delivery systems because of their distinctive characteristics. Hyaluronic acid (HA) is a naturally occurring biopolymer present in the extracellular matrix, recognized for its biocompatibility, biodegradability, low toxicity, and non-immunogenic properties. These characteristics provide HA an exemplary choice for the development of tailored drug delivery systems, particularly in oncological treatment. HÀ-based nanocomposites can be engineered to specifically target tissues or cells, as HA interacts with the CD44 receptor, which is overexpressed in numerous cancer cells. This focused strategy enhances drug delivery efficacy and minimizes off-target effects. Hyaluronic acid can be altered with several functional groups, including carboxyl and hydroxyl groups, facilitating the development of various nanocarriers such as micelles, liposomes, nanoparticles, and hydrogels (Fu et al. 2023).

Recent studies indicate that HA-based nanocomposites can improve the stability and release kinetics of encapsulated pharmaceuticals, rendering them excellent

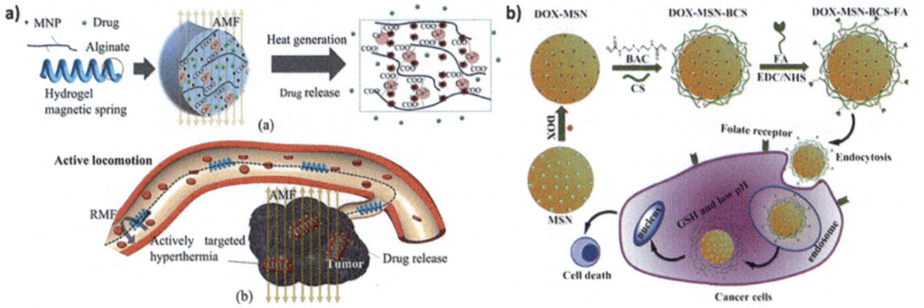

Fig. 8.4 **a** Alginate-based hydrogel used for targeted drug delivery systems, Reproduced with permission from Iravani and Varma (2022). **b** carbon nanocomposites used as targeted drug delivery for cancer cells, Reproduced with permission from Mikušová and Mikuš (2021)

carriers for anticancer agents. These nanocomposites exhibit enhanced therapeutic efficacy, decreased toxicity, and substantial tumor growth suppression (Salari et al. 2021).

8.3.3.4 Collagen-Based Nanocomposites

Collagen-based nanocomposites are another commonly utilized material for medication delivery due to their distinctive features. Collagen is a key structural protein in the extracellular matrix, renowned for its biocompatibility, biodegradability, low antigenicity, and exceptional safety profile. These attributes render collagen an exemplary prospect for the development of drug delivery systems, tissue engineering scaffolds, wound dressings, and gene therapies (Verma et al. 2023).

Collagen-based nanocomposites can be developed into multiple formats, including hydrogels, nanoparticles, microspheres, microneedles, liposomes, and composite gels. These formulations can augment drug stability, provide regulated and sustained release, and promote cellular absorption. The superior encapsulation capacity, robust structural integrity, and effective mass transfer characteristics of collagen-based nanocomposites render them ideal for the targeted delivery of biomolecules with enhanced efficacy (Verma 2022).

Recent advancements indicate that collagen-based nanocomposites are applicable in cancer treatment, wound healing, and colon-targeted medication administration. Nonetheless, issues such as insufficient mechanical strength and elevated degradability must be resolved for clinical applications (Arun et al. 2021) (Fig. 8.4).

8.3.4 Diagnostic Imaging

Biocompatible nanocomposites have become an emerging field in theranostic which refers to a system that can be a combined application of therapeutics and imaging agents (Qi et al. 2020). Nanotheranostics are based on carriers at submicron or nanoscale. These nanotheranostic systems can be developed from various materials, including dendrimers (Fu et al. 2014), liposomes (Wang et al. 2020), polymeric nanoparticles (Shao et al. 2019), carbon-based nanomaterials (Costa et al. 2018), metal or inorganic nanocarriers (Mardhian et al. 2020), and hybrid systems that combine these categories, such as polymer-coated nanocarriers (Li et al. 2020; Guo et al. 2019). Carbon-based nanomaterials, such as carbon nanotubes (either standalone or functionalized with other materials), graphene oxide, fullerenes, and carbon quantum dots, have been widely utilized for detecting drugs in biological samples and for their imaging capabilities (Govindasamy et al. 2018; Keerthi et al. 2018; Sung et al. 2018).

An ideal nanotheranostic should possess longer circulation time in body, efficient release properties, high tissue specificity and penetration, strong imaging capability, and a favorable target-to-background ratio (Ferber et al. 2014). These systems can localize diagnostic and therapeutic agents to disease sites, minimizing undesired side effects. Their nanometric size contributes to extended blood circulation. For tumor diagnosis and therapy, nanosized particles can extravasate into tumor tissues and remain there due to poor lymphatic drainage, offering advantages over smaller multifunctional molecules or functionalized macromolecules (Dasgupta et al. 2020).

The following are some of the diagnostic applications of biocompatible nanocomposites:

- Biosensing and bioimaging
- Disease diagnostics
- Early detection of cancer

However, in recent years nanomedicine has introduced innovative applications in disease diagnostics and invasive therapies. Developing novel tools with enhanced imaging capabilities remains critical for the early detection of diseases, particularly cancer, which would greatly benefit the medical field. Alongside imaging, nanotherapeutics play a vital role in treating severe illnesses. While early nanotheranostic applications primarily focused on cancer, their use has expanded to include conditions like neurological disorders and cardiovascular diseases (CVD). Preclinical studies have shown promising results for theranostics in CVD (Zhao et al. 2018; Sharma et al. 2019; Mog et al. 2019).

Furthermore, the creation of biodegradable polymers and nanocomposites made from renewable resources is accelerating in the field of materials research. Prominent biodegradable polymers and nanocomposites hold great promise for use in packaging, biosensing, and biomedicine. However, because of its stiffness and low molecular weight, its use is frequently restricted (Nukala et al. 2022). Many studies have been conducted on the use of ligands in the treatment of many incurable diseases. The significance of enzymes with possible inhibitory effects has been highlighted by prior research that has shown the anti-scab activity of particular biodegradable polymers and nanocomposites (Maciel-Flores et al. 2024). Furthermore, the surfaces of montmorillonite clay may be altered by the pharmacokinetic characteristics of the recently created biodegradable polymers and nanocomposites.

Researchers have looked at a number of methods to improve the characteristics of nanocomposites made from renewable resources in order to overcome these obstacles, such as using peroxides and low-weight plasticisers (Sardar 2023; Turco et al. 2021). Enhancing physical, thermal, and biological degradation properties is the goal of these changes. The characteristics of materials can be improved by using biodegradable polymers and nanocomposites made by mixing organic clays with polymers (Karabo Innocent Matumba 2024). According to research, mechanical and thermal stability can be considerably increased by adding reasonable amount of organoclay (Nazrun et al. 2024; Kalendova et al. 2024). Research is still being done to improve their qualities by coupling agents or cross-linking, although these techniques frequently have issues with cost and environmental impact (Shafiee et al. 2022). We anticipate that employing different synthesis techniques to enhance the characteristics of nanocomposites made from renewable resources will open up new avenues for investigation in a variety of materials science fields.

8.4 Conculsion

To summarize, the growing interest in bio-based materials is a result of the imperative need to address sustainability within material science, which is driven by global environmental concerns such as pollution, climate change, and resource scarcity. The potential of biocompatible nanocomposites as alternatives to non-renewable resources, specifically those derived from renewable and residual sources, is the subject of this book chapter. The emphasis on structure–property guidelines is the result of systematic investigations into the complex architecture and interactions present in biopolymers and biocompatible nanocomposites. Biopolymers, non-toxic compounds, and additives were employed to develop lightweight and high-performance materials in the investigation of biocompatible nanocomposites. The ultimate objective is to develop bio-based materials that exhibit exceptional performance, including functionality, high rigidity and strength, and lightweight, all while maintaining sustainability and a competitive cost.

References

Abbasinia S, Monfared-Hajishirkiaee R, Ehtesabi H (2024) Nanocomposite sponge based on sodium alginate, polyvinyl alcohol, carbon dots, and woolly hedge nettle antibacterial extract for wound healing. Ind Crops Prod 214:118554

Abbaspoor S, Ashrafi A, Salehi M (2018) Synthesis and characterization of ethyl cellulose micro/nanocapsules using solvent evaporation method. Colloid Polym Sci 296:1–6

Acik G, Karatavuk AO (2020) Synthesis, properties and biodegradability of cross-linked amphiphilic Poly (vinyl acrylate)-Poly (tert-butyl acrylate) s by photo-initiated radical polymerization, 127:109602

Aguilar A et al (2019) Application of chitosan in bone and dental engineering. Molecules, 24(16)

Ahmed ST, Leferink NG, Scrutton NS (2019) Chemo-enzymatic routes towards the synthesis of bio-based monomers and polymers, 467:95–110
Ahsan W et al (2023) Biodegradation of different types of bioplastics through composting—a recent trend in green recycling. Catalysts 13(294):2023
Alexandre M, Dubois P (2000) Polymer-layered silicate nanocomposites: preparation, properties and uses of a new class of materials. Mater Sci Eng R Rep 28(1):1–63
Alsayed A, Ashraf M (2020) Modified nanofiltration membrane treatment of saline water: a review. Desalination Water Treat 187:93–105
Armentano I et al (2018) Nanocomposites based on biodegradable polymers, 11(5):795
Arun A et al (2021) Collagen nanoparticles in drug delivery systems and tissue engineering. Appl Sci 11(23):11369
Ashammakhi N et al (2006) Biodegradable nanomats produced by electrospinning: expanding multifunctionality and potential for tissue engineering. J Nanosci Nanotechnol 6(9–10):2693–2711
Bakhshandeh B et al (2017) Tissue engineering; strategies, tissues, and biomaterials. Biotechnol Genet Eng Rev 33(2):144–172
Barra G et al (2023) A comprehensive review on the thermal stability assessment of polymers and composites for aeronautics and space applications, 15(18):3786
Bellier N et al (2022) Recent biomedical advancements in graphene oxide- and reduced graphene oxide-based nanocomposite nanocarriers. Biomater Res 26(1):65
Bhatia S (2016) Nanoparticles types, classification, characterization, fabrication methods and drug delivery applications. Natural polymer drug delivery systems: nanoparticles, plants, and algae. Springer International Publishing, Cham, pp 33–93
Boro U, Kashyap N, Moholkar VS (2022) Sonochemical synthesis of Poly(lactic acid) nanocomposites with ZnO nanoflowers: effect of nanofiller morphology on physical properties. ACS Eng Au 2(1):46–60
Cakmak OK (2024) Biodegradable polymers—a review on properties, processing, and degradation mechanism. Circ Econ Sustain 4(1):339–362
Cardoso AR et al (2022) An ultra-sensitive electrochemical biosensor using the Spike protein for capturing antibodies against SARS-CoV-2 in point-of-care, 16:100354
Chen M et al (2014) Antitumor efficacy of a PLGA composite nanofiber embedded with doxorubicin@ MSNs and hydroxycamptothecin@ HANPs, 4(95):53344–53351
Costa PM et al (2018) Functionalised carbon nanotubes enhance brain delivery of amyloid-targeting Pittsburgh compound B (PiB)-derived ligands, 2(2):168
Darder M et al (2021) Gentamicin-montmorillonite intercalation compounds as an active component of hydroxypropylmethylcellulose bionanocomposite films with antimicrobial properties, 69(5):576–588
Dasgupta A et al (2020) Imaging-assisted anticancer nanotherapy, 10(3):956
Ding Y-W et al (2022) Advances in modified hyaluronic acid-based hydrogels for skin wound healing, 10(13):3393–3409
Dong Y et al (2009) Degradation behaviors of electrospun resorbable polyester nanofibers. Tissue Eng Part B Rev 15(3):333–351
Doshi J, Reneker DH (1995) Electrospinning process and applications of electrospun fibers. J Electrostat 35(2):151–160
Duarah R et al (2016) High performance bio-based hyperbranched polyurethane/carbon dot-silver nanocomposite: a rapid self-expandable stent. Biofabrication 8(4):045013
Ferber S et al (2014) Polymeric nanotheranostics for real-time non-invasive optical imaging of breast cancer progression and drug release, 352(1):81–89
Ferrari R et al (2011) ε-Caprolactone-based macromonomers suitable for biodegradable nanoparticles synthesis through free radical polymerization, 44(23):9205–9212
Flórez M, Cazón P, Vázquez MJP (2023) Selected biopolymers' processing and their applications: a review, 15(3):641

Fong H, Chun I, Reneker DH (1999) Beaded nanofibers formed during electrospinning. Polymer 40(16):4585–4592

Fu C-P et al (2023) Hyaluronic acid-based nanocarriers for anticancer drug delivery. Polymers 15(10):2317

Fu F et al (2014) Multifunctional lactobionic acid-modified dendrimers for targeted drug delivery to liver cancer cells: investigating the role played by PEG spacer, 6(18):16416–16425

Fu L et al (2017) Bilayer Poly(Lactic-co-glycolic acid)/nano-hydroxyapatite membrane with barrier function and osteogenesis promotion for guided bone regeneration, 10(3):257

Gaviria MI et al (2022) Highly sensitive fluorescent biosensor based on acetylcholinesterase and carbon dots–graphene oxide quenching test for analytical and commercial organophosphate pesticide detection, 10:825112

George A, Shah PA, Shrivastav PS (2019) Natural biodegradable polymers based nano-formulations for drug delivery: a review. Int J Pharm 561:244–264

Govindasamy M et al (2018) Determination of neurotransmitter in biological and drug samples using gold nanorods decorated f-MWCNTs modified electrode, 165(9):B370

Guo F et al (2019) Multi-responsive nanocarriers based on β-CD-PNIPAM star polymer coated MSN-SS-Fc composite particles, 11(10):1716

Gupta RC et al (2019) Hyaluronic acid: molecular mechanisms and therapeutic trajectory, 6:458280

Howard D et al (2008) Tissue engineering: strategies, stem cells and scaffolds. J Anat 213(1):66–72

Iravani S, Varma RS (2022) Alginate-based micro- and nanosystems for targeted cancer therapy. Mar Drugs 20(10):598

Jamróz E, Kulawik P, Kopel PJP (2019) The effect of nanofillers on the functional properties of biopolymer-based films: a review, 11(4):675

Juang R-S et al (2023) Electrochemical detection of parathyroid hormone using graphene-incorporated conductive polymer decorated with MoS2 nanoparticles, 666:131336

Kai D, Liow SS, Loh XJ (2014) Biodegradable polymers for electrospinning: towards biomedical applications. Mater Sci Eng C Mater Biol Appl 45:659–670

Kalayil N, Budar AA, Dave RK (2024) Nanofibers for drug delivery: design and fabrication strategies. BIO Integr 5(1):978

Kalendova A, Kupkova J, Urbaskova M, Merinska D (2024) Applications of clays in nanocomposites and ceramics. Minerals 14:93

Matumba KI, Mokhena TC, Ojijo V, Sadiku ER, Ray SS (2024) Morphological characteristics, properties, and applications of Polylactide/Poly(ε-caprolactone) blends and their composites—a review. Macromol Mater Eng, 2400056:1–24

Kasturi S et al (2021) Highly sensitive electrochemical biosensor based on naturally reduced rGO/Au nanocomposite for the detection of miRNA-122 biomarker, 93:186–195

Katti DS et al (2004) Bioresorbable nanofiber-based systems for wound healing and drug delivery: optimization of fabrication parameters. J Biomed Mater Res-Part B Appl Biomater 70(2):286–296

Keerthi M et al (2018) One-pot biosynthesis of reduced graphene oxide/prussian blue microcubes composite and its sensitive detection of prophylactic drug dimetridazole, 165(2):B27

Khane Y et al (2014) Fabrication, characterization and application of biocompatible nanocomposites: a review. Eur Polym J 214:113187

Khane Y et al (2024) Fabrication, characterization and application of biocompatible nanocomposites: a review, p 113187

Ki CS et al (2005) Characterization of gelatin nanofiber prepared from gelatin–formic acid solution. Polymer 46(14):5094–5102

Kim S, Chung H (2024) Biodegradable polymers: from synthesis methods to applications of lignin-graft-polyester. Green Chem 26:10774–10803

Kim Y et al (2023) Chitosan-based biomaterials for tissue regeneration. Pharmaceutics, 15(3)

Kizilbey K (2019) Optimization of rutin-loaded PLGA nanoparticles synthesized by single-emulsion solvent evaporation method. ACS Omega 4:555–562

Kwon SG et al (2018) Recent advances in stem cell therapeutics and tissue engineering strategies. Biomater Res 22(1):36

Lai C et al (2016) Synthesis of surface molecular imprinted TiO2/graphene photocatalyst and its highly efficient photocatalytic degradation of target pollutant under visible light irradiation. Appl Surf Sci 390:368–376

Lai J et al (2024) Alginate-based encapsulation fabrication technique for drug delivery: an updated review of particle type, formulation technique, pharmaceutical ingredient, and targeted delivery system. Pharmaceutics 16(3):370

Lee J-S et al (2020) Porphyrin nanofiber/single-walled carbon nanotube nanocomposite-based sensors for monitoring hydrogen peroxide vapor, 306:127518

Lemarchand C et al (2006) Influence of polysaccharide coating on the interactions of nanoparticles with biological systems. Biomaterials 27(1):108–118

Li B et al (2023) Flexible enzymatic biosensor based on graphene sponge for glucose detection in human sweat, 36:102525

Li G et al (2020) pH/Reduction dual-responsive comet-shaped PEGylated CQD-DOX conjugate prodrug: Synthesis and self-assembly as tumor nanotheranostics, 110:110653

Li K et al (2022) Graphene oxide nanosheet-mediated fluorescent RPA "turn-on" biosensor for rapid RNAi transgenic plant detection, 1189:339222

Lin W et al (2020) The role of hyaluronic acid in cartilage boundary lubrication, 9(7):1606

Liu G et al (2016) In-situ hydrothermal synthesis of molecularly imprinted polymers coated carbon dots for fluorescent detection of bisphenol A. Sensors and actuators B: Chemical, 228

Liu Y et al (2022) Ultrasmall Fe-doped carbon dots nanozymes for photoenhanced antibacterial therapy and wound healing. Bioact Mater 12:246–256

Luo Y et al (2015) Green synthesis of silver nanoparticles in xylan solution via Tollens reaction and their detection for Hg 2+, 7(2):690–700

Maciel-Flores CE, Lozano-Alvarez JA, Bivián-Castro EY (2024) Recently reported biological activities and action targets of Pt(II)- and Cu(II)-based complexes. Molecules 29:1066

Malaki M et al (2019) Advanced metal matrix nanocomposites, 9(3):330

Mardhian DF et al (2020) FGF2 engineered SPIONs attenuate tumor stroma and potentiate the effect of chemotherapy in 3D heterospheroidal model of pancreatic tumor, 4(1):26

Meghana M et al (2023) A road map on synthetic strategies and applications of biodegradable polymers, 80(11):11507–11556

Mikušová V, Mikuš P (2021) Advances in chitosan-based nanoparticles for drug delivery. Int J Mol Sci 22(17):9652

Mittal V (2009) Polymer layered silicate nanocomposites: a review, 2(3):992–1057

Mog B et al (2019) Nano-antagonist alleviates inflammation and allows for MRI of atherosclerosis, 3(4):342

Mukherjee C et al (2023) Recent advances in biodegradable polymers–properties, applications and future prospects 192:112068

Nandhini J, Karthikeyan E, Rajeshkumar SJ (2024) Eco-friendly bio-nanocomposites: pioneering sustainable biomedical advancements in engineering, 19(1):86

Nazrun T, Hassan MK, Hossain MD, Ahmed B, Hasnat MR, Saha S (2024) Application of biopolymers as sustainable cladding materials: a review. Sustainability 16:27

Nukala SG, Kong I, Patel VI, Kakarla AB, Kong W, Buddrick O (2022) Development of biodegradable composites using polycaprolactone and bamboo powder. Polymers 14:4169

Orasugh JT et al (2018) Jute cellulose nano-fibrils/hydroxypropylmethylcellulose nanocomposite: a novel material with potential for application in packaging and transdermal drug delivery system. Ind Crops Prod 112:633–643

Ospina-Villa JD et al (2019) Encapsulation of proteins from Leishmania panamensis into PLGA particles by a single emulsion-solvent evaporation method. J Microbiol Methods 162:1–7

Ozcelikay G et al (2021) Biocompatible nanopolymers in drug delivery systems and their recent electrochemical applications in drug assays. In: Hussain CM, Thomas S (eds) Handbook of polymer and ceramic nanotechnology. Springer International Publishing, Cham, pp 875–894

Panchal SS, Vasava D (2020) Biodegradable polymeric materials: synthetic approach, 5(9):4370–4379
Penczek S, Pretula J, Slomkowski SJ (2021) Ring-opening polymerization, 3(2):33–57
Prabha G, Raj V (2016) Preparation and characterization of polymer nanocomposites coated magnetic nanoparticles for drug delivery applications. J Magn Magn Mater 408:26–34
Qi B et al (2020) Tuned near infrared fluorescent hyaluronic acid conjugates for delivery to pancreatic cancer for intraoperative imaging, 10(8):3413
Qiu K et al (2013) Doxorubicin-loaded electrospun poly (L-lactic acid)/mesoporous silica nanoparticles composite nanofibers for potential postsurgical cancer treatment, 1(36):4601–4611
Rachmawati H et al (2016) Curcumin-loaded PLA nanoparticles: formulation and physical evaluation, 84(1):191–202
Ramesh Kumar S et al (2020) Bio-based and biodegradable polymers-State-of-the-art, challenges and emerging trends, 21:75–81
Raso J et al (1999) Influence of different factors on the output power transferred into medium by ultrasound. Ultrason Sonochem 5(4):157–162
Reakasame S, Boccaccini AR (2018) Oxidized alginate-based hydrogels for tissue engineering applications: a review. Biomacromol 19(1):3–21
Reddy YVM et al (2022) Fine-tuning of MXene-nickel oxide-reduced graphene oxide nanocomposite bioelectrode: sensor for the detection of influenza virus and viral protein, 214:114511
Ressler A (2022) Chitosan-based biomaterials for bone tissue engineering applications: a short review, 14(16):3430
Sahoo DR, Biswal T (2021) Alginate and its application to tissue engineering. SN Appl Sci 3(1):30
Said RAM et al (2020) Review—insights into the developments of nanocomposites for its processing and application as sensing materials. J Electrochem Soc 167(3):037549
Salari N et al (2021) Hyaluronic acid-based drug nanocarriers as a novel drug delivery system for cancer chemotherapy: a systematic review. DARU J Pharm Sci 29(2):439–447
Sardar H (2023) Drug like potential of Daidzein using SwissADME prediction: In silico Approaches. PHYTO Nutrients 02:02–08
Satchanska G, Davidova S, Petrov PD (2024) Natural and synthetic polymers for biomedical and environmental applications, 16(8):1159
Sedira S, Mendaci B (2019) Hydrothermal synthesis of spherical carbon nanoparticles (CNPs) for supercapacitor electrodes uses. Mater Renew Sustain Energy 9(1):1
Shafiee S, Bazli L, Karrabi M, Ghoreishy MHR, Bazli M (2022) Effect of organoclay addition on rheological, thermal, and mechanical properties of nitrile rubber/phenolic resin blend. Polymers 14:1463
Shaikh AA et al (2024) Biopolymer-based nanocomposites for application in biomedicine: a review, 44(2):83–116
Shao L et al (2019) Auto-fluorescent polymer nanotheranostics for self-monitoring of cancer therapy via triple-collaborative strategy, 194:105–116
Shariatinia Z (2020) Biopolymeric nanocomposites in drug delivery. In: Nayak AK, Hasnain MS (eds) Advanced biopolymeric systems for drug delivery. Springer International Publishing, Cham, pp 233–290
Sharififard H et al (2018) A novel natural chitosan/activated carbon/iron bio-nanocomposite: Sonochemical synthesis, characterization, and application for cadmium removal in batch and continuous adsorption process. Biores Technol 270:562–569
Sharma R et al (2023) A critical review on classified excipient sodium-alginate-based hydrogels: modification. Charact Appl Soft Tissue Eng 9(5):430
Sharma M et al (2019) Nanotheranostics, a future remedy of neurological disorders, 16(2):113–128
Shin M et al (2024b) Carbon-based nanocomposites for biomedical applications. RSC Adv 14(10):7142–7156
Shin M et al (2024) Carbon-based nanocomposites for biomedical applications, 14(10):7142–7156
Shinoka T et al (1995) Tissue engineering heart valves: valve leaflet replacement study in a lamb model. Ann Thorac Surg 60:S513–S516

Sinha Ray S et al (2002) New polylactide/layered silicate nanocomposites. 1. Preparation, characterization, and properties. Macromolecules 35(8):3104–3110

Soares JC et al (2017) Immunosensor for pancreatic cancer based on electrospun nanofibers coated with carbon nanotubes or gold nanoparticles, 2(10):6975–6983

Soman V, Vishwakarma K, Poddar MK (2023) Ultrasound assisted synthesis of polymer nanocomposites: a review, 30(11):406

Soppimath KS et al (2001) Biodegradable polymeric nanoparticles as drug delivery devices. J Control Release 70(1–2):1–20

Soufi GJ, Iravani SJ (2020) Eco-friendly and sustainable synthesis of biocompatible nanomaterials for diagnostic imaging: current challenges and future perspectives, 22(9):2662–2687

Su T et al (2014) Glucose oxidase triggers gelation of N-hydroxyimide–heparin conjugates to form enzyme-responsive hydrogels for cell-specific drug delivery. Chem Sci 5(11):4204–4209

Sung S-Y et al (2018) Graphene quantum dots-mediated theranostic penetrative delivery of drug and photolytics in deep tumors by targeted biomimetic nanosponges, 19(1):69–81

Suresh K et al (2017) Enhanced mechanical and thermal properties of polystyrene (PS) nanocomposites prepared using organo-functionalized Ni-Al layered double hydroxide (LDH) via melt intercalation technique. J Sci: Adv Mater Devices 2

Thomas D, Latha MS (2023) Alginate based nanocarriers for controlled drug delivery applications. In: Jana S, Jana S (eds) Alginate biomaterial: drug delivery strategies and biomedical engineering. Springer Nature Singapore, Singapore, pp 61–83

Tice TR, Gilley RM (1958) Preparation of injectable controlled-release microcapsules by a solvent-evaporation processs, 2:343–352

Torati SR et al (2023) Gold deposited laser-induced graphene electrode for detection of miRNA-141

Turco R, Santagata G, Corrado I, Pezzella C, Di Serio M (2021) In vivo and post-synthesis strategies to enhance the properties of phb-based materials: a review. Front Bioeng Biotechnol. 14(8):619266

Vaia RA, Ishii H, Giannelis EP (1993) Synthesis and properties of two-dimensional nanostructures by direct intercalation of polymer melts in layered silicates. Chem Mater 5(12):1694–1696

Velasco-Mallorquí F et al (2020) New volumetric CNT-doped gelatin-cellulose scaffolds for skeletal muscle tissue engineering. Nanoscale Adv 2(7):2885–2896

Verma AK (2022) Collagen-based biomaterial as drug delivery module. In: Collagen biomaterials. IntechOpen

Verma AK (2023) An insight into collagen-based nano biomaterials for drug delivery applications. In: Malviya R, Sundram S (eds) Engineered biomaterials: synthesis and applications. Springer Nature Singapore, Singapore, pp 397–428

Wang C et al (2016) Enhanced cancer immunotherapy by microneedle patch-assisted delivery of anti-PD1 antibody. Nano Lett 16(4):2334–2340

Wang JT et al (2020) Organ biodistribution of radiolabelled γδ T cells following liposomal alendronate administration in different mouse tumour models, 4(2):71

Yang M et al (2021) Polypyrrole/sulfonated multi-walled carbon nanotubes conductive hydrogel for electrochemical sensing of living cells, 418:129483

Yassue-Cordeiro PH et al (2018) 1—Chitosan-based nanocomposites for drug delivery. In: Inamuddin AM, Asiri, Mohammad A (eds) Applications of nanocomposite materials in drug delivery. Woodhead Publishing, p 1–26

Yi Q et al (2022) All-3D-printed, flexible, and hybrid wearable bioelectronic tactile sensors using biocompatible nanocomposites for health monitoring, 7(5):2101034

Yuan Z et al (2016) Doxorubicin-loaded mesoporous silica nanoparticle composite nanofibers for long-term adjustments of tumor apoptosis, 27(24):245101

Zhai P et al (2020) The application of hyaluronic acid in bone regeneration, 151:1224–1239

Zhang Y et al (2018) Injectable sustained release PLA microparticles prepared by solvent evaporation-media milling technology. Drug Dev Ind Pharm 44(10):1591–1597

Zhang W et al (2023) Reduced graphene oxide-embedded nerve conduits loaded with bone marrow mesenchymal stem cell-derived extracellular vesicles promote peripheral nerve regeneration. Neural Regen Res 18(1):200–206

Zhang Z et al (2015) Electrospun PLA/MWCNTs composite nanofibers for combined chemo-and photothermal therapy 26:115–123

Zhao Y et al (2018) In vivo therapeutic evaluation of polymeric nanomedicines: effect of different targeting peptides on therapeutic efficacy against breast cancer, 2(4):360

Zhuang X, Wang F, Hu XJ (2024) Biodegradable polymers: a promising solution for green energy devices, 204:112696

Zohri et al (2009) Polymeric NanoParticles: Production, Applications and Advantage. Internet J Nanotechnol 3:1

Chapter 9
Biocompatible Nanocomposites for Vaccine Delivery

R. Rajakumari, G. Sangeetha, Nandakumar Kalarikkal, and Sabu Thomas

Abstract Biocompatible nanocomposites for vaccine delivery systems enable the sustained and targeted drug release. The vaccine delivery procedure protects, stabilizes and controls the immunizing agent which serves as adjuvants to improve the efficacy of vaccines. This chapter provides the complete review about the importance of nanobiomaterials in the development of vaccines and its delivery. Firstly, this chapter emphasizes about the lipid-based delivery vehicles, polymer-based delivery, hydrogels and microneedles-based delivery systems. Furthermore, the different delivery routes of these nanobiomaterials which influences the immune responses is also reviewed. In addition, the biocompatible nanocomposites used for the development of cancer vaccine, covid-19 vaccine, malaria vaccine, tuberculosis and hepatitis B vaccines. Throughout the chapter, the importance of biomaterials in vaccine delivery, its efficacy, limitations and advancements was highlighted. This chapter provides a different type of nanocomposites used in the delivery of vaccines, its mechanism of action, applications, challenges and future prospects in this evolving field.

Keywords Nanocomposites · Vaccine · Biomaterials · Immune response · Cancer · Covid-19

Authors R. Rajakumaria, G. Sangeetha have equal contribution.

R. Rajakumari (✉) · G. Sangeetha
Krupanidhi College of Pharmacy, Bengaluru, Karnataka, India
e-mail: rajimpharm@gmail.com

N. Kalarikkal
School of Pure and Applied Physics, Mahatma Gandhi University, Kottayam, Kerala, India
e-mail: nkkalarikkal@mgu.ac.in

S. Thomas
School of Chemical Sciences, Mahatma Gandhi University, Kottayam, Kerala, India

R. Balakrishnan et al. (eds.), *Biocompatible Nanocomposites*, Advanced Structured Materials 239, https://doi.org/10.1007/978-981-96-7450-3_9

9.1 Introduction

Vaccines have been instrumental in preventing and controlling the infectious diseases. However, in spite of its success, there are several challenges in the development of vaccine with respect to its targeted controlled delivery, efficacy and stability. Nanotechnological advancements helps in the development of biocompatible nanocomposites for improving the targeted delivery of vaccines. Moreover, these nanocomposites improve the bioavailability, antigen uptake, stability, activation of immune response and targeted distribution (Bookstaver et al. 2018). These nanocomposites contribute for the innovations for the improvement of conventional and development of novel vaccines with enhanced immune system interaction. The nanomaterials are very small in size with adaptable surface properties and it has the ability to encapsulate the adjuvants which provides controlled release of the vaccine, protecting the active components in the vaccine preventing from degradation (Shichinohe and Watanabe 2023).

Biomaterial nanocomposites are the materials which consists of nanoparticles and developed into tunable properties in order to attain the desired characteristics. For the delivery of vaccine, nanocomposites are used as a carrier for the delivery of antigens and to improve the immune response as well. The blend of nanomaterials and biocompatible matrix helps to efficiently interact with the immune system without producing toxic reactions. The main purpose of the biocompatible nanocomposites in the delivery of vaccines is to;

1. Improve the stability of vaccine and thereby protects from the degradation.
2. Enhance the immune response by activation of the antigens and targeting the particular cell.
3. Deliver controlled and sustained release of the antigens.
4. Providing the targeted delivery to the specific cell to improve the efficacy and thereby decreasing the side effects.

Biocompatible nanocomposites are developed to encapsulate a diverse range of ingredients for vaccine formulation like antigens/adjuvants, nucleic acids, proteins, peptides for the formulation of vaccines (Lei et al. 2022; Ju et al. 2023; Yang et al. 2023; Yu et al. 2023).

The excellent potential of nanobiomaterials involved in the wide-ranging developmental plans in the realm of biomedical engineering. The challenges pertaining to the delivery of vaccines, targeting the cells, safety and efficacy of the vaccines that biomaterial nanocomposites must overcome in the field of delivery of vaccines. Therefore, a systematic approach in the recent developments in engineering the delivery of vaccines is very important. This chapter reviews the different biomaterials used for the preparation which includes polymer-based nanoparticles, liposome-based nanoparticles, hydrogels and microneedles. Additionally, this chapter explains the morphology, structure, size and properties of these materials which alters the immune reactions. Moreover, this also provides a brief summary of the materials as adjuvants and its addition to the biomaterials for the vaccine development. This chapter also summarizes the progress in the biomedical research for vaccines which improves the research and also enables the design and application of various biomaterials for the progress in vaccine development.

9.2 Types of Nanocomposites Used for Vaccine Delivery

Biocompatible nanocomposites for the delivery of vaccines comes in different forms based on the properties of matrices and the nanoparticles. The main types of biocompatible nanocomposites used in the preparations of vaccines are lipid-based, polymer-based, ceramic-based and metallic-based delivery systems. There are various forms of nanocomposites and it provides diverse advantages for the manufacturing of vaccines. The most widely used biocompatible nanocomposites used in the delivery of vaccines are (Fig. 9.1);

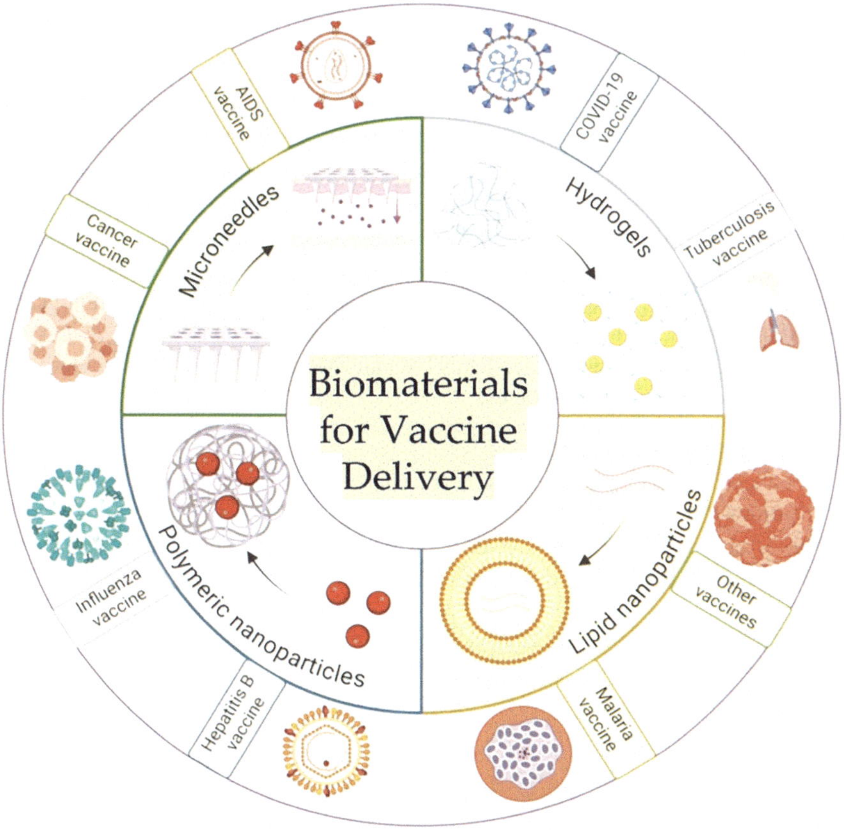

Fig. 9.1 Biocompatible nanocomposites for vaccine delivery adapted from (Zhuo et al. 2024)

9.2.1 Lipid-Based Nanocomposites (Liposomes and Nanoparticles)

Liposomes are widely used as nanodrug carriers for vaccines because of its excellent biocompatibility, biodegradability and non-toxic properties. Moreover, liposomes have high targeting capability in the specific site, encapsulate the drug with protection, release the active moiety in a controlled fashion, improve the therapeutic activity by reducing the side effects. They have the capability to deliver the antigens and adjuvants to the antigen presenting cells simultaneously which enhances the immune response. In addition, they prevent the degradation of antigens, enhancing the antigen presenting cells uptake, and aid in the cytoplasmic release and immunostimulatory activity (Tenchov et al. 2021; Agallou et al. 2023).

There are neutral, cationic and anionic liposomes and they are used in nucleic acid drug delivery. Cationic liposomes containing lipids and nucleic acids are shown to be the promising candidate for delivery of vaccines and possess strong immune response. Moreover, these cationic liposomes containing vaccines were effectively targeted the dendritic cells in mice. These discoveries suggest that liposome-based delivery systems are used for targeting the cancer cells and infections (Amin et al. 2022).

The lipid-based nanocomposites are the widely used system for the delivery of vaccines with the help of liposomes and lipid nanoparticles. The liposomes are made up of spherical vesicles consists of lipid bilayers with hydrophobic and hydrophilic antigens providing a unique and versatile release of the components (Fig. 9.2). They look exactly the structure of cell membranes which offers to facilitate the interactions with the cell membrane and provides the endocytosis process for the release of the encapsulated antigen (Nikolova et al. 2022; Firdaus et al. 2023). In the recent years, the lipid nanoparticles have been emerged as the pioneer because of the successful delivery of mRNA vaccines (Pfizer-BioNTech and Moderna Covid-19 vaccines) and it provides the following advantages;

1. Improved stability of mRNA
2. Enhanced cellular uptake because of the charge on the lipid nanoparticles enables the union of cell membranes.
3. Protecting the RNA from nucleases for the controlled delivery into cells where the RNA will be translated into proteins to activate the immune system.

The lipid nanoparticles are useful in the optimizing the size of the particles, surface charge and lipid contents for the enhancement of vaccines delivery.

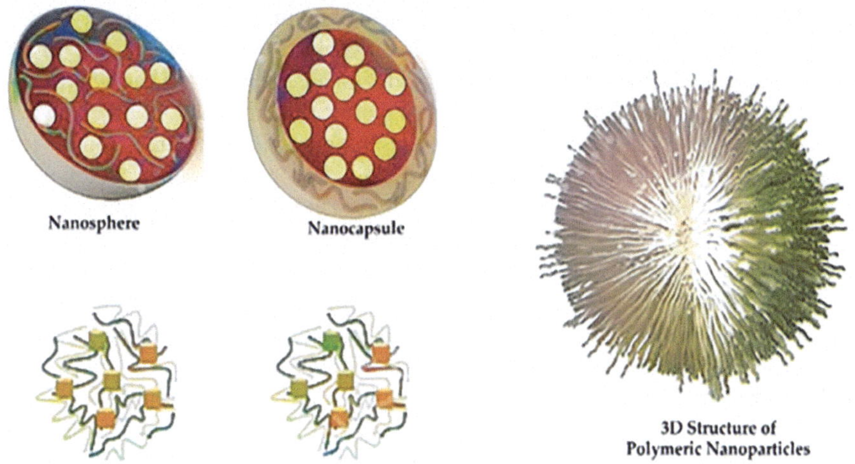

Fig. 9.2 Delivery of vaccines from the polymeric matrix. Polymeric nanocarriers such as nanosphere and nanocapsule are represented in the diagram. The drug releases from the core which is enclosed by the oily core where the drug dissolves

9.2.2 *Polymer-Based Nanocomposites*

The polymer-based nanocomposites are important class of nanocomposites for the delivery of vaccines. These polymer-based nanocomposites consist of biodegradable polymers such as poly (lactic-co-glycolic acid) (PLGA), polycaprolactone (PCL) and chitosan which are some of the few polymers used for vaccines formulation. These polymers are biocompatible and biodegradable which provides the sustained and controlled release of the vaccines for a period of time which reduces the multiple doses of vaccine administration (Fig. 9.2) (Backlund et al. 2023).

PLGA nanoparticles were widely used to encapsulate the various components like antigens, proteins and nucleic acid for the formulation of vaccines. The release of the encapsulated components in the vaccine can be controlled by varying the ratio of polymer and monomer and its molecular weight. The PLGA nanoparticles are widely used and it is formulated to target the specific cell through the tunable surface properties with the addition of ligands and antibodies (Gao et al. 2022; Tuomela and Levings 2023).

PLGA nanocomposites have recently garnered significant interest in the cancer therapy. PLGA nanocomposites undergo hydrolysis in in vivo conditions which breaks down into lactic acid and glycolic acid which is metabolized through tricarboxylic acid cycle. In addition, the studies suggest that there is a close link between the metabolic environment and immunosuppressive microenvironment of tumor. The regulatory T cells promote the growth and differentiation in the presence of lactic acid under acidic conditions which plays a complex role in maintaining the immunosuppressive state. Especially, PLGA nanocomposites-based formulations are approved for the clinical use in the treatment of cancer thereby providing valuable insights for the development of vaccines for cancer. The hydrolytic degradation of PLGA may lead to the accumulation of lactic and glycolic acids, which can potentially cause inflammation (Wu et al. 2022; Horvath and Basler 2023).

2.3 Chitosan-Based Nanoparticles have gained increased attention because of its mucoadhesive properties which improves the efficacy of vaccines administered through the mucosal routes like nasal or oral vaccines. This polymer is used to improve antigen uptake by the dendritic cells which provides an immune boosting effect. The presence of amino and hydroxyl groups in the chitosan molecule allows the inclusion of different functional groups via chemical and molecular modifications. After the chemical and molecular modifications, the chitosan derivatives improve the permeation and adhesion for the delivery of vaccines. The glycosylated chitosan and trimethyl chitosan have been extensively used for vaccine delivery (Wu et al. 2022). The recent findings reveal that chitosan-based nano-vaccines are used in nasal delivery systems which have adhesive properties representing immunostimulatory abilities. The challenges in ensuring safety and effectiveness in the vaccines created a major drawback in the development of vaccine. Hao et al. revealed the encapsulation of the material in the chitosan gel matrix and this gel formulation generates antibody as well as the activation of the immune cells. In the in vivo studies, they have identified that the vaccinated mice was protected from the fatal effects of Zika virus. This innovation provides a substantial advancement in the development of safe and effective vaccines. However, the intricate manufacturing and purification process associated with the chitosan possess challenges by limiting its effectiveness (Dmour and Islam 2022; Gaglio et al. 2023).

Self-Assembling **Peptides.**

The peptides are made from the amino acids which can easily self-assemble into nanostructures by non-covalent interactions and they are considered as the promising material for delivery of vaccines. These peptide materials have excellent biocompatibility, easy preparation method and adjustable functionality to improve its characteristics. In a study, Zhu et al. developed a peptide-based hydrogel nanoformulation with improved immune activation and anti-inflammatory effects. Wang et al. in another study, he used glycopeptide conjugates as the vaccine adjuvants where it binds to the specific receptors by producing immune stimulatory effects (Huo et al. 2023; Wu and Wang 2023).

9.2.3 *Silica Nanocomposites*

The mesoporous silica nanoparticles are used for vaccine delivery because they have the capacity to load large amounts of active molecules and high surface area. These composites are biocompatible and these molecules can be functionalized to improve the targeting of the cells and to load immunomodulatory agents. These silica nanocomposites are useful in providing controlled and targeted release of antigens thereby prolonging the immune system activation. Because of the porous nature, these silica nanocomposites can encapsulate mRNA, antigens and adjuvants (Guo et al. 2022; Huang et al. 2022).

9.2.4 *Metallic Nanocomposites*

The metallic nanoparticles such as gold, silver and iron are most widely used for biomedical applications because of their functionalization and stability (Chue-Gonçalves et al. 2021a, b). The studies revealed that gold nanoparticles have the ability to interact with the immune cells such as macrophages and dendritic cells. These metallic nanoparticles enhance the cellular uptake and immunogenicity of the vaccine. Moreover, the gold nanoparticles interact with the receptors of the immune cells by enhancing cytokine production and stimulates the immune system. The gold nanocomposites are used for the delivery of DNA and RNA vaccines for enabling the penetration of these materials into the cells (Abdalhamed et al. 2022; Sengupta et al. 2022; Hanna et al. 2023).

9.3 Hydrogels and Microneedles-Based Vaccine Delivery

Hydrogels are three-dimensional, hydrophilic polymer networks capable of holding large amounts of water or biological fluids making them ideal for drug and vaccine delivery (Fig. 9.3). Hydrogels, composed of biocompatible and degradable polymers such as natural polymers like alginate, chitosan, hyaluronic acid, and gelatin, provide biocompatibility and stability, while synthetic polymers such as polyethylene glycol (PEG), poly(vinyl alcohol) (PVA), and polylactic-co-glycolic acid (PLGA) offer controlled degradation and mechanical properties. Additionally, zwitterionic polymers, such as modified dextran, reduce immune side effects and improve antigen delivery (Chung et al. 2023). Hydrogels can act as adjuvants by stimulating the immune system and enhancing antigen-specific immune responses. Their biocompatibility reduces the likelihood of tissue toxicity, making them safer alternatives to traditional delivery methods. They also present an alternative to needle-based vaccine administration, which is particularly advantageous for improving patient compliance and reducing associated risks. Despite these advantages, challenges like achieving uniform degradation, optimizing antigen loading and ensuring consistent immune responses remain areas of active investigation. Nonetheless, hydrogels represent a transformative approach to vaccine delivery, contributing to more effective and accessible immunization strategies globally (Basu et al. 2021). Research areas in hydrogel include stimuli-responsive hydrogels, which release antigens in response to specific triggers within the body, such as changes in pH (Wei et al. 2017) or temperature (Pastor et al. 2020; Wu et al. 2012). Bioadhesive hydrogels adhere to mucosal surfaces like the nasal or oral cavity, providing sustained antigen delivery and potentially stimulating mucosal immunity (Mei et al. 2023). Combining hydrogels with nanomaterials can enhance antigen delivery by improving cellular uptake and immune response. Many research studies are ongoing on nanogel integration in vaccine delivery. It holds promise for developing more effective vaccines (Pérez et al. 2024).

Microneedles are tiny, needle-like structures, typically made of materials like silicon, polymer or sugar, that are designed to penetrate the skin's outer layer without reaching nerve endings, enabling painless administration of vaccines (Fig. 9.4). Microneedles offer efficient vaccine delivery by targeting antigen presenting cells like Langerhans and dendritic cells, optimizing immune responses. Their design facilitates self-administration, proving especially useful in pandemics. Additionally, microneedles provide a pain-free alternative by avoiding nerve-dense regions, improving patient compliance, particularly among pediatric and geriatric individuals. These features make microneedles a revolutionary tool for accessible and effective vaccination (Bajwa et al. 2022; Menon et al. 2021).

There are three types of microneedles for vaccine delivery including dissolving microneedles, which are made of biodegradable materials that dissolve after delivering the vaccine through a 'poke and release' type mechanism, reducing the risk of infection and injuries due to needle-stick post-application. Coated microneedles have a coating of vaccine that dissolves upon insertion into the skin and acts as a 'coat and poke' mechanism. It has limitations in the efficiency and uniformity of coating. And, solid microneedles create micropores in the skin, allowing the vaccine to be applied topically. It is composed of materials and metals such as silicon, titanium, stainless steel, and in some cases, polymers. Solid microneedles act as 'poke and patch' mechanism (Sartawi et al. 2022). There has been notable research on the dissolving microneedle arrays for the delivery of Hepatitis B vaccine adjuvanted by APC-targeted poly (lactic-co-glycolic acid) nanoparticles (Zheng et al. 2023). Coated microneedles were prepared with sulforhodamine, ovalbumin, and two HIV antigens to induce systemic IgG and secretory IgA in saliva, and confirmed that coated microneedles could offer a versatile technique for oral mucosal vaccination (Ma et al. 2014). Solid microneedles using porous silicon and Poly (ethylene

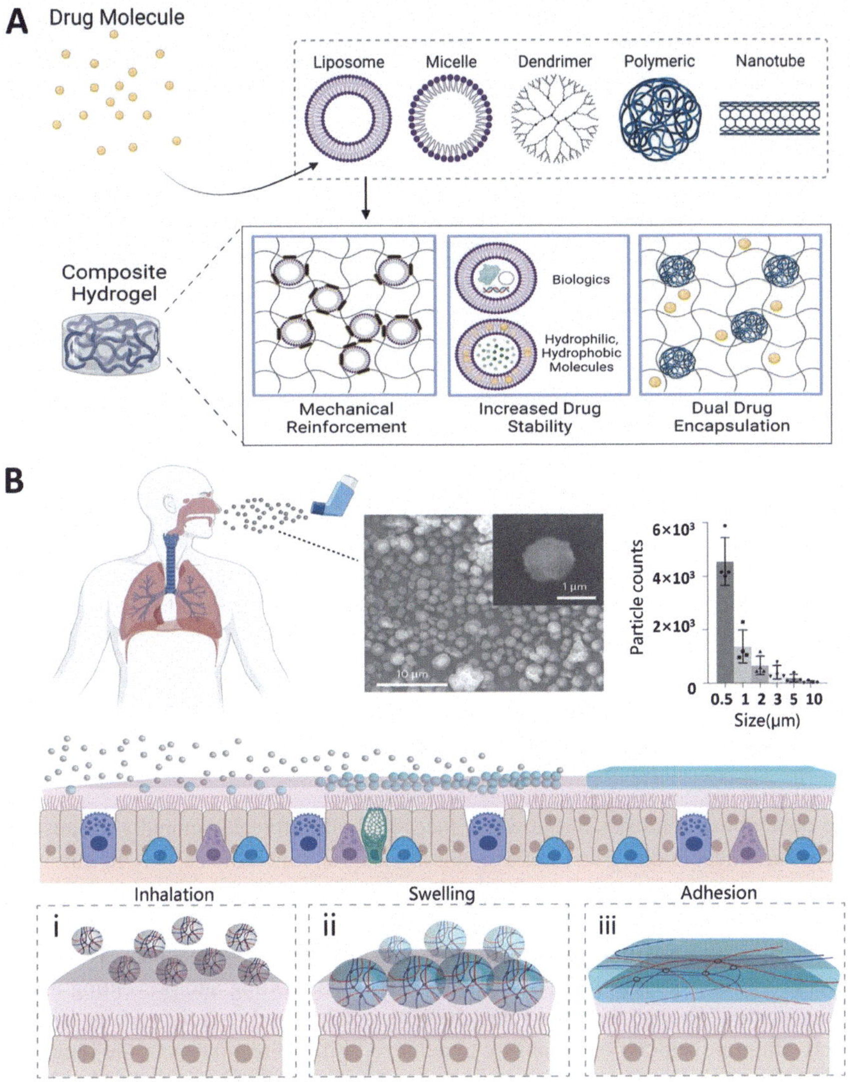

Fig. 9.3 Hydrogel. **a** Nanocarrier-hydrogel composites for vaccine delivery. Reproduced from (Kass and Nguyen 2022) with permission from WIREs Nanomedicine and Nanobiotechnology. **b** Changes in the contact process of the inhalable bioadhesive hydrogel with the mucus surface. Dry SHIELD particles spheres) are inhaled and they become swollen (blue spheres) once they are in contact with the mucus layer (pink layer). Finally, it forms a layer of hydrogel (blue layer) and adheres to the mucus layer. The process includes inhalation (i), swelling (ii) and adhesion (iii). The representative SEM image showing the morphology of SHIELD particles before swelling. The inset shows a zoomed-in view of one particle. Aerodynamic diameter of SHIELD particles. Reproduced with permission from Nature Materials (Mei et al. 2023)

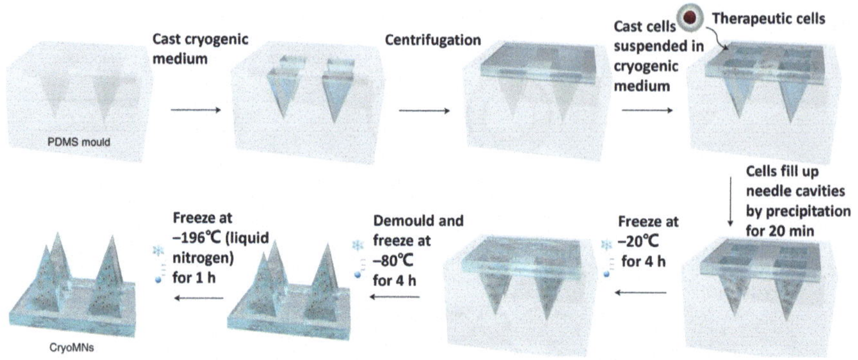

Fig. 9.4 Frozen microneedles. A Preparation of frozen microneedles. Reproduced with permission from Nature Biomedical Engineering (Chang et al. 2021)

glycol) diacrylate (PEGDA) are particularly biocompatible and desirable for vaccine delivery by the transdermal vaccine delivery method (Chandbadshah et al. 2022). It played a key role in the treatment of a Covid-19 pandemic. Ongoing research in both domains continues to expand their applications and improve their efficacy, making them central to next-generation vaccine delivery systems.

9.4 Biocompatible Nanocomposites in Vaccine Development

The developing countries in the world faces several outbreaks annually because of the communicable diseases such as malaria, dengue, tuberculosis, AIDS, influenza and numerous respiratory infections. These infectious diseases remain danger to the health of public accentuating the necessity for the prevention strategies of the diseases along with the development of medications and vaccines. Therefore, vaccines are considered as the important tools for controlling and preventing infectious diseases. Though there are some challenges associated with the vaccine such as safety, immunogenicity and stability to achieve immune responses in the long run. Currently, numerous researches focus on the use of nanobiomaterials for the development of vaccines. The nanobiomaterials support the rational design of adjuvants and antigens for improving the efficacy, tolerability and targeted delivery. In the recent years, the nanobiomaterials are typically used for the wide range of diseases. This chapter examines the developments in nanobiomaterial delivery systems for infectious diseases vaccines and cancer vaccines. The following table reviews the biomaterials used for the vaccine's formulation, its delivery forms and its therapeutic effects (Table 9.1).

Table 9.1 Biocompatible nanocomposites in vaccine development

Vaccine	Nanobiomaterial	Delivery method	Therapeutic effects
Vaccine for cancer	Hydrogels and nanoparticles	Nanovaccine and Hydrogel complex NvlH	It improves the accumulation of tumor antigens, stimulates the anti-tumor immune response, and suppresses the growth of large tumors
Vaccine for AIDS	Liposomes and nanoparticles	Liposomes containing HIV envelope proteins and immune modulatory substances	Enhanced antigen capture of cells
Vaccine for tuberculosis	Chitosan-based nanoparticles	Chitosan-coated with BCG and CFP-PLGA combination with vaccine components	Makes a strong immune response Controls host resistance to *Mycobacterium tuberculosis*
Vaccine for influenza	Nanoparticles and microneedles	Nanoparticles loaded with antigens	Stimulates the immune response and protect against various subtypes of influenza
Vaccine for malaria	Hydrogels, nanocapsules and nanoemulsion	API vector to target vaccine delivery	Decreases parasitic resistance and enhances antimalarial activity
Vaccine for hepatitis B	Self-assembling peptides and nanoparticles	Ferritin NP-preS1 nanovaccine	Enhances the cellular immune response to infections
Vaccine for Covid-19	Liposomes	LNPs cover around and protect mRNA	Helps uptake of mRNA in the body and encourages endosomal disruption for the delivery of nucleic acids in the host cell and thereby triggers the immune response

9.4.1 Vaccine for Cancer

Cancer immunotherapy is considered as the most potential therapy in preventing cancer growth and metastasis by immune system activation to detect and destroy the cancer cells. This process has become basis of clinical cancer treatment with vaccines developing as the most useful strategy. Using the vectors, these cancer vaccines deliver antigens, adjuvants to the lymph nodes. Then the antigen presenting cells process these vectors to improve immune responses and immune memory by immune cycle (Chao et al. 2020). Because of the low immunogenicity, poor targeting and immunosuppressive tumor microenvironment the efficacy of the cancer vaccines is limited. To overcome these limitations, advanced vaccine delivery systems incorporating nanocarriers and micelles have been developed, offering improved precision and effectiveness. In spite of its benefits, these vaccine delivery systems have its own impediments like complex preparation methods, ineffective targeting and antigen encapsulation (Fan et al. 2021).

Hydrogels are considered as the most useful technique to address these challenges due to their exceptional properties. Presently, the most widely used vaccine delivery systems is hydrogels wherein it permits the activation of antigens, adjuvant, cytokines, drugs and nanoparticles. The most potential system is the injectable hydrogels where they show highly efficient in immune system activation (Qin et al. 2021; Mei et al. 2022). These hydrogels are very easily formulated via *in-situ* method and administered using syringe in the targeted pathological site. This method has been very beneficial for the comfort of the patients and tolerability. A recent study presented an innovative hydrogel containing single dose nanovaccine injectable medication and immune barrier complex such as NvIH. This injection is given as single intralesional injection which alleviates immunosuppression, enhances antigen in the tumor site and activates the anti-tumor immune response. This study also revealed that immune barrier complex NvIH effectually suppress and reduce the large tumor cells (Chao et al. 2020; Yang et al. 2020; Ji et al. 2021; Kirtane et al. 2021).

Despite the advantages and advancements, the development of cancer vaccines is still facing lots of setbacks. The development of cancer vaccines is a complex process with inadequate treatment exposure and extended observation periods required to assess the therapeutic results (Guo et al. 2025). However, the incorporation of biomaterials in cancer vaccines offers as evidence that the clinical trials are there to target the progressive solid tumors and various cancers. These trials emphasize the possibility of in-situ cancer vaccines based on biomaterials and biological material. The development of vaccines incorporated biological material is widely useful for clinical applications (Saxena et al. 2021; Liu et al. 2022).

9.4.2 Vaccine for AIDS

The human immunodeficiency virus (HIV) is extremely dangerous infectious disease with high mortality and morbidity rates. The treatment for HIV is mired by the absence of effective medication but the nanobiomaterials display the potential to enhance the efficiency of alternate HIV therapies. Moreover, researchers developed the liposomes to encapsulate the proteins and drug, improve the uptake of antigens, prolong the antigens by lymphocytes and enhance the antigen capture by antigen presenting cells. Another evolving approach is that attaching immunomodulatory agents to the antigen presenting cells or T-cells during immunotherapy which modifies its immunostimulatory function (Freeling et al. 2014; Jones et al. 2017).

The recent studies shows that the nanotechnological applications in the prevention and treatment of HIV have been widely studied and examined in the animal trials and clinical trials. Currently, the development of long-acting antiretroviral injectable medications for the treatment of HIV enhances the patient compliance and reduces the frequency of dosing which signifies the major clinical nanotechnological advancement. The first long-acting caboterivir and rilpivirine antiretroviral drugs with controlled-release profile were approved and accepted by the European Union for patients in 2020. The lipid nanoparticles contain three to four antiretroviral drug combinations which are tested in the non-human primates reveals that nanoencapsulation prolongs the drug circulation time and increases the uptake of drugs by monocytes in the blood and macaques lymph nodes. This is particularly apparent in patients who receives antiretroviral drugs as HIV can persist in macrophages in the typical antiretroviral therapy. These outcomes recommend that long-acting antiretroviral injectable medications could become a promising treatment option for patients with HIV. Therefore, more research is required to determine the ideal treatment strategies for the humans (Freeling et al. 2015; Hanson et al. 2015).

9.4.3 Vaccine for Tuberculosis

Mycobacterium tuberculosis is the bacterium responsible for tuberculosis which is the deadliest bacterial infection globally and creates a significant problem on the public health domain. There are challenges associated with the management of tuberculosis because of the frequent occurrence of HIV-tuberculosis co-infections and rise of drug-resistant *M. tuberculosis* which extremely worsens the diagnosis and treatment of tuberculosis. Thus, the therapeutic vaccination is considered as the promising alternative treatment for tuberculosis which reduces the pathology and prevent degeneration after the medication unlike the traditional prophylactic vaccines. The Bacillus Calmette-Guerin (BCG) vaccine is one of the protective vaccines for the children against tuberculosis but it is not effective for the adults (Fatima et al. 2020; Hatherill et al. 2020; Milián-Suazo et al. 2022).

The research findings reveals that several tuberculosis vaccines are currently under preclinical trials and clinical development and its main purpose is to enhance the efficacy of BCG Vaccine. A study by Magallanes et al. showed that BCH vaccine coated with the chitosan along with PLGA confirmed that enhances the prevention of tuberculosis. Moreover, their study indicates that combining BCG vaccination with protein enrichments helps to decrease the transmission of tuberculosis in the in vivo studies. Also, this combination helps in reduction of lesions, infections and bacillary counts which leads to give a strong immune response. Furthermore, the exosomes have shown to be promising choice for detecting and treating tuberculosis. Noncoding RNA exosomes (ncRNAs) plays a vital role in the immune response of *M. tuberculosis* and offers promising opportunities as a novel biomarker for the next-generation diagnostic approaches (Contreras-Magallanes et al. 2021). In addition, this novel biomarker helps in combating bacterial infections and displays significant contribution in the development of advanced diagnostic tools. *M. tuberculosis* has progressed to destroy the human immune responses and continues to be within the host cells thereby obscuring the action of innate and adaptive immunity. Nevertheless, the developments in the vaccine research and the significant role of exosomes in the immune system modulation help in improving the treatment outcomes of tuberculosis (Baker et al. 2014; Bouzeyen and Javid 2022).

9.4.4 Vaccine for Influenza

The development of vaccines for influenza is well established and the current research is about the nanobiomaterials for manufacturing innovative vaccines for the influenza virus. These advances could enhance the patient compliance and decrease the need for yearly influenza vaccines. Formulating a common influenza vaccine remains vital for addressing the public health hazards by the influenza strains and pandemics (Rouphael et al. 2017).

The most significant approach in the current research is microneedle vaccine delivery system which is under the clinical trials. The clinical studies suggests that the use of soluble microneedles patches for the influenza vaccination shows that they can maintain the stability for a year and induces immune responses when compared to the traditional injectable vaccines. Pan et al. developed a mucosal nanoparticle vaccine HMNF which contains the influenza antigenic epitopes and when administered through the nasal mucosa, it induces potent antibody and immune responses and protect against multiple influenza A and B vaccine subtypes (Mistilis et al. 2017).

In a study, the intranasal HMNF vaccination in mice results in a strong immune response, increased levels of specific antigens and the activity of T-cell responds actively to the various antigen mutations. In addition, the HMNF vaccine was produced by *E. coli* system which supports the trans-nasal self-administration of vaccines and expected to develop a broad-spectrum influenza vaccine in the future (Mistilis et al. 2017).

9.4.5 Vaccine for Malaria

Malaria remains a global health burden, particularly in regions with limited healthcare infrastructure. The disease causes a significant public health challenge, with over 247 million cases and approximately 619,000 deaths reported in 2021, according to the World Health Organization. While vector control strategies such as insecticide-treated nets (ITNs) and antimalarial treatments have significantly reduced malaria-related deaths, vaccine development has faced considerable hurdles (WHO 2022).

Malaria vaccines face significant challenges due to the parasite's complex lifecycle and immune evasion strategies. Efforts targeting different lifecycle stages—pre-erythrocytic (sporozoites), erythrocytic (merozoites) and transmission (gametocytes)—have achieved limited success. Key barriers include rapid antigen degradation during storage, suboptimal immune activation with traditional adjuvants, and inefficient delivery of antigens to lymph nodes and antigen presenting cells hindering effective immunization (Krishna et al. 2019). To overcome the conventional delivery technology challenges, nanocomposites are hybrid systems composed of multiple components, such as polymers, lipids, and inorganic materials, designed to encapsulate and deliver vaccine antigens. Biocompatible nanocomposites have emerged as a transformative platform for malaria vaccine delivery, offering distinct advantages in enhancing stability, targeting, immune activation, and efficacy (Sanders et al. 2021).

Xu et al. 2022a designed a layer-by-layer (LbL) antigen delivery platform specifically engineered for malaria vaccines. A biocompatible modified chitosan nanoparticle (LBL NP) was synthesized and utilized for LbL loading and release of multiple malaria antigens from pre-erythrocytic and erythrocytic stages. LbL NP served as an antigen/protein delivery vehicle and was demonstrated to target the highest *Plasmodium falciparum* Circumsporozoite Protein (PfCSP) specific T-cell responses in mice compared to multiple controls. These nanoparticles, primarily composed of biocompatible materials such as poly (lactic-co-glycolic acid) (PLGA) and chitosan, provide a stable environment for encapsulating antigens and adjuvants. They protect vaccine components from degradation, ensuring antigen integrity during storage and transport. The nanoparticles are engineered to enhance antigen delivery to antigen presenting cells (APCs), leveraging targeted mechanisms such as ligand functionalization to improve dendritic cell uptake and antigen presentation. Furthermore, the controlled-release properties of these nanocomposites enable sustained antigen exposure, enhancing the immune response and potentially reducing the need for booster doses.

Laurenson and Laurens (2024) discussed the recent advancements in malaria vaccine development, focusing on the groundbreaking WHO recommendations for two vaccines: RTS,S and R21/Matrix-M. The first malaria vaccine recommended in 2021, demonstrated its feasibility, safety, and effectiveness in pilot programs conducted in Ghana, Kenya, and Malawi, reducing severe malaria cases and all-cause mortality by 13%, even with moderate coverage. Two years later, the WHO approved R21/Matrix-M, which similarly targets the *Plasmodium falciparum* circumsporozoite protein. Together, these vaccines represent a transformative step forward, expected to reduce malaria morbidity and mortality significantly. Many malarial vaccines are under development and have been put into clinical studies at various stages. These vaccines were developed by an innovative approach including the use of live attenuated sporozoite inoculations, advanced RNA-based platforms, and combinatory vaccines that integrate the PfCSP with antigens targeting other stages of the parasite's lifecycle. These strategies aim to enhance immune protection by addressing multiple facets of the parasite's biology, thereby increasing vaccine efficacy and durability (Luke et al. 2003).

9.4.6 *Vaccine for Hepatitis B*

Hepatitis B vaccines are designed to prevent chronic liver infections caused by the hepatitis B virus (HBV). HBV is an important agent of liver disease including cirrhosis, hepatocellular carcinoma, and hepatic failure. Most traditional vaccines use recombinant hepatitis B surface antigen (HBsAg) combined with aluminum-based adjuvants to stimulate an immune response (Fabrizi et al. 2021 and Pattyn et al. 2021). These vaccines are effective but have limitations such as stability issues, the need for multiple doses, and sometimes suboptimal immune responses in certain populations (www.spice.unc.edu). Advances in vaccine technology, such as the use of polymeric nanoparticles (Gupta et al. 2007), solid lipid nanoemulsion (Sahu et al. 2019), and nanoemulsion (Baker et al. 2014), have been employed to enhance the delivery and efficacy of the hepatitis B surface antigen (HBsAg). By encapsulating the antigen, nanoparticles protect it from degradation during storage and transport, ensuring its stability even in suboptimal conditions. They enable targeted delivery to antigen presenting cells (APCs) like dendritic cells, ensuring efficient antigen uptake and a robust immune response. Moreover, nanoparticles can co-deliver HBsAg with potent adjuvants, such as toll-like receptor (TLR) agonists, enhancing both humoral and cellular immunity (Pattyn et al. 2021). Intranasal and oral formulations of nano-vaccines provide needle-free administration options, stimulating mucosal, systemic immunity and simplifying the transport of vaccines. Nanotechnology in vaccine delivery provides multifunctional platforms that also hold promise for combination vaccines, integrating antigens for hepatitis B with other co-infections like HIV or hepatitis C (Lindstrand et al. 2021).

Chitosan is a biodegradable polymer extensively utilized in nanotechnology due to its biocompatibility and ability to encapsulate antigens efficiently. Recent research demonstrated the potential of chitosan nanoparticles loaded with recombinant hepatitis B surface antigen to enhance immune responses. These nanoparticles exhibited controlled antigen release, high stability, and significant adjuvant activity, inducing strong anti-HBsAg IgG (Immunoglobulin G) responses in vivo. Additionally, mannosylated chitosan nanoparticles improved targeting of immune cells, promoting higher immunogenicity compared to non-mannosylated systems (Mehrabi et al. 2018). Nano-adjuvant based on silk fibroin nanoparticles has emerged as promising carriers for hepatitis B vaccines. These nanoparticles demonstrated sustained antigen release, maintaining stability for up to six months. Immunization studies revealed that physically loaded HBsAg-SF nanoparticles significantly enhanced IgG production, outperforming conventional vaccines in eliciting immune responses (Rezaei et al. 2021). Hepatitis B surface antigen vaccine fabricated chitosan–polyethylene glycol nanocomposite (HBsAg—CS-PEG- NC) has shown enhanced stability and reduced toxicity while maintaining high antigen encapsulation efficiency. These systems demonstrated controlled antigen release and heightened immune responses (Nishanth et al. 2020). The above research study on nanocomposites in the Hepatitis B vaccine evidenced the integration of nanocomposites in hepatitis vaccine development marks a significant advancement, addressing challenges related to delivery, stability and immune activation.

9.4.7 Vaccine for Covid-19

The Covid-19 pandemic has brought unprecedented challenges to global health systems, requiring rapid innovation in vaccine development to combat the SARS-CoV-2 virus and its evolving variants. Nanotechnology has emerged as a critical enabler, offering innovative solutions for vaccine design, delivery, and efficacy (Xu et al. 2022b). Nanocomposites—materials engineered at the nanoscale combining two or more components—have proven particularly transformative in the development of Covid-19 vaccines. These attributes make nanocomposites an ideal platform for modern vaccine technologies, including mRNA, DNA, and protein subunit vaccines. The success of mRNA vaccines for COVID-19, such as Pfizer-BioNTech's BNT162b2 and Moderna's mRNA-1273, owes significantly to nanotechnology. Lipid nanoparticles (LNPs) played a pivotal role as delivery systems. These nanoscale carriers protect fragile mRNA from enzymatic degradation, facilitate cellular uptake, and enable precise delivery to target cells (Khurana et al. 2021). The self-assembled protein nanoparticles have been employed to display the SARS-CoV-2 spike receptor-binding domain (RBD) in a multivalent format, significantly enhancing immunogenicity. Self-assembled nanostructures such as Ferritin and I53-50 platforms elicited strong neutralizing antibody responses, demonstrating potential as versatile carriers for subunit vaccines (Kang et al. 2021).

Additionally, chitosan, a biocompatible and biodegradable polymer, demonstrates antimicrobial and anticancer properties, making it a versatile material for therapeutic applications. In the context of Covid-19, chitosan-based systems provide a promising platform for developing pulmonary vaccines due to their mucoadhesive properties, permeability enhancement, and targeted delivery capabilities (Tousian and Khosravi (2023). Nanocovax, a protein subunit vaccine based on the SARS-CoV-2 spike protein, encapsulated in nanoparticles, has shown high immunogenicity and safety in preclinical studies. It elicited strong IgG responses and protected against viral challenges in animal models, offering a promising platform for human application (Tran et al. 2021). Nanocomposites have revolutionized Covid-19 vaccine development by enhancing antigen stability, delivery efficiency, and immunogenicity. Their versatility and scalability make them vital tools in addressing current and future global health challenges, paving the way for innovative vaccine technologies.

9.5 Other Infectious Diseases

Emerging infectious diseases, including those caused by dengue, chikungunya, zika virus, typhoid fever, pneumococcal infections, rabies and ebola have spurred interest in novel vaccine strategies. Infectious diseases cause a substantial global health challenge, leads to significant morbidity and mortality and burdening healthcare systems and economies worldwide. Biomaterials-based vaccines offer a promising avenue for combating infectious diseases. While this field is relatively early in development of vaccines, recent developments showed by the successful development of liposomal Covid-19 vaccines led to more research on material-based vaccine development that provides robust protection against a diverse range of pathogens (Polack et al. 2020). Nano-vaccines offer several advantages, such as targeted antigen delivery, enhanced bioavailability and reduced side effects. Researchers are developing various nanoparticle-based vaccines to harness these benefits, including virus-like particles, liposomes, polymeric nanoparticles, nanogels, lipid nanoparticles, emulsion vaccines, exosomes and inorganic nanoparticles (Zhong et al. 2017).

A new strategy of encapsulating the immunodominant EDIII domain of Envelop protein of all the four serotypes of dengue virus (1–4) into polymeric (PLGA) nanoparticles. It enhances antigen stability, bioavailability and immunogenicity, promoting neutralizing antibodies and T-cell activation. Preclinical studies show broad protection and potential to prevent severe complications like dengue hemorrhagic fever (Khan et al. 2022). Research on producing Chikungunya virus-like particles (VLPs) and subunit vaccines in insect cells highlights this method as a scalable and effective approach for vaccine development: Chikungunya virus, a mosquito-borne pathogen causing severe arthritic disease. With no licensed treatments, researchers have developed scalable glycoprotein subunit and virus-like particle (VLP) vaccines using baculovirus technology in insect cells. VLPs have shown strong immunogenicity and protection in animal models, making them a promising vaccine candidate (Stefan et al. 2016). Nanocomposites represent a versatile and promising tool for vaccine development, particularly for addressing complex infectious diseases. Their ability to enhance immune responses and their adaptability to various pathogens makes them a focal point of next-generation vaccine technologies.

9.6 Conclusion and Future Prospects

Nanobiomaterials that enables the precise control of delivery of vaccines and the spatial and temporal interactions with the cells were effective in vaccine localization and delivery. This involves combination of the chemicals, material science, biotechnological and medical disciplines and possess huge potential applications. Extensive efforts have been made to improve the role of nanobiomaterials in the development of vaccines and immunomodulation to prevent infections against various infectious diseases, auto-immune diseases, regenerative medicine and cancer treatment.

The development of vaccines using nanobiomaterials is progressing at a remarkable pace with new breakthroughs every year. Nanobiomaterials play a vital role in the delivery of vaccines which enhances the efficacy of vaccines with immunostimulatory properties. In addition to its stimulatory effects, these materials minimize the pain, improve the systemic effects and stability of antigens. They can easily target the particular cells, tissues or organ system by delivering the specific signals to the multiple immune cell types and pathways. The development of vaccine is a complex process which involves different scientific fields and technologies depends on the research objectives, type of pathogen and preparation methodology. These nanobiomaterials enhance the immune responses and these materials have the tendency to send signals to the specific tissues or cells. Moreover, these nanobiomaterials can be developed for wide variety of delivery systems like lipid nanoparticles, polymeric nanoparticles, microneedles and hydrogels. These delivery systems (lipid-polymer hybrid nanoparticles, nanoparticle composite hydrogels, and hydrogel forming microneedles) provide a targeted controlled release of vaccines for the treatment of variety of different infection and cancers. These nanobiomaterials vaccine delivery systems display the most influential synergetic advantages and effects.

The nanobiomaterials hold great potential in transforming the development of vaccines but the challenges associated with the delivery should not be ignored for the long-term growth. One major complication is that transportation and storage of vaccine but then the silica nanoparticles shown to protect the vaccines from environmental conditions guides to improve the manufacturing process and its stability. In spite of the efforts taken to improve the delivery methods using polymer-based, peptide-based delivery systems to enhance the in vivo T-cell responses, efficacy of the system should be thoroughly evaluated. Furthermore, the safety and long-term effects of nanobiomaterials on human health are the critical factors to be addressed. Additionally, the future direction of the section of the nanobiomaterials would be based on unique immune response of the individuals. There are several unsolved topics and the research areas remains to permit comprehensive investigation to ensure that the biomaterials are used very effectively in the clinical settings.

Our understanding of the immune system develops by the way the biomaterials functions as vaccine adjuvants are becoming clearer. Further research paves the way for the biomaterials to develop innovative adjuvants. In addition, the biomaterials have the natural immunogenic properties which is used to develop carrier systems for vaccine and control immune cell activity in the body. The adjuvants were used widely in the development of vaccines for enhancing the immune response, safety and efficacy. By the continuous advancements and the rational development, it is expected that more effective activation of immune response, pathogen coverage and addressing the challenges associated with transportation and storage can be done. The innovations in medicine and biotechnology pave the way for the biomaterials for the delivery of vaccines drives the development of safe, effective and specific novel vaccine adjuvants.

References

Abdalhamed AM, Naser SM, Mohamed AH, Zeedan GSG (2022) Development of gold nanoparticles-lateral flow test as a novel field diagnostic assay for detecting foot-and-mouth disease and lumpy skin disease viruses. Iran J Microbiol 14:574

Agallou M, Margaroni M, Tsanaktsidou E et al (2023) A liposomal vaccine promotes strong adaptive immune responses via dendritic cell activation in draining lymph nodes. J Control Release 356:386–401

Amin M, Seynhaeve ALB, Sharifi M et al (2022) Liposomal drug delivery systems for cancer therapy: the rotterdam experience. Pharmaceutics 14:2165

Backlund C, Jalili-Firoozinezhad S, Kim B, Irvine DJ (2023) Biomaterials-mediated engineering of the immune system. Annu Rev Immunol 41:153–179

Bajwa IK, Kaur N, Dsouza JM, Mathew JL (2022) Evaluation of efficacy and safety of intradermal delivery of vaccines through microneedle (s) in human beings: a protocol for a systematic review. Syst Rev 11(1):170

Baker JL, Chen L, Rosenthal JA et al (2014) Microbial biosynthesis of designer outer membrane vesicles. Curr Opin Biotechnol 29:76–84

Basu P, Saha N, Saha T, Saha P (2021) Polymeric hydrogel-based systems for vaccine delivery: a review. Polymer 230:124088

Bookstaver ML, Tsai SJ, Bromberg JS, Jewell CM (2018) Improving vaccine and immunotherapy design using biomaterials. Trends Immunol 39:135–150

Bouzeyen R, Javid B (2022) Therapeutic vaccines for tuberculosis: an overview. Front Immunol 13:878471

Chandbadshah SBVJ, Mannayee G (2022) Structural analysis and simulation of solid microneedle array for vaccine delivery applications. Mater Today: Proc 65:3774–3779

Chang H, Chew SW, Zheng M et al (2021) Cryomicroneedles for transdermal cell delivery. Nat Biomed Eng 5(9):1008–1018

Chao Y, Chen Q, Liu Z (2020) Smart injectable hydrogels for cancer immunotherapy. Adv Funct Mater 30:1902785

Chue-Gonçalves M, Pereira GN, Faccin-Galhardi LC et al (2021a) Metal nanoparticles against viruses: possibilities to fight SARS-CoV-2. Nanomaterials 11:3118

Chue-Gonçalves M, Pereira GN, Faccin-Galhardi LC, Kobayashi RK, Nakazato G (2021b) Metal nanoparticles against viruses: possibilities to fight SARS-CoV-2. Nanomaterials 11(11):3118

Chung JT, Lau CML, Chung CH, Rafiee M, Yao S, Chau Y (2023) Vaccine delivery by zwitterionic polysaccharide-based hydrogel microparticles showing enhanced immunogenicity and suppressed foreign body responses. Biomater Sci 11(14):4827–4844

Contreras-Magallanes YG, Durán-Aguilar M, Sosa-Gallegos SL et al (2021) Prime vaccination with chitosan-coated phipps bcg and boosting with cfp-plga against tuberculosis in a goat model. Animals 11:1046

Dmour I, Islam N (2022) Recent advances on chitosan as an adjuvant for vaccine delivery. Int J Biol Macromol 200:498–519

Fabrizi F, Cerutti R, Dixit V, Ridruejo E (2021) Hepatitis B virus vaccine and chronic kidney disease. The advances. Nefrología (English Edition) 41(2):115–122

Fan M, Jia L, Pang M et al (2021) Injectable adhesive hydrogel as photothermal-derived antigen reservoir for enhanced anti-tumor immunity. Adv Funct Mater 31:2010587

Fatima S, Kumari A, Das G, Dwivedi VP (2020) Tuberculosis vaccine: a journey from BCG to present. Life Sci 252:117594

Firdaus FZ, Bartlett S, Hussein WM et al (2023) Liposomal formulations of a polyleucine–antigen conjugate as therapeutic vaccines against cervical cancer. Pharmaceutics 15:602

Freeling JP, Koehn J, Shu C et al (2014) Long-acting three-drug combination anti-HIV nanoparticles enhance drug exposure in primate plasma and cells within lymph nodes and blood. AIDS 28:2625–2627

Freeling JP, Koehn J, Shu C et al (2015) Anti-HIV drug-combination nanoparticles enhance plasma drug exposure duration as well as triple-drug combination levels in cells within lymph nodes and blood in primates. AIDS Res Hum Retroviruses 31:107–114
Gaglio SC, Perduca M, Zipeto D, Bardi G (2023) Efficiency of chitosan nanocarriers in vaccinology for mucosal immunization. Vaccines (Basel) 11:1333
Gao L, Li J, Song T (2022) Poly lactic-co-glycolic acid-based nanoparticles as delivery systems for enhanced cancer immunotherapy. Front Chem 10:973666
Guo J, De May H, Franco S et al (2022) Cancer vaccines from cryogenically silicified tumour cells functionalized with pathogen-associated molecular patterns. Nat Biomed Eng 6:19–31
Guo Y, Wang Z, Li G, Zhan M, Xiao T, Wang J, Shen M (2025) A polymer nanogel-based therapeutic nanovaccine for prophylaxis and direct treatment of tumors via a full-cycle immunomodulation. Bioactive Mater 43:129–144
Gupta PN, Khatri K, Goyal AK, Mishra N, Vyas SP (2007) M-cell targeted biodegradable PLGA nanoparticles for oral immunization against hepatitis B. J Drug Target 15:701–713
Hanna SJ, Thayer TC, Robinson EJS et al (2023) Single-cell RNAseq identifies clonally expanded antigen-specific T-cells following intradermal injection of gold nanoparticles loaded with diabetes autoantigen in humans. Front Immunol 14:1276255
Hanson MC, Abraham W, Crespo MP et al (2015) Liposomal vaccines incorporating molecular adjuvants and intrastructural T-cell help promote the immunogenicity of HIV membrane-proximal external region peptides. Vaccine 33:861–868
Hatherill M, White RG, Hawn TR (2020) Clinical development of new TB vaccines: recent advances and next steps. Front Microbiol 10:3154
Horvath D, Basler M (2023) PLGA Particles in Immunotherapy. Pharmaceutics 15:615
https://spice.unc.edu/2018/heplisav-b-a-new-hepatitis-b-vaccine-that-can-be-used-for-pre-and-post-exposure-prophylaxis-of-healthcare-personnel-hcp/. Accessed on 21 Nov 2024
https://www.who.int/teams/global-malaria-programme/reports/world-malaria-report-2022
Huang Y, Li P, Zhao R et al (2022) Silica nanoparticles: Biomedical applications and toxicity. Biomed Pharmacother 151:113053
Huo Y, Hu J, Yin Y et al (2023) Self-assembling peptide-based functional biomaterials. ChemBioChem 24:e202200582
Ji G, Zhang Y, Si X et al (2021) Biopolymer immune implants' sequential activation of innate and adaptive immunity for colorectal cancer postoperative immunotherapy. Adv Mater 33:2004559
Jones RB, Mueller S, Kumari S et al (2017) Antigen recognition-triggered drug delivery mediated by nanocapsule-functionalized cytotoxic T-cells. Biomaterials 117:44–53
Ju Y, Hu Y, Yang P et al (2023) Extracellular vesicle-loaded hydrogels for tissue repair and regeneration. Mater Today Bio 18:100522
Kang Y, Sun C, Zhen Z, Yuan R, Zheng Q, Li J, Zhou P, Chen X, Liu Z, Zhang X, Yu X, Kong X, Zhu Q, Zhong Q, Xu M, Zhong N, Zeng Y, Feng G, Ke C, Zhao J, Zeng M (2021) Rapid development of SARS-CoV-2 spike protein receptor-binding domain self-assembled nanoparticle vaccine candidates
Kass LE, Nguyen J (2022) Nanocarrier-hydrogel composite delivery systems for precision drug release. WIREs Nanomed Nanobiotechnol 14(2):e1756
Khan R, Ahmed F, Afroz S, Khan N (2022) Tetravalent formulation of polymeric nanoparticle-based vaccine induces a potent immune response against dengue virus. Biomater Sci
Khurana A, Allawadhi P, Khurana I, Allwadhi S, Weiskirchen R, Banothu A, Chhabra D, Joshi K, Bharani K (2021) Role of nanotechnology behind the success of mRNA vaccines for COVID-19. Nano Today 38:101142–101142
Kirtane AR, Verma M, Karandikar P et al (2021) Nanotechnology approaches for global infectious diseases. Nat Nanotechnol 16:369–384
Krishna S et al (2019) Nanoparticle-based delivery of malaria vaccines: advances and challenges. Adv Drug Deliv Rev 147:57–76
Laurenson AJ, Laurens MB (2024) A new landscape for malaria vaccine development. PLoS Pathog 20(6):e1012309

Lei L, Bai Y, Qin X et al (2022) Current understanding of hydrogel for drug release and tissue engineering. Gels 8:301
Lindstrand A, Cherian T, Chang-Blanc D, Feikin D, O'Brien KL (2021) The world of immunization: achievements, challenges, and strategic vision for the next decade. J Infect Dis 224(Supplement_4):S452–S467
Liu J, Liew SS, Wang J, Pu K (2022) Bioinspired and biomimetic delivery platforms for cancer vaccines. Adv Mater 34:2103790
Luke TC, Hoffman SL (2003) Rationale and plans for developing a non-replicating, metabolically active, radiation-attenuated plasmodium falciparum sporozoite vaccine. J Exp Biol 206(21):3803–3808
Ma Y, Tao W, Krebs SJ, Sutton WF, Haigwood NL, Gill HS (2014) Vaccine delivery to the oral cavity using coated microneedles induces systemic and mucosal immunity. Pharm Res 31:2393–2403
Mehrabi M, Dounighi N, Sorkhabadi S, Doroud D, Amani A, Khoobi M, Ajdary S, Pilehvar-Soltanahmadi Y (2018) Development and physicochemical, toxicity and immunogenicity assessments of recombinant hepatitis B surface antigen (rHBsAg) entrapped in chitosan and mannosylated chitosan nanoparticles: as a novel vaccine delivery system and adjuvant. Artif Cells, Nanomedicine, Biotechnol 46:230–240
Mei H, Cai S, Huang D et al (2022) Carrier-free nanodrugs with efficient drug delivery and release for cancer therapy: From intrinsic physicochemical properties to external modification. Bioact Mater 8:220–240
Mei X, Li J, Wang Z, Zhu D, Huang K, Hu S, Cheng K (2023) An inhaled bioadhesive hydrogel to shield non-human primates from SARS-CoV-2 infection. Nat Mater 22(7):903–912
Menon I, Bagwe P, Gomes KB, Bajaj L, Gala R, Uddin MN, Zughaier SM (2021) Microneedles: a new generation vaccine delivery system. Micromachines 12(4):435
Metz S, Pijlman G (2016) Production of chikungunya virus-like particles and subunit vaccines in insect cells. Methods Mol Biol 1426:297–309
Milián-Suazo F, González-Ruiz S, Contreras-Magallanes YG et al (2022) Vaccination strategies in a potential use of the vaccine against Bovine tuberculosis in infected herds. Animals 12:3377
Mistilis MJ, Joyce JC, Esser ES et al (2017) Long-term stability of influenza vaccine in a dissolving microneedle patch. Drug Deliv Transl Res 7:195–205
Nikolova MP, Kumar EM, Chavali MS (2022) Updates on responsive drug delivery based on liposome vehicles for cancer treatment. Pharmaceutics 14:2195
Nishanth AN, Namasivayam KR, Arvind Bharani RS, Nivedh K, Syed NH, Samuel R (2020) Hepatitis B-surface antigen (HBsAg) vaccine fabricated chitosan-polyethylene glycol nanocomposite (HBsAg-CS-PEG-NC) preparation, immunogenicity, controlled release pattern, biocompatibility or non-target toxicity. Int J Biol Macromol 144:978–99
Pastor Y, Ting I, Martínez A, Irache J, Gamazo C (2020). Intranasal delivery system of bacterial antigen using thermosensitive hydrogels based on a Pluronic-Gantrez conjugate. Int J Pharm 119154
Pattyn J, Hendrickx G, Vorsters A, Van Damme P (2021) Hepatitis B Vaccines. J Infect Dis 224(Supplement_4):S343–S351
Pérez MLS, Montironi I, Funes JA, Margineda C, Campra N, Cariddi LN, Alustiza F (2024) Nanogel-Mediated antigen delivery: biocompatibility, immunogenicity, and potential for tailored vaccine design across species. Vaccine 42(17):3721–3732
Polack FP, Thomas SJ, Kitchin N, Absalon J, Gurtman A, Lockhart S, Perez JL, Pérez Marc G, Moreira ED, Zerbini C, Bailey R, Swanson KA, Roychoudhury S, Koury K, Li P, Kalina WV, Cooper D, Frenck RW Jr, Hammitt LL, Gruber WC (2020) Safety and efficacy of the BNT162b2 mRNA Covid-19 vaccine. N Engl J Med 383:2603–2615
Qin L, Zhang H, Zhou Y et al (2021) Nanovaccine-based strategies to overcome challenges in the whole vaccination cascade for tumor immunotherapy. Small 17:2006000
Rezaei F, Keshvari H, Shokrgozar MA, Doroud D, Gholami E, Khabiri A, Farokhi M (2021) Nano-adjuvant based on silk fibroin for the delivery of recombinant hepatitis B surface antigen. Biomater Sci 9(7):2679–2695

Rouphael NG, Paine M, Mosley R et al (2017) The safety, immunogenicity, and acceptability of inactivated influenza vaccine delivered by microneedle patch (TIV-MNP 2015): a randomised, partly blinded, placebo-controlled, phase 1 trial. The Lancet 390:649–658

Sahu KK, Kaurav M, Pandey RS (2019) Chylomicron mimicking solid lipid nanoemulsions encapsulated enteric minicapsules targeted to colon for immunization against Hepatitis B. Int Immunopharmacol 66:317–329

Sanders PR et al (2021) A peptide-based malaria vaccine delivered via nanoparticle systems. Vaccine 39(25):3439–3450

Sartawi Z, Blackshields C, Faisal W (2022) Dissolving microneedles: applications and growing therapeutic potential. J Control Release 348:186–205

Saxena M, van der Burg SH, Melief CJM, Bhardwaj N (2021) Therapeutic cancer vaccines. Nat Rev Cancer 21:360–378

Sengupta A, Azharuddin M, Al-Otaibi N, Hinkula J (2022) Efficacy and immune response elicited by gold nanoparticle-based nanovaccines against infectious diseases. Vaccines (Basel) 10:505

Shichinohe S, Watanabe T (2023) Advances in adjuvanted influenza vaccines. Vaccines (Basel) 11:1391

Tenchov R, Bird R, Curtze AE, Zhou Q (2021) Lipid nanoparticles– from liposomes to mRNA vaccine delivery, a landscape of research diversity and advancement. ACS Nano 15:16982–17015

Tousian B, Khosravi AR (2023) Chitosan-based pulmonary particulate systems for anticancer and antiviral drug carriers: a promising delivery for COVID-19 vaccines. Results Chem 6:101146

Tran T, May B, Ung T, Nguyen M, Nguyen T, Dinh et al (2021) Preclinical immune response and safety evaluation of the protein subunit vaccine nanocovax for COVID-19. Front Immun 12

Tuomela K, Levings MK (2023) Acidity promotes the differentiation of immunosuppressive regulatory T cells. Eur J Immunol 53:2350511

Wei J, Xue W, Yu X, Qiu X, Liu Z (2017) PH Sensitive phosphorylated chitosan hydrogel as vaccine delivery system for intramuscular immunization. J Biomater Appl 31(10):1358–1369

Wu B, Wang H (2023) Peptide assemblies as promising tumor vaccines: current platforms and progress. BIO Integr 4:45–51

Wu Y, Wei W, Zhou M, Wang Y, Wu J, Ma G, Su Z (2012) Thermal-sensitive hydrogel as adjuvant-free vaccine delivery system for H5N1 intranasal immunization. Biomaterials 33(7):2351–2360

Wu J, Sahoo JK, Li Y et al (2022) Challenges in delivering therapeutic peptides and proteins: a silk-based solution. J Control Release 345:176–189

Xu Y, Zhou Z, Brooks B, Ferguson T, Obliosca J, Huang J, Kaneko I et al (2022) Layer-by-layer delivery of multiple antigens using trimethyl chitosan nanoparticles as a malaria vaccine candidate. Front Immunol 13:900080

Xu C, Lei C, Hosseinpour S, Ivanovski S, Walsh LJ, Khademhosseini A (2022) Nanotechnology for the management of COVID-19 during the pandemic and in the post-pandemic era. Nat Sci Rev 9(10):nwac124.

Yang F, Shi K, Jia Y et al (2020) A biodegradable thermosensitive hydrogel vaccine for cancer immunotherapy. Appl Mater Today 19:100608

Yang P, Ju Y, Hu Y et al (2023) Emerging 3D bioprinting applications in plastic surgery. Biomater Res 27:1

Yu Y, Gao Y, He L et al (2023) Biomaterial-based gene therapy. MedComm (Beijing) 4:e259

Zheng X, Zhu J, Zheng C et al (2023) Dissolving microneedle arrays as a Hepatitis B vaccine delivery system adjuvanted by APC-targeted Poly (Lactic-co-Glycolic Acid) (PLGA) nanoparticles. AAPS PharmSciTech 24:42

Zhong M, Zhuang G (2017) Construction and immunological evaluation of immunostimulatory nanocomposites with vaccine potential. J Infect Dis Ther 5:1–4

Zhuo Y, Zeng H, Su C, Lv Q, Cheng T, Lei L (2024) Tailoring biomaterials for vaccine delivery. J Nanobiotechnol 22(1):480

Chapter 10
Biocompatible Nanocomposites for Diagnostic Imaging

Lakshmi Manoj, Suman M. Subair, Sandra Varghese, and K. R. Rakhimol

Abstract Biocompatible nanocomposites are hybrid organic–inorganic materials synthesized at the nanoscale, with the aim of improving their structural, functional and biological properties. They are about everything pertaining to new developments caused by diagnostic imaging. Nanocomposites can offer high mechanical strength, enhanced thermal and electrical properties, biocompatibility, as well as biodegradability at abandonment, and hence, they have a wide application field in the biomedical world, from tissue engineering to drug delivery and including bioimaging: In this chapter, synthesis, characterization, and application of biocompatible nanocomposites and their revolutionary promises in diagnostic imaging will be highlighted. The book also emphasizes some benefits associated with combining nanoparticles with polymers, particularly magnetic nanoparticles and quantum dots, in order to form advanced imaging agents. Promising nanocomposites have also gone ahead by assuring very good resolution and contrast in MRI, CT, and optical imaging modalities while offering possibilities for functionalized targeting and imaging with reduced cytotoxicity. In addition, organic polymer-inorganic hybrids broaden the horizon for finer imaging and therapeutic delivery. Some of the key challenges facing this domain today include the difficult concepts of biocompatibility and toxicity coupled with regulatory types of issues. Identification of such strategies might pave the way toward opening new avenues for clinical translation of these materials and pave the path toward future directions in this regard. This chapter describes the latest trends in innovative synthesis methods, including sol–gel processes, surface modification strategies, and functionalization, for improved drug delivery and imaging performance.

Keywords Biocompatible · Synthesis · Diagnostic imaging · MRI · Contrast agent

L. Manoj · S. M. Subair · S. Varghese
School of Nanoscience and Nanotechnology, Mahatma Gandhi University, Kottayam, Kerala, India

K. R. Rakhimol (✉)
School of Polymer Science and Technology, Mahatma Gandhi University, Kottayam, Kerala, India
e-mail: rakhimolkr@gmail.com

R. Balakrishnan et al. (eds.), *Biocompatible Nanocomposites*, Advanced Structured Materials 239, https://doi.org/10.1007/978-981-96-7450-3_10

10.1 Introduction to Biocompatible Nanocomposites

Biocompatible nanocomposites are composite materials of organics and inorganics engineered to interact safely with biological systems. Their hybrid structure comprises mechanical strength and stability of inorganic materials along with flexible and functional organic compounds, which make them superior. Such nanocomposites are endowed with unique features, namely enhanced imaging sensitivity, targeted functionality, multimodal capabilities, and biocompatibility. They will give a good improvement in contrast for imaging modalities like MRI, CT, and fluorescence imaging, enable targeted delivery, and can be combined into multiple imaging agents. Proper and accurate selection of biocompatible polymers or natural materials minimizes cytotoxicity and immune reactions. Biocompatible nanocomposites have horizons for advancement in medical diagnostics, as they enhance the resolution of images and provide real-time visualization, thus early and accurate detection of diseases such as cancer, cardiovascular diseases, or neurodegenerative disorders. These special composite materials, both inorganic and organic, have currently taken a shape by becoming a nanocomposite. What is done is to take a number of such materials, at least one of which has a dimension between 1 and 100 nm, and then combine them. The basic thing to achieve using the nanocomposites is to get interfaces between the nanocomponents and the matrix that are much higher when compared to the ones that can work at a much smaller level. Nanocomposites may be combined into biocompatible materials such as hydrogels, ceramics, or polymers. The plethora of techniques, such as sol–gel, self-assembly, chemical coprecipitation, and hydrothermal procedures, can be employed for producing the various kinds of NCs, including metal oxide polymers, metal–carbon nanomaterials, and metal oxide-metal nanohybrid matrices. Some of the many properties it has include electrical conductivity, flame resistance, redox reactivity, chemical resistance, greater optically clarity, and catalytic activity. The other new functions are added with polarity change on the surface. Further, in contrast to the bulk materials, nanocomposite materials provide several benefits which are found to be good for sensing applications, conversion, batteries, remediation, and energy storage. Due to flexibility concerning biologic functionalization, nanofeatures, and many other features, nanocomposites may be applied to a number of biomedical applications. Therefore, nanocomposites can be applied in medical devices, tissue engineering, stem cell regulation, imaging, targeted medication delivery, artificial implants, and the detection of pathogens and malignant cells (Khane et al. 2024). Bionanocomposites are classified based on application and composition; schematic representation of classification is given below.

Biocompatibility is the most critical factor in medical application because the materials and devices cannot cause irritation or inflammation in the patients, which defeats the objective of the treatment (Gobena and Woldeyonnes 2024; Omar et al. 2022). The important properties of bionanocomposites which suits them for biomedical applications are antimicrobial properties, biodegradability, biocompatibility, and mechanical properties. A schematic representation is given below (Figs. 10.1, 10.2 and 10.3).

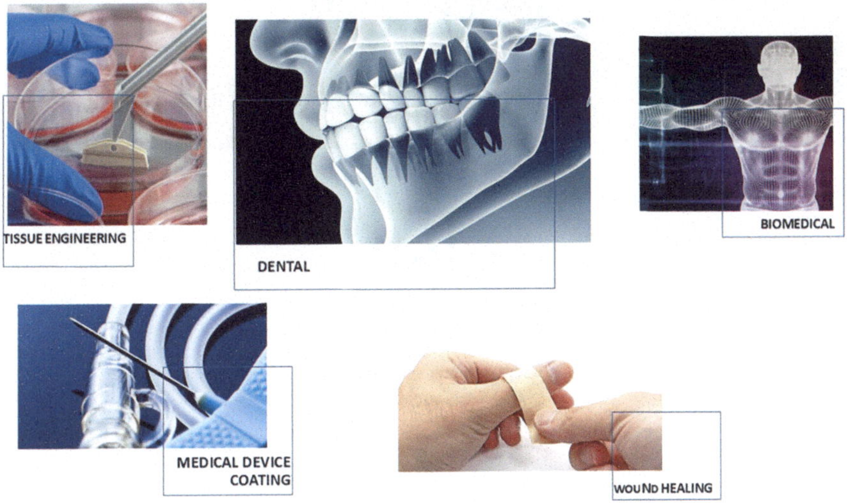

Fig. 10.1 Applications of nanocomposites

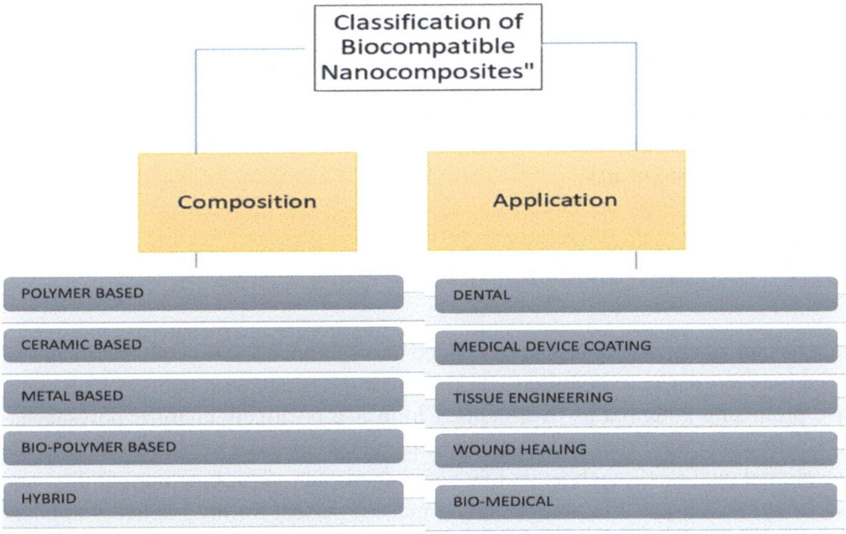

Fig. 10.2 Classification of nanocomposites

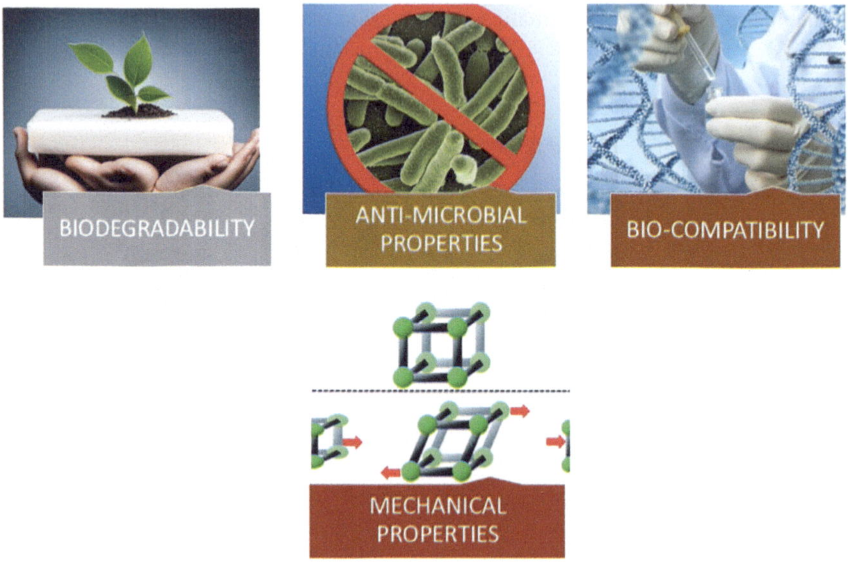

Fig. 10.3 Properties of bionanocomposites

10.2 Materials Used in Biocompatible Organic, Inorganic, and Hybrid Materials

The materials used in biocompatible nanocomposites can be classified into-

- Organic nanocomposites
- Inorganic nanocomposites
- Hybrid nanocomposites.

10.2.1 Organic Nanocomposites

Organic nanocomposites can also be divided into natural and synthetic. Natural products arise everywhere to be used to produce new drugs and treatments, which could have promising potential as treatments. Chitin is a positively charged polysaccharide from crustaceans that was partially deacetylated in order to obtain chitosan. Its exceptional qualities—such as its low cost, mucus adhesion, biocompatibility, and lack of cytotoxicity or immunogenicity—make it perfect for drug delivery applications. Silk fibroin has also been investigated recently as a natural biomaterial for tissue-engineered bone, cartilage, ligaments, and nerves, as well as a drug delivery substrate. Silk is a naturally occurring polymer material with a high resistance to deformation, and silk fibroin is a structural protein made from silk threads that have been separated from the cocoons of the silkworm Bobymexmori (Justina et al. 2016).

Because the properties of natural materials are different in material produced or processed at one time under consistent conditions can differ from batch to batch in its composition, structure, or properties, with such influences being environmental conditions at sourcing (temperature, humidity). Variations between raw materials (e.g., plant, animal, or mineral sources). Processing methods or differences in handling, their complete exploitation has been limited. However, the production of synthetic polymers provides the flexibility and reproducibility required for nanomedical applications. In dentistry, one of the most studied synthetic organic molecules is poly-lactic-co-glycolic acid co-polymer. Because of its excellent compatibility, Poly-Lactic-Co-Glycolic Acid (PLGA) has been approved by the U.S. Food and Drug Administration for use in drug administration, diagnostics, and other medical applications. Most importantly, when Poly-Lactic-Co-Glycolic Acid (PLGA) decomposes, it releases carbon dioxide and water (Justina et al. 2016).

Dendrimers are synthetic, highly branched polymers that behave like micelles and show promise in medicinal applications. They can be used as ligands for a variety of medical purposes, targeted carriers for antiviral or chemotherapeutic medications, gene delivery systems, and tissue healing. Dendrimers have few uses in dentistry, nevertheless, and the results vary depending on the dental area and polymer selection.

Solid lipid nanoparticles are indeed excellent drug delivery systems because of the physical stability, small size, great drug loading, huge surface area, very low toxicity, and enhanced transport of lipophilic active chemicals yet gelation proved a problem over the time as well as lack of scientific study on the nexus between these nanoparticles and the biology of the mouth cavity pose serious hindrances in the related use in the medical domain. Liposomes are spherical vesicles, shaped like small bowls that can have a diameter ranging from about 20 nm to several micrometers, and are biodegradable, non-toxic, and non-immunogenic. Their possible applications in periodontics are of interest, as they have already been used in studies to encapsulate medications for the treatment of intracellular infections. However, liposomes—especially those of a greater size—are not thought to be very stable. Of all the nanomaterials, some are more likely to be applied in clinical settings than others. For instance, chitosan and silk fibroin-based nanoparticles are more suited for tissue engineering, bone development, vascular tissue formation, and wound healing in dentistry, whereas Poly-Lactic-Co-Glycolic Acid (PLGA) nanoparticles can release minocycline over a 5 day period (Justina et al. 2016).

10.2.2 Inorganic Nanocomposites

The inorganic nanocomposites are the nanoparticle or nanostructured inorganic components which are dispersed in an inorganic or organic matrix. Such materials exhibit significantly enhanced properties of differing nature, including physical, chemical, thermal, and mechanical, due to their nanoscale dimensions, high surface area, and interfacial interactions. Their main attributes are thermal stability, better mechanical strength, optical and electronic property improvement, and chemical stability. Examples of this kind of composite include oxide-based nanocomposites, such as silica, titanium, alumina, silver, gold, and copper; ceramic; carbon-based nanocomposites such as graphene oxide and carbon nanotube composites; and chalcogenide-based nanocomposites. Iron oxides, cobalt, nickel, and clay are only some of the examples of magnetic nanocomposites. There are also several clay-based nanocomposites which are particularly used as packaging and barrier materials and other interesting perovskite nanocomposites used in solar cells and light-emitting devices. These can also be adopted in construction for high-temperature applications, biosensors, biosensors devices, and energy devices.

Magnetic nanoparticles (mNPs) are a very special class of inorganic nanomaterials and are made from magnetite or maghemite as the core material. They can process extra iron particularly kept in the human body's protein, ferritin. Iron oxide mNPs are beneficial to iron homeostasis through adsorption, excretion, and storage. The high superparamagnetic property of the superparamagnetic iron oxide nanoparticles causes significant magnetic moments when they are exposed to the external magnetic fields, which enhance the signal change per unit of particles. Inorganic iron "cores" have biocompatible surface modifications, give a stable performance under physiological conditions, and prevent aggregation. These modifications can be further functionalized by attaching a part of proteins, peptides, DNA or by attaching antibodies magnetic resonance imaging, or MRI, is an important imaging modality that produces very high spatial resolution contrast differences between tissues. In terms of the ability to optimize image intensity, mNP-based contrast agents obtained the best performance in virtue of their high magnetic moment. Iron oxide mNPs have been approved by the U.S. Food and Drug Administration for their use as MRI signal enhancers.

Research on the next generation of active targeted contrast compounds is still under progress, offering fascinating new possibilities for diagnosis, imaging, and therapy. The initial targeting agents were monoclonal antibodies, but because of their size, they may not diffuse well past normal biological barriers. A lot of research has been done on MRI agents that use mNPs to help diagnose solid tumors. One technique created with mNPs, specifically SPIOs, to treat malignancies is intracellular hyperthermia. In order to heat up the cell and cause necrosis, this treatment uses a delivery agent that functions as a nanoscale heater. An efficient treatment is offered by magnetic materials with a temperature range of 42–60 °C. mNPs can be directly injected into tumors to cause tumor regression, or they can be passively targeted through general biodistribution. Another method that affects the delivery of nucleic acids is magnetic transfection, which involves a magnetic field acting on conjugated nucleic acid vectors. Colloidal gold, also known as gold nanoparticles (AuNPs), is easily manufactured, multipurpose indicators, and detection probes. Because of their capacity to visualize based on the interaction between the particles and light, AuNPs are very appealing for use in biological imaging techniques. They are used in dental materials, textile fabrics, coating stainless steel materials, water treatment, and burn therapy medications because of their strong antibacterial activity against bacteria, viruses, and other eukaryotic pathogens. Since AuNPs have a high atomic number, they have been utilized as contrast agents in electron microscopy for many years. They can be applied to X-ray scattering, fluorescence microscopy, photothermal coherence tomography, optical microscopy in phase contrast or differential interference contrast mode, and multiphoton SPR microscopy. Immunostaining, in which antibody-conjugated AuNPs are intended to bind against antigens on fixed and permeabilized cells before being observed via TEM or light microscopy, is the most well-known use of AuNPs in bioimaging. Additionally, AuNPs are used as cell transport vehicles, such as when conjugating NPs with cancer-targeting moieties to send them into cancerous tissue. Gene guns, specific or non-specific endocytosis, cellular uptake, and nuclear translocation of AuNPs coated with nucleoplasm, are examples of delivery methods. More specifically, AuNPs are used in plataron-related sensing to register the presence of analyte molecules and provide a concentration readout. One of the most dependable intrinsic distinguishing characteristics of AuNPs that has been used for sensing is their plasmon resonance frequency. Under intense light, AuNPs show an anti-photo bleaching effect and possess flawless fluorescent properties. When AuNPs are stained on imaging cells, the fluorescence of the AuNPs is recorded while the cells are illuminated. QDs are colloidal crystals that are nanometers in size and contain elements from groups II to VI of the periodic table. The majority of them are bio-applications, including employing CdSe and being shielded from hazardous cadmium by a ZnS wall. Because QDs have strong, sustained light emissions, controllable fluorescence, and are less susceptible to photobleaching, they are useful for long-term cell monitoring (Khalid et al. 2020).

10.2.3 Hybrid Nanocomposites

Hybrid nanocomposites are created at the nanoscale by combining organic and inorganic components with the aim of gaining a different spectrum of synergistic properties. The inorganic material mainly imparts mechanical strength and thermal stability and/or specific optical or electrical properties, while organic provides flexibility, processability, or biocompatibility. Characterized mainly with tailored properties, high interfacial interaction, multifunctionality, and versatility, hybrid nanocomposite is preferred mostly when multi-qualities are needed. Some types of hybrid nanocomposites would include polymer-based, like polymer–clay; metal-based, like metal–organic frameworks; ceramic-based, like ceramic-polymer; and biomaterial-based, like chitosan-silica. Examples are perovskite-polymer, quantum dot-polymer, and magnetic hybrids, which are some of the well-known optical and electronic hybrid nanocomposites. They are commonly applied in biomedicine, energy, catalysis, and electronics, hybrid nanocomposites preferably suitable in optoelectronics, biomedicine, energy, environment, and construction material. Biomedicine applications will include drug delivery, tissue engineering, biosensing, energy-related applications in solar cells, fuel cells, supercapacitors, and batteries, and environmental uses of the applications. Next-generation energy storage will include solar-polymer hybrids in lithium-ion batteries and graphene-metal oxide hybrids in batteries and supercapacitors. Due to its ability to mix interesting features from different components, hybrid organic–inorganic nanomaterials have caught the interest of researchers working on the creation of novel and intelligent nanocomposite materials. These consist of a range of physical characteristics, high catalytic activity, and exciting optical qualities. It is still necessary to add an organic component or polymer matrix to improve the mechanical, thermal, and chemical properties. This method maximizes the interface between the components, shapes the NP filler's percolation threshold, and enhances the processability of the integrated polymer-based component. However, the synthesis of these nanomaterials is hampered by the incompatible properties that prevent these components from integrating (Khalid et al. 2020).

10.3 Synthesis of Biocompatible Nanocomposites

The previously mentioned idea states that biocompatible nanocomposites can be employed as a nanofiller, additive, or matrix platform, or both. In order to improve their performance and tailor them for specific applications, these nanocomposites are commonly modified chemically or mechanically.

Here, we'll briefly review some of the processes utilized to create biocompatible. The ultimate properties of nanocomposites are greatly influenced by the technique or processing method employed, regardless of other variables.

10.3.1 Synthesis Methods of Biocompatible Nanocomposites

10.3.1.1 Solution-Based Methods

Dissolving metal alkoxides or precursors into solution is a process of making biocompatible ceramics nanocomposites by sol–gel. Hydrothermal/solvothermal synthesis of high-temperature, high-pressure synthesis has produced nanostructures having well-defined morphology and crystallization properties. Electrochemical deposition is a technique of adding inorganic materials on substrate for the purpose of hybrid nanocomposites. Precipitation/co-precipitation causes nanoparticles to get embedded in polymers or biocompatible matrices.

10.3.1.2 Melt-Based Methods

In melt blending, polymers and nanoparticles are heated to high temperatures before mixing to obtain nanocomposites. Melt blending has an extensive application for thermoplastics. Hot-melt extrusion incorporates extrusion as well as blending to achieve a well-uniformed distribution of nanoparticles into its ideal polymer matrix.

10.3.1.3 In Situ Methods

In Situ Polymerization—Polymerization of monomers in the presence of nanoparticles, thus ensuring their uniform distribution within the polymer matrix. In Situ Growth—Growth of nanoparticles directly in a biocompatible matrix, oftentimes an improved particle–matrix interaction is reached.

10.3.1.4 Biotemplating and Biomimetic Applications

Biotemplating uses biological structures (e.g., plants, bacteria, or natural fibers) to template nanocomposites with a hierarchical architecture. Biomineralization mimics natural processes of mineralization in such a manner that biocompatible nanocomposites can be synthesized, e.g., hydroxyapatite for bone repair.

10.3.1.5 Physical Methods

Physical Vapor Deposition (PVD) evaporates materials onto a substrate to create ultrathin nanocomposite films. Laser Ablation produces nanoparticles through laser pulse ablation of a target material in a liquid medium. Mechanical Milling uses high energy ball milling to mix nanoparticles and matrix material for nanocomposites synthesis.

10.3.1.6 Advanced Techniques

3D Printing and Additive Manufacturing: This technique is meant to construct layers gradually, making it possible for precise control in architecture and composition of nanocomposites.

Microfluidics: Develop nanocomposites through controlled micro-environmental conditions which can be in full reproducibility and scalability.

Self-Assembly: Capitalizes on molecular interactions to assemble automatically nanostructures into hierarchical nanocomposites.

10.3.2 Sol–Gel Method

Template synthesis was one of the earliest bottom-up techniques for producing nanoparticles. Roy initially proposed it as a "gel route" in 1969 when he developed diphasic materials, like ceramic-ceramic nanocomposites, using inorganic gels as templates at temperatures far lower than their melting points. This greatly improved method is used to create hybrid organic–inorganic nanocomposites and other multi-component nanomaterials. A common technique for producing polymer-based nanocomposites is sol–gel. After hydrolysis, an inorganic substance is dispersed across an organic medium during the poly-condensation phase. The ultimate result, a polymer-based nanocomposite, is made by drying the mixers in a hot environment. The combined powder is then heated to form a nanocomposite. The sol–gel technique may be used to create porous aerogels, layered densified materials, and spun fibers that are lightweight and mechanically flexible. This environmentally benign method transformed the biopolymer back into ethanol by dissolving chitin in 1-butyl-3-imidazolium acetate using supercritical fluid technology. The mesoporous material ultimately exhibited significant porosity and low density. Because of the intricate solubility and chemical composition of bio-based components that undergo solubilization and regeneration, the sol–gel process can only be used on a laboratory scale to manufacture bio-renewable nanocomposites, despite being suitable for producing a range of nanocomposites (Khane et al. 2024).

10.3.3 Mixing the Solution

In fact, this procedure is quite similar to the sol–gel process; however, in this case, the dispersed matrix is first transformed into composite by coating, dipping, spraying, casting, or molding once it has been combined with the additives. However, sustainable development advocates argue that this method is not environmentally friendly because solvents are used in the recovery and evaporation processes. Because of this, there is now more interest in finding solvent-free or environmentally friendly substitutes (Khane et al. 2024).

10.3.4 Intercalation of Melt

Materials for nanocomposite construction are frequently made using this straightforward method. Usually, the process begins with melting the matrix phase, then fillers and additives are added, and finally the mixture is molded or cast into the required shape. This approach has many advantages over solution intercalation and intercalation polymerization. This technique involves combining melted matrix components, such as polymers, with reinforcements, such as nanoparticles. After that, extrusion and other secondary processing methods are applied to these polymer composites. This method has been used to create a few bio-renewable nanocomposites. One example of a bio-renewable composite material for biomedical purposes is PU thermoplastic, which is made up of polymer/halloysite nanotubes (HNTs) and long-chain polyols generated from fatty acids like linoleic acids. In another study, a twin-screw extruder and the melt mixing method were used to create nanocomposite materials. These materials demonstrated great potential for use in the electrical, electronic, and automotive sectors. Among these were carbon nanotubes and bioderived polyamides 4, 10. For the production of bio-renewables, the melt mixing method leaves a smaller chemical footprint than the solution mixing method since it is more ecologically friendly and utilizes less solvents (Khane et al. 2024).

10.3.5 Composite Processing Based on Dispersion

Several studies used simple synthesis techniques, such dispersion methods, to produce polymer composites with the required properties. These kinds of dispersion techniques combine reinforcing ingredients in a liquid using ultrasonic cation, assisted by a heated compression process (Khane et al. 2024).

10.3.6 Infiltration of Liquid

The first phase in this procedure is to combine the base matrix material with the metal reinforcing particles. Then, utilizing liquid infiltration technology, the base material and reinforcing particles are heated to a predetermined temperature. To eliminate the natural porosity of the composite, the generated materials are thermally treated at temperatures below the melting point of the underlying matrix material (Khane et al. 2024).

10.3.7 Spray Pyrolysis

The following are the five steps involved in making composite material: First, dissolve the matrix and reinforcing material precursors in an organic or inorganic solvent. Next, create a mist or fog out of the liquid composite material using an ultrasonic atomizer. The composite mists are then heated in a hot air oven. The composite materials in oxide form are produced by the vaporization process, which breaks down the heated sample (Khane et al. 2024).

10.3.8 Stir Casting

Adding hard reinforcing particles, such as SiC and TiC, to melted metal through stir-casting is an economical way to create metal matrix nanocomposites. Reinforcing particles are uniformly distributed by a vortex created by an impeller or stirrer covered in graphite. Metal oxides, carbon-based nanomaterials, and ceramic particles are frequently combined using this technique. But the primary problem is that nanoparticles aggregate because of their enormous surface area and poor wettability, and the mixer's effect causes porosity to form (Khane et al. 2024).

10.4 Surface Modification Techniques for Biocompatibility

Some of the methods applied in order to improve the biocompatibility of materials that are used in medical devices and implants include Plasma Treatment, Chemical Grafting, Thin Film Coating, Protein Immobilization, Physical–Mechanical Methods, Bioconvergence Modification, and Functional Composite Method. Plasma Treatment changes the surface properties through the application of ionized gases. The functional groups or molecules get attached by Chemical Grafting for better interaction with the biological tissues. In biocompatible materials, it would deposit a very thin film onto the substrate surface or attach specific proteins that may help in cell adhesion and growth. This reduces the immune response to this very small amount, allowing the better integration of tissues, as well as more perfect functions of the appliances used medically or in prosthetic devices (Roh et al. 2023).

10.5 Functionalization of Targeted Drug Delivery and Improved Imaging Performance

Functionalization of nanoparticles enhances drug delivery efficacy and targeting. Strategies include ligand conjugation, polymer coating, and pH-sensitive coatings. Ligand conjugation targets specific molecules on cancerous cells, while polymer coating uses biocompatible polymers for blood circulation and stability. pH-sensitive coatings react to tumor acidic environments, delivering medication precisely. Improved imaging performance leads to better patient outcomes and precise diagnoses. Key elements include High Resolution Resolve for capturing small details and improved camera contrast, and hybrid imaging, which uses multiple imaging modalities like PET-CT and PET-MRI. Advanced algorithms: Very sophisticated software which processes and analyzes imaging data in much more advanced ways. The face of medical diagnosis and treatment is rapidly changing to ensure better delivery of targeted drugs and unparalleled clarity in the internal structures (Saurwein et al. 2021).

10.6 Role of Biocompatible Nanocomposites in Diagnostic Imaging

The necessity for the new disease recognition and diagnosis stimulates the developments of imaging methods and contrast agents. Currently, existing imaging techniques remain inadequate in enabling rapid and reliable visualization of microanatomy of tissues and lesion characterization. Meeting this need is the creation of non-toxic contrast media with extended post-injection retention times. Nanoparticles are a promising technology for breaking some existing barriers in imaging and diagnostics (Andoh et al. 2024). The great size of nanoparticles is believed to provide prospects of mass interaction with a host of different molecules.

This property makes them suitable as excellent carriers for diagnostic agents that need to be delivered and interact efficiently. The nanoscale dimensions of the nanoparticles also help optimize the features of the nanocomposites, improving their strengths, gas barrier properties, catalytic activities, and drug delivery efficiencies. Some of these advances highlight the applicability of nanoparticles in various aspects including diagnosis, energy storage as well as in medicine (Han et al. 2019).

Nanocomposites are a distinct form of nanomaterial but nanomaterials can also be defined as an engineered material possessing distinct nanoscale features.

10.6.1 MRI Contrast Agents

About 10 million MRI examinations using contrast agents are performed on a global scale annually. In modern medicine, MRI with contrast agents is an important diagnostic tool (Wahsner et al. 2019). Magnetic Resonance Imaging of patients with malignancies was done around 5 million times last year in the United States, alone.

MRI contrast agents are a class reliable within the industry. MR images are improved through the use of these agents designed to distribute in the tissue and change the water proton relaxation time constant, which displays organ, degree of blood perfusion. Classification system for MRI examinations with contrast agents mostly.

Abstract composition classified by: metal composition, method of investigation, method of introduction, magnetic properties, effects have on the diagnosis, distribution in the body and clinical use.

1. *Chemical structure*: Metal presence/absence (e.g., gadolinium, iron)
2. *Administration route*: Intravenous, oral, intramuscular, or intra-arterial
3. *Magnetic behavior*: Paramagnetic, superparamagnetic, or ferromagnetic
4. *Imaging effect*: T1 enhancement, T2 enhancement, or dual-contrast
5. *Biodistribution*: Blood-pool, liver-specific, tumor-specific, or lymphographic
6. *Clinical application*: CNS, oncology, cardiovascular, hepatobiliary, or musculoskeletal imaging (Xiao et al. 2016).

Most MRI contrast agents incorporate elements containing lanthanide, notably gadolinium.

10.6.1.1 Superparamagnetic Fe_2O_3/Fe_3O_4 Nanoparticles

Superparamagnetic Iron (III) Oxide Nanoparticles (SPIONs) consist of tiny synthetic particles, typically 10–100 nm in diameter, composed of maghemite (γ-Fe_2O_3) or magnetite (Fe_3O_4) (Wahajuddin and Arora 2012).

Iron oxide-based nanoparticles exhibit remarkable potential for MRI contrast enhancement because of:

1. Intrinsic magnetic properties
2. Excellent biocompatibility.

10.6.1.2 Gadolinium-Based Agents (GBCAS)

With their unique water proton relaxation properties, GBCAs have established a prominent position in clinical MRI applications, regarded as safe and effective. The use of GBCA is vital for ensuring prompt and accurate diagnostic outcomes (Ranga et al. 2017). Currently majority enhancing agents are centered around chelates of Gd. Gd contains toxicity in its free form, and therefore, the synthesis of GBCAs involves forming chelates to facilitate the elimination of their toxic effects. The chelating process also regulates their pharmacokinetics (Parillo et al. 2024). Gadolinium's (Gd III) paramagnetic properties make it an indispensable tool for medical imaging and treatment monitoring, with widespread applications across various clinical specialties.

GBCAs are chemical substances, injected into the body during the magnetic resonance imaging (MRI) scans that enhances the clarity of the diagnostic studies. Additionally, this enables the medical imaging doctor to generate reports with greater precision (Parillo et al. 2024).

10.6.2 CT Contrast Agents

Computed tomography (CT) is a clinical application tool used for diagnosis; those are available at low cost in hospitals all over the world. CT is a clinical diagnostic tool that leads for 3D reconstruction. CT imaging faces difficulties in discriminating between soft tissues, normal organs, and tumors due to the overlap in their mass attenuation coefficients, leading to reduced contrast resolution.

Based on the components CT enhancing agents are:

Gold (Au)-nanomaterials.
Bismuth (Bi)-nanomaterials (Lai et al. 2024).

10.6.2.1 Gold (AU)-Based Nanomaterials

Au-based nanomaterials, boasting excellent biocompatibility and other desirable properties, have become a staple in medical practices. The optical, biological, and therapeutic behavior of Au nanoparticles is sensitive to their sizes; therefore, size optimization is crucial to a specific application (Lai et al. 2024). To further comprehend the relationship between the sizes of AuNP with their pharmacokinetics as well as the generation of contrast in CT, six different AuNP formulations, ranging from 4 to 152 nm, were prepared, characterized, and investigated here (Lai et al. 2024).

10.6.2.2 Bismuth-Based Agents Nanomaterials

Bismuth nanoparticles can be designed into advanced nanoagents that support multimodal imaging and therapy through strategic complexation with many functional agents, including small molecule drugs, fluorescent labels, and supporting chemicals. Metallic bismuth nanoparticles are biodegradable into non-toxic compounds called Bi (III), that are then excreted through the kidneys following renal excretion after completion of CT analysis (Yu et al. 2022).

10.6.3 Optical Imaging

Bioluminescence and fluorescence and other optical techniques are already beginning to emerge as very powerful tools for both diagnosis and therapy. The method of bioimaging in its foundational form—Optical imaging—provides a visualization power that cannot be paralleled. Its benefits are multifaceted: (1) It is non-invasive with a natural intuitive nature. (2) it covers 15, 12, and 8 orders, respectively, of temporal, intensity, and spatial scales, (3) at an extremely high-resolution level.

10.6.3.1 Fluorescent Nanoparticles

A photon hitting a molecule excites it to a more vibrational excited state, rapidly losing off excess energy as heat. Then, returning to the unexcited state, it fluoresces by emitting a photon that has a reduced energy in a process that takes approximately 10^{-8} s and stops right after an excitation source is taken away (Kirti 2017). Fluorescent NPs are usually used for localized molecule marking or processes outlining. Best optimized florescent NPs should have wavelengths between 700–750 and 750–800 nm in case of their excitation and emission wavelength respectively (Sun et al. 2010). Highly efficient nanoparticles must be used to develop the involvement regarding good signal-to-noise ratio.

Though it is one of the most versatile, it also consists of limitations those are which that, optical imaging (OI) alone is inadequate for accurate localization of NPs, requiring supplementary techniques to obtain high spatial resolution, etc. (Boschi and De Sanctis 2017).

10.6.4 Ultrasound Contrast Agents

They present different properties like speed of sound and density of the reflections generated upon reflection once an ultrasonic wave has been transmitted to the body. This morphological analysis of soft tissues has often been applied in the field of radiology as well as cardiology in conjunction with ultrasound imaging.

Ultrasound Contrast Agents are intravascular contrast media but also have applications in urinary bladder instillation for detecting ureteric reflux and uterine instillation to assess tubal patency (Cosgrove 2006).

10.6.4.1 Microbubble Contrast Agents

Microbubble contrast agents have opened up new possibilities for ultrasound, signaling the start of a new era. Minuscule 3 μm-diameter gas bubbles are injected directly into the veins. The safety of microbubble injection into the circulation, despite initial concerns, has been substantiated through extensive clinical experience, demonstrating that volumes below 200 μl are harmless and comparable to radiography and MRI contrast agent (Blomley 2001).

10.6.4.2 Polymer-Based Nanocomposites

The polymer-based nanomaterials are the most flexible and chemically resistant among all the types of nanoparticles (Solangi et al. 2024). Engineered polymeric microbubbles (MB) and nanomaterials, including nanocups and microcapsules, are known to show high drug loading, margination, and acoustic responses. This review provides recent progress in the chemistry of polymeric material designing, chemical tuning, and emerging applications for US-sensitive materials in diagnosis and therapy (Barmin et al. 2023).

10.7 Diagnostic Imaging Mechanism of Action

The diagnostic tool used is medical imaging, which produces pictorial images of the physiological processes or tissues and internal organs with normal and abnormalities (Hussain et al. 2022). For the accurate diagnosis, with the aid of electromagnetic radiation and certain other technologies, diagnostic imaging produces images of internal structures of the body. Medical image analysis involves applying various methods including image classification and segmentation on the diverse medical images, for example, Diagnostic x-ray scanning, CT imaging, Retinographies, and Ultrasound images.

Diagnostic X-ray imaging makes images through interactions between an X-ray beam and a patient. As X-ray beam travels through the body, its photons experience absorption and scattering by interaction with tissues of the human body; the amount absorbed depends on the density of the material in its passage.

(CT) Imaging is an imaging modality that uses X-rays to produce cross-sectional images, or 'slices,' of the body, reconstructed from measurements of X-ray beam attenuation. CT scanners use detectors to measure X-ray transmission, not image formation, with a thin beam (1–10 mm) scanning the body (Bell and Nadrljanski 2010).

10.7.1 Interaction of Nanocomposites with Biological Tissues and Cells

Nanotechnology gives a transformative boost to personalized medicine, offering unparalleled potential for improving the treatment of diseases. Biointeractions of nanoparticles can have unpredictable effects. Nanoparticles ($\leq$200 nm) are much easier to enter cells than micron-sized particles, which may still passively diffuse into cells. Nanoscale matters such as proteins that natural tissues consist of also interact with cells directly. The reason nanoparticles are highly taken up by cells is because of their numerous applications in nanomedicine, including tracking and imaging cells, targeting diseases, drug/gene delivery, disease diagnosis, and therapy, which are considerable due to functionalization and precise localization within cells.

10.7.2 Mechanisms Behind Enhanced Imaging Resolution and Contrast

Biology and medicine use various sources of energy, light, X-rays, ultrasound, etc., in imaging modalities. The merits and demerits of each modality lie in the physical and biological principles that govern the interaction between energy and tissue. This set of complementary methodologies makes for a very powerful diagnostic and research toolkit.

- *MRI* In this technique, a strong magnetic field is used in association with highly advanced pulse sequences that offer much more contrast due to tissue characteristics. Examples include diffusion-weighted imaging and functional MRI, which offer much greater resolution in some areas (Yusefi and Helfield 2022).

- *CT (Computed Tomography)* Multi-slice CT technology allows faster image acquisition and finer slices, thus improving spatial resolution. Techniques such as iterative reconstruction enhance the quality of images by reducing noise (Barentsz et al. 2006) 5.3. Targeting strategies for specific organs, tissues, or cells.

Delivering therapeutic agents to precise locations within the body, such as specific organs, tissues, or cell types, poses a major challenge in the treatment of various human diseases (Attia et al. 2019).

Nanomedicines

Aims to targeting specific cells, tissues, or organs, thereby minimizing systemic toxicity and side effects, and maximizing therapeutic efficacy. Also enables direct delivery of active therapeutic molecules to the site of action, reducing dosage requirements and adverse effects, and improving treatment outcomes (Himri and Guaadaoui 2018).

Gene Therapy

It refers to the introducing of a therapeutic gene into host cells that will be expressed and result in a desirable biological response.

Gene therapy, or the replacement of the abnormal gene by a healthy one, is considered to be the best approach toward cancer therapy that overcomes chemotherapy's deficiencies since it acts specifically but decreases non-specific toxicity.

10.8 Biocompatibility and Toxicity Considerations

The ability of a biomaterial to safely and effectively perform its intended biomedical function is known as biocompatibility, instead of eliciting a beneficial cellular or tissue response (Li et al. 2012). On the other hand, NP toxicity involves harmful effects on physiology and structure of organs/tissues, which is determined by many physiochemical parameters (Laurencin et al. 1990).

10.8.1 In Vivo and in Vitro Studies on Cytotoxicity, Immunogenicity, and Long-Term Effects

Cytotoxicity basically defines the damaging effect caused by an agent on the cells, irrespective of where they come from or whatever their type may be (Di Nunzio et al. 2017). In vitro toxicity describes laboratory experiments conducted on a substance outside the host organism and which examines its harmful effects due to some chemicals or drugs. At the moment, primary cells and co-cultures also become more frequently used by scientists for mechanistic action and toxicity studies that give a prognosis of a possible dangerous influence. In vivo studies these are tested in living organisms (Suhorutsenko et al. 2011).

Researchers assessed the effectiveness of an In Vitro and In Vivo Comparative Immunogenicity Assessment assay in predicting the immunogenicity of biotherapeutic products (Anderson 1999). This biotherapeutic assay detected sequence variants and single-point mutations, and its principles have been applied to in vitro assays using immune cells to evaluate the risk of rejection in transplant patients (Suhorutsenko et al. 2011).

10.8.2 Strategies to Improve Safety Profiles

10.8.2.1 Biodegradable Polymers

Products, particularly plastics, engineered for certain periods before a degradation process occurs that degrades the product into less harmful byproducts. Generally, biodegradable polymers provide environmental benefits that include raw material regeneration, biodegradation by microorganisms with reduced CO_2 emissions, and introduction of harmless byproducts to H_2O, CO_2, and CH_4 (Samir et al. 2022).

10.8.2.2 Surface Coating

This technique provides a straightforward and efficient way to alter the surface characteristics of a membrane. Surface coating enhances the performance of LIB cathodes by boosting capacity retention, thermal stability, and rate capability (Sánchez-Bodón et al. 2023).

10.9 Applications in Diagnostic Imaging

In medical science, diagnostic imaging is a technique of displaying the anatomy along with physiology of various organs and tissues from humans for medical purposes as well as in-depth research of both normal and abnormal anatomy and physiology (Hendee 1999). In addition to diagnosing anomalies and treating illnesses, diagnostic imaging system is employed to diagnose anomalies, cure illnesses, and reveal underlying structures beneath the skin and bones (1). A sector of medical imaging is the bedrock of healthcare science (Hendee 1999). Radiology, which make use of imaging devices such as X-ray radiography, X-ray computed tomography (CT), endoscopy, nuclear magnetic resonance imaging (MRI), magnetic resonance spectroscopy (MRS), positron emission tomography (PET), thermography, medical photography, electrical source imaging (ESI), digital mammography, tactile imaging, magnetic source imaging (MSI), medical optical imaging, single-photon emission computed tomography (SPECT), and ultrasonic and electrical impedance tomography are all part of this crucial aspect of biological imaging in general (Endo 2005; Feng et al. 2020).

10.10 Cancer Imaging: Early Detection and Tumor Targeting

Cancer imaging plays a pivotal role in diagnosing and treating multiple kinds of cancer. Doctors can see tumors and metastases in previously uncommon resolution due to advanced imaging techniques which involves MRI, CT, and PET scans. These technologies significantly enhance patient outcomes by enabling earlier diagnosis of malignant growths, frequently before symptoms arise (Farwell et al. 2014).

10.10.1 Computed Tomography (CT) Scan

CT (X-ray CT) is another name for computed tomography. Diagnostician, biologists, archeologists, as well as and many other scientists use CT scans to generate tomographic images of multiple objects that examine. Computed Tomography (CT) utilizes X-rays that traverse across the body. In contrast to traditional radiography, the latter employs a more sensitive X-ray detection equipment, develops images by dissecting sections of the body, and uses a computer to alter the data. The detectors and X-ray tube circulate around the patient. The most astounding feature of CT is its knack to visualize the most minute variations in X-ray absorption stages. The range of densities noted is approximately ten times greater than with traditional radiography. It is feasible to identify density gradients within soft tissues in addition to differentiating fat from other soft tissues. Sir Godfrey Hounsfield, developed the first computed tomography (CT) scanner in 1972 while as an electrical engineer. Hounsfield and Allan McLeod Cormack shared the 1979 Nobel Prize in Physiology either Medicine when Cormack, a physicist, created a comparable method at the same time (Andoh et al. 2024; Nicholls 2019).

10.10.2 Nuclear Magnetic Resonance Imaging (NMRI)

Nuclear Magnetic Resonance Imaging (NMRI) is a highly effective and non-contact tool for detecting and assessing cancer. In order to generate cross-sectional photographs concerning your anatomy from many perspectives, it uses radio waves and magnets. Unlike some other imaging modalities, it enables physicians to examine soft tissues in greater detail. There are quite many different types of cancer that can be detected with MRIs. Their capacity to identify cancers throughout different bodily sections can assist medical professionals in determining the tumor's size along with location. However, those kinds of cancer that an MRI can identify are limited (Andoh et al. 2024; Haris et al. 2015).

10.10.3 Positron Emission Tomography (PET) Scan

PET scan detects initial signs of neurological conditions, cardiac disorders, and cancer. In order to locate sick cells, a safe radioactive tracer is injected. 18F-fluorodeoxyglucose (FDG) has been employed in Positron Emission Tomography (PET) to quantitatively measure local glucose metabolism. Quantification of FDG uptake by PET facilitates the distinction between benign and malignant cancers, a determination of these degree of malignancy, the verdict of the effectiveness of chemotherapy or radiation treatment, and the prediction of prognosis because malignant tumors show enhanced glucose metabolism. PET examining is a potent diagnostic visualizing method that may identify a broad variety of cancers in the body. It functions by drawing attention to regions with elevated metabolic activity, which is frequently an indication of the spread of cancer. Hypermetabolic have also been demonstrated using PET. Malignant lymphoma, melanoma, breast, pulmonary, colon cancer, the pancreatic, and throat and head cancers can all be detected using PET imaging's high sensitivity. Additionally, it works well for patients whose primary tumor is unknown. As an outcome, PET imaging may be able to identify several malignancies in their early stages in a single investigation (Andoh et al. 2024; Yasuda et al. 2000; Zhu et al. 2011).

10.11 Cardiovascular Imaging: Vascular Imaging, Plaque Detection

Cardiovascular imaging is used in approximately 10 million patients annually, accounting for a significant portion of the almost 40 million people who are affected by cardiovascular disease. It serves as crucial in the process of determining therapeutic decisions. Cardiorespiratory visioning is significant because of the diagnosis, assessment, as well as cardiology treatment conditions, such as coronary artery disease, heart failure, heart valve complications, and hereditary cardiac defects. Physicians can evaluate heart function, blood flow, artery integrity, and the presence of plaque or clots using cardiovascular imaging techniques, which gives them vital information for managing and preventing disease (Manoel et al. 2024; Shaw et al. 2010).

10.11.1 Vascular Imaging

Vascular imaging is a subfield of diagnostic visualization that centers on assessing the form and function of capillaries, veins, and arteries. In order to diagnose and track vascular diseases such aneurysms, blood clots, atherosclerosis (plaque accumulation), and other conditions that affect blood circulation and artery integrity, this imaging is essential. Heart attacks, strokes, peripheral artery disease, and other conditions that might cause major consequences can be identified and treated with the use of vascular imaging (Shaw et al. 2010).

10.11.2 Intravascular Ultrasound (IVUS)

IVUS imaging is kind of interventional imaging that results in high-definition photographs of vascular structures via reflected ultrasound waves. IVUS images are generated using the varied electromagnetic resistance at the junction of various tissue complexes, just like many other types of ultrasound imaging. The three primary components of an IVUS visualization system are usually a pullback device to remove the catheter in models with an auto-pullback system, an imaging console with the hardware and software essential to convert the collected IVUS signals to images, and a catheter with a small ultrasound transducer or array. An IVUS catheter with a tiny transducer at its distal tip is positioned over a guidewire beyond the target lesion during IVUS imaging. The transducer that detects ultrasound is electrically stimulated while the catheter is inserted, and high-frequency ultrasonic waves—usually with a core frequency of 20–60 MHz—are generated and spread across various tissues. A pullback of the transducer tip through the lesion or segment of interest is then performed at a speed of 0.5–2.0 mm/s using the automatic pullback device. The continuously acquired reflection signals from the arterial wall will be utilized to generate a series of tomographic images of the vessel wall (Peng et al. 2021; Suh et al. 2011).

10.11.3 Optical Coherence Tomography (OCT)

OCT is a cutting-edge visioning methodology that generates cross-sectional images of living tissue's interior microstructure with great resolution. Although its initial medical uses were documented less than ten years ago, the technique's roots may be traced back to initial studies on white-light interferometry, which produced the one-dimensional (1-D) visual ranging method known as optical coherence-domain reflectometry (OCDR). OCT is capable of imaging minute blood veins and other structures as far underneath the surface as 1–2 mm in the skin and other tissues where they are widely distributed. The manageable simplicity and affordability of the apparatus that OCT complex is constructed on give it an advantage in comparison to high-frequency ultrasonic imaging, another technology that penetrates deep penetrating levels although has inferior. Notwithstanding these appealing characteristics, OCT requires to resolve a number of technical challenges before it may substitute existing diagnostic imaging modalities (Kwong and Yucel 2003; Shaw et al. 2010).

10.11.4 Plaque Detection

The three layers that make up the healthy coronary artery wall consist of the muscular tunica medium, the outer tunica adventitia, and the inner tunica intima. The inner tunica intima, that possesses lower echo genicity than its outer tunica adventitia, which is highly echo reflecting, generates atherosclerotic plaque. Nevertheless, the majority of atherosclerotic lesions observed in thin cap fibroatheromas (TCFA) demonstrate moderate echo genicity. The three layers can be easily identified because the tunica media, that which primarily composed of smooth muscle fibers, which looks dark in the black-and-white sonogram image since ultrasound waves are not deflected by it. Plaque detection is the process of identifying and measuring plaque accumulation in the arteries (Shaw et al. 2010; Suh et al. 2011; Xie et al. 2017).

10.11.5 Cardiac Catheterization

Cardiac catheterization is a medical visualizing technique given to visualize the inside concerning heart's blood vessels (coronary arteries). Calcium, cholesterol, lipids, and other materials build up on the artery walls to form this plaque. In this procedure, a thin tube called a catheter is put into a blood vessel in the wrist either groin and directed toward the heart arteries. An injection of contrast dye makes the arteries visible on an X-ray image. This allows doctors to identify regions of reduced blood circulation, resulting from plaque buildup, that could lead to heart conditions like angina or heart attacks. Coronary angiography helps guide treatment decisions, such as whether to perform angioplasty or surgery (CT or Invasive Coronary Angiography in Stable Chest Pain 2022; Shaw et al. 2010).

10.11.6 Neurological Imaging

Neurological imaging encompasses a range of techniques for seeing the physique and the physiology of the Central Nervous System. These techniques are vital for discovering and tracking anomalies, injuries, and neurological illnesses. Over the preceding ten years, an assortment of neuro visualizing procedures have been used to better understand the pathophysiology of challenging neurological illnesses and to give us biological signs for early diagnosis and therapy response (Yen et al. 2023).

10.11.7 Brain Targeting for Alzheimer's

Over the past forty years, imaging has been used in many ways to research Alzheimer's disease (AD). Neuroimaging is used to observe white matter, gray matter, and cerebrospinal fluid; assess unique brain activity; and detect aberrant protein accumulations. Positron Emission Tomography (PET) and Nuclear Magnetic Resonance Imaging (NMRI) are two neuroimaging procedures that are now in use globally (Yen et al. 2023).

10.11.8 Positron Emission Tomography (PET)

Positron emission tomography (PET) assess specific cerebrum activity or detects aberrant accumulations of Tau and amyloid protein via a tracer, a little, safe quantity of a radioactive compound. PET is utilized less often in clinical settings but more often in dementia research. PET scans for tau and amyloid are used to identify who participates part in research studies, estimate the effects of innovative medications intended to alter tau or amyloid pathways, and determine who could have been most in danger of developing Alzheimer's disease.

Different types of PET scans measure different proteins in the brain.

- Amyloid PET scans quantify aberrant beta-amyloid protein accumulation. The appearance of neuritic plaques, a defining feature of Dementia disease, is correlated with higher beta-amyloid levels.
- The protein tubulin associated unit, that causes Neuronal tangles in Alzheimer's disease and multiple other dementias, can be abnormally accumulated and detected by tau PET scans.

However, in PET imaging, radiation exposure is a frequent concern. There have been attempts to lower the quantity of radiotracer used in PET imaging in order to reduce this potential risk. But low-dose PET has low image quality and is naturally noisy (Zhu et al. 2019).

10.11.9 Nuclear Magnetic Resonance Imaging (MRI)

Nuclear Magnetic Resonance Imaging (MRI) is a method that is not invasive that lets us examine the brain without putting patients at risk for exposure to radiation. MRI is a sophisticated medical procedure that creates precise images of the anatomy of the cerebrum from various perspectives using a strong magnet, radio waves, an antenna, and a computer. Certain specialized MRI scans not only provide images of the anatomy of the brain but also images of its chemical science (MR spectroscopy) and function (functional MRI). Volumetric MRI (vMRI) scans allow experimenter to focus on specific brain regions and structures that are renowned for being crucial to memory enlargement and recall. In order to make an accurate diagnosis and gain a superior understanding of the replacements taking place in this particular cerebrum location, people can be contrast over time with either by themselves or in collaboration with others in different diagnostic classes.

However, those who have metallic implants, including pacemakers or cochlear implants, cannot get an MRI scan due to the size of the magnet (Yen et al. 2023).

10.12 Stroke Imaging

Stroke is one of the primary causes of disability and death in advanced countries, with an estimated 2.5% prevalence in adults overall. The rate of stroke escalates with age, surpassing an estimated 45% among individuals over 85 if silent infarcts are also taken into consideration. One stroke arises for every 15–18 casualties, rendering arterial ischemic stroke (AIS) one of the most common three causes of death in affluent nations. Nuclear magnetic resonance imaging and computed tomography are two methods for imaging strokes. The imaging mainstay that most hospitals have available constantly is CT. Acute cerebral hemorrhage could be promptly checked out via CT imaging. Additional benefits to this procedure are provided by the speed and recent technological developments in CT. In stroke CT, radiation exposure continues to be noteworthy, especially if contrast-angiography and perfusion-CT datasets are obtained (El-Koussy et al. 2014).

10.13 Current Challenges and Future Outlook

Along a broad range of ways and a great many scrutinies there is not even a feeling of being under any technology constraint concerning Biocompatible imaging techniques that can be accessed today. Thus, pre-clinical discovery precedes in pace more than the clinical translation of such discoveries. Generally, in U.S., PET radio-tracers are, by regulatory definition, considered new drugs and thus require filing an IND requisition alongside the U.S. Food and Drug Administration. Parallel radio-tracers abide by the same criteria as pharmaceuticals that are administered underneath IND. Novel Biocompatible imaging approaches will thus be continued to be adopted, as an avenue toward improving cancer diagnosis and treatment, independent of the regulatory structure in the United States throughout the world. Specifically, theranostics are expected to experience rapid growth in the near future. The increasing availability of theranostic drugs in cancer treatment will enhance AI's use in a number of essential applications (Andoh et al. 2024).

Medical image analysis has only recently started using the technology of DL to solve some of the key issues in CV visualizing. Anyhow, this transition from an artificially manufactured system to a trained for learning attributes from statistics is progressive. Remarkably, it is noticed that for most medical application studies, the techniques based on DL tend to achieve considerable performance. However, there are many challenges for AI experts in applying DL technology to the new tool for the diagnosis of vital CVD. To be specific, the dominant feature evulsion of deep learning brings new methods into being. The present application of this high technology in clinical practice still faces critical dilemmas (Manoel et al. 2024).

- Models perhaps overfitting or hasty generalized due to inadequacy of high-quality labeled drilling examples. Because of this, it is essential to increase generalization of the model across various situations, which is arduous and time-consuming.
- Clinical images involve ethical and legal concerns when used in commercial systems, so DL model performance could be impacted if this reliable information is not used (Manoel et al. 2024).

Deep learning is just one representative in diverse academies of artificial intelligence, which is unneeded does not mean it is the optimal algorithm. The maximum capabilities of deep learning should be appraised more reasonably. While looking to the future, most of these challenges are issues that will improve with the evolution of AI and ML. AI can support standardization in image analysis to reduce variability and improve diagnostics. Machine learning algorithms can also analyze large imaging data sets for pattern recognition and outcome prediction, potentially revolutionizing personalized medicine approaches. New imaging techniques are also being developed. Improved image quality and effectiveness could result from safer and more efficient contrast agents made possible by nanotechnology and novel materials (Manoel et al. 2024)

Within the last ten years, various computational techniques and appliances have been established with a view to detecting neurological disorders as early as possible. Precisely, Artificial Intelligence and more particularly Deep Learning is given to obtain detailed and computerized strokes detection outcome. Such findings are not a stand-alone feature of the identifying process. There are specific issues while implementing the computer-enhanced technique to acquire valid results:

- Detailed images are vital. The instruments which are to acquire these high-resolution brain images is very costly.
- Moreover, variance in neural tissues enlarge the trouble in recognizing the ischemic stroke.
- The existence of any older stroke area provides another challenge. This makes it difficult to distinguish between the old and new stroke regions (El-Koussy et al. 2014; Shaw et al. 2010).

Neuroimaging is integrating with other technologies like BCI and neurostimulation devices. Such integrations can enhance our ability in analyzing and treating neurological disorders. Techniques, namely fMRI and MEG are becoming more advanced; thus, the mapping of brain activity and connectivity are improved. Neuroimaging will tend toward personalized medicine as data acquired through imaging inform patient-specific treatments. As such, this will create an effective and targeted form of treatment for diseases, for instance, Alzheimer's, stroke, and epilepsy. In the development of non-invasive imaging technologies, patients are spared a variety of invasive procedures; instead, they are relaxed with greater comfort. Innovations in ultrasound, optical imaging, and wearable devices facilitate making neuroimaging much less cumbersome and accessible for the patients (Zhu et al. 2019).

10.14 Regulatory Aspects to Take into Account for Clinical Nanocomposites

While clinical translation of therapeutic nanocomposites creates novel possibilities for tailored because of the particular properties and possibility of harm affiliated with nanoparticles, medicine and improved patient results also require serious thought about many regulatory hurdles. Safety and efficacy come first. Concerns revolve around assessing the nanocomposite's potential toxicity and adequately analyzing it, and mechanism of pharmacokinetics and disposal procedures are important to be considered. Different regulatory agencies, including the US Food and Drug Administration (FDA) and the European Medicines Agency (EMA), have defined precise criteria for assessing safety and efficacy of nanomedicine, often demand intensive details on the production proceedings, quality control procedures and the description of the nanocomposite's physicochemical characteristics. Production and quality assurance are also crucial. The possibility of scalable and dependable creating nanocomposites for clinical research and potential market is condemning. To ensure consistency between batches and minimize any variability in the pharmaceutical product, regulatory agencies require strict quality control procedures. Additionally, it is important to establish the nanocomposite's exact dimension, structure, exterior feature, and arrangement to ensure optimal standard and observe latent dangers. The regulatory agency may require full testimony of the manufacturing and batch release criteria.

A nanocomposite-based therapy must balance the potential benefits with reported hazards, which a regulatory body does analyzing features including the extent of the illness, available treatments, and any adverse effects. Another topic is risk-reduction tactics to look at. It is very important for seeking regulatory approval to determine and reduce any risks associated with the nanocomposite, including environmental impact, immunological reaction, and off-target effects. Additionally, honorable conduct is a crucial factor to take into account. Participation in clinical trials requires informed consent involving nanocomposites. The sick should be knowledgeable about the therapy's possible threats and welfare including unreliability regarding nanomaterials. Transparency and transmission with sick person, medical executives, and the community concerning the growth and hazards of medications based on nanocomposite technology are necessary in order to build trust and confidence. Experimenter and originator can increase the possibility of outstanding clinical translation and bring the advantages of medications based on nanocomposite technology to patients by working with regulatory bodies and aggressively resolving these problems (Endo 2005).

10.15 Recent and Prospective Developments in Biocompatible Imaging

Sonography, Radiography, nuclear magnetic resonance imaging (NMRI), SPECT Scan, PET Scan, and light imaging have major current limitations on the basis of using numerous contrast media and radio-tracers for cancer imaging. This means that patients will greatly benefit from new technologies that are currently under development, which include PET Scanning powered by ion beams, cancer surgery directed by near-infrared fluoroscence, hyper-polarization NMRI and time-of-flight PET for rendition, positron emission mammography, and spectroscopy-intensified sigmoidoscopy for cancer screening. Nanocomposites may be able to enhance the domain of cancer diagnosis and medication. The current standpoint of fact-finding render important information about the potential applications of nanocomposites in the therapeutic evolution of this area. Additional development of those procedures, upgrading of the delivery system, and research into symbiotic partnerships with current and future clinical trials will target alternative therapy techniques. Future research is aimed at creating nanocomposites for particular cancer treatments in order to escalate therapeutics precision and reduce the consequence that accompany them.

Recent advancements have pushed magnetic resonance imaging significantly ahead, at least, and continue in the evolutionary trend of cardiovascular imaging. The central focus in advancing the field is myocardial perfusion imaging techniques and also 4D flow imaging, among other advances in spatial and temporal resolution. Trade-off in spatial and temporal resolution had thus far been the Achilles' heel of cardiac MRI. This difficulty has been resolved recently, improving both spatial and temporal resolution at the same time. In cardiac MRI, the incorporation of 4D flow imaging marks a paradigm change. This method not only makes it possible to visualize blood flow in three dimensions but also records the flow patterns' temporal change over time-the fourth dimension. Valvular pathologies, congenital anomalies in the heart, and vascular abnormalities have been quite successfully aided by the deployment of 4D flow imaging to gain in-depth insight into blood flow profiles. Enhancements ensure the assessment of myocardial perfusion imaging can now better distinguish cases of ischemic heart disease. Advanced techniques and first-pass perfusion sequencing also help in making adequate determinations of myocardial blood flow, as also facilitated by stress perfusion imaging with arterial spin tagging. These developments enable clinicians to detect minor perfusion abnormalities and, therefore help in early detection and intervention of coronary artery disorders. Non-invasive multimodalities are the way of the future for cardiovascular diagnosis. High temporal resolution and high precision for detecting coronary artery stenosis are the results of innovations in coronary CT angiography. This technology will keep expanding and provide a quicker way to identify substantial coronary lesions than invasive alternatives. The simulated computation of fractional flow reserve in plaque ascertain with CCTA is granting another growing innovation that complements CCTA, which includes CCTA FFR. Calculating FFR, formerly, was reserved for invasive coronary angiography. Anyhow, with the above innovations, it is now possible to detect flow-limiting coronary lesions with non-invasive techniques. This makes it possible to refer a specific group of patients for invasive coronary angiography in the event that the most effective medicinal treatment is not available.

Much has developed in fMRI technology, scanning faster and crisper. Methods such as Direct Imaging of Neuronal Activity, which provides millisecond-level temporal resolution, can better capture a more accurate brain activity. EEG has enhancements in the analysis of brain activity and neural oscillations. New techniques and algorithms increase the resolution as well as the accuracy of EEG recordings, making it a more potent tool for diagnosis of neurological disorders. In order to improve image analysis, automate activities, and produce quantitative results, AI and machine learning are fastly being incorporated into neuroimaging. In cases of abundant visioning modalities such as CT, PET, and MRI, the better picture of the structural and functional features of the cerebrum can be obtained for improved diagnostic precision and thus can help in establishing a personalized treatment plan for the patient. Non-invasive imaging techniques require less intra-operative procedure. It increases the comfort of the patient. Advances in wearable technology, ultrasound, and optical imaging are making neuroimaging less strenuous and more accessible to patients.

10.16 Conclusion

Biocompatible nanocomposites are promising agents to revolutionize diagnostic imaging and deal with the limitations traditionally related to resolution, contrast, and specificity. Combinations of organic and inorganic components at the nanoscale allow for up-to-date enhanced modalities of imaging, like MRI and CT, as well as targeted delivery with minimal toxicity. These advances open doors not only to precision diagnosis but also synergistic therapeutic applications, thereby underpinning the tremendous potential of nanocomposites in personalized medicine.

Despite these great promises, there are still substantial challenges, including scalability, regulatory hurdles, and long-term biocompatibility barriers toward clinical translation. The resolution of such issues would require an integrated approach toward material science, bioengineering, and medical research. Characterization, functionalization, and biocompatibility assessment protocols will be critical toward accelerating adoption in the clinical space.

They envision smarter, multifunctional nanocomposites that can have a capability for real-time diagnostics and therapy. Enhanced imaging agents, hybrid systems, and bio responsive materials are expected to continue redefining medical imaging. Since scientists, clinicians, and policymakers continue working closely together, biocompatible nanocomposites are poised to change healthcare and enhance the efficacy of early detection and treatment outcomes worldwide.

References

Anderson J (1999) Issues and perspectives on the biocompatibility and immunotoxicity evaluation of implanted controlled release systems. J Control Release 57(2):107–113. https://doi.org/10.1016/S0168-3659(98)00178-3

Andoh V, Ocansey D, Naveed H, Wang N, Chen L, Chen K, Mao F (2024) The advancing role of nanocomposites in cancer diagnosis and treatment. Int J Nanomed 19:6099–6126. https://doi.org/10.2147/IJN.S471360

Attia MF, Anton N, Wallyn J, Omran Z, Vandamme TF (2019) An overview of active and passive targeting strategies to improve the nanocarriers efficiency to tumour sites. J Pharm Pharmacol 71(8):1185–1198. https://doi.org/10.1111/jphp.13098

Barentsz J, Takahashi S, Oyen W, Mus R, De Mulder P, Reznek R, Oudkerk M, Mali W (2006) Commonly used imaging techniques for diagnosis and staging. J Clin Oncol 24(20):3234–3244. https://doi.org/10.1200/JCO.2006.06.5946

Barmin RA, Moosavifar M, Dasgupta A, Herrmann A, Kiessling F, Pallares RM, Lammers T (2023) Polymeric materials for ultrasound imaging and therapy. Chem Sci 14(43):11941–11954. https://doi.org/10.1039/D3SC04339H

Bell D, Nadrljanski M (2010) Computed tomography. Radiopaedia.org. https://doi.org/10.53347/rID-9027

Blomley MJK (2001) Science, medicine, and the future: Microbubble contrast agents: a new era in ultrasound. BMJ 322(7296):1222–1225. https://doi.org/10.1136/bmj.322.7296.1222

Boschi F, De Sanctis F (2017) Overview of the optical properties of fluorescent nanoparticles for optical imaging. Euro J Histochem 61(3). https://doi.org/10.4081/ejh.2017.2830

Cosgrove D (2006) Ultrasound contrast agents: an overview. Eur J Radiol 60(3):324–330. https://doi.org/10.1016/j.ejrad.2006.06.022

CT or Invasive Coronary Angiography in Stable Chest Pain (2022) N Engl J Med 386(17):1591–1602. https://doi.org/10.1056/NEJMoa2200963

Di Nunzio M, Valli V, Tomás-Cobos L, Tomás-Chisbert T, Murgui-Bosch L, Danesi F, Bordoni A (2017) Is cytotoxicity a determinant of the different in vitro and in vivo effects of bioactives? BMC Complement Altern Med 17(1):453. https://doi.org/10.1186/s12906-017-1962-2

El-Koussy M, Schroth G, Brekenfeld C, Arnold M (2014) Imaging of acute ischemic stroke. Eur Neurol 72(5–6):309–316. https://doi.org/10.1159/000362719

Endo M (2005) Recent progress in medical imaging technology. Syst Comput Jpn 36(3):1–17. https://doi.org/10.1002/scj.20154

Farwell MD, Pryma DA, Mankoff DA (2014) PET/CT imaging in cancer: current applications and future directions. Cancer 120(22):3433–3445. https://doi.org/10.1002/cncr.28860

Feng K, Wang Z, Yang Y (2020) Development of medical imaging sensors. Int J Distrib Sens Netw 16(1):155014772090360. https://doi.org/10.1177/1550147720903607

Gobena ST, Woldeyonnes AD (2024) A review of synthesis methods, and characterization techniques of polymer nanocomposites for diverse applications. Discov Mater 4(1):1–18. https://doi.org/10.1007/S43939-024-00119-0

Han X, Xu K, Taratula O, Farsad K (2019) Applications of nanoparticles in biomedical imaging. Nanoscale 11(3):799–819. https://doi.org/10.1039/c8nr07769j

Haris M, Yadav SK, Rizwan A, Singh A, Wang E, Hariharan H, Reddy R, Marincola FM (2015) Molecular magnetic resonance imaging in cancer. J Transl Med 13(1):313. https://doi.org/10.1186/s12967-015-0659-x

Hendee WR (1999) Physics and applications of medical imaging. Rev Mod Phys 71(2):S444–S450. https://doi.org/10.1103/RevModPhys.71.S444

Himri I, Guaadaoui A (2018) Cell and organ drug targeting. In: Nanostructures for the engineering of cells, tissues and organs. Elsevier, pp 1–66. https://doi.org/10.1016/B978-0-12-813665-2.00001-6

Hussain S, Mubeen I, Ullah N, Shah SSUD, Khan BA, Zahoor M, Ullah R, Khan FA, Sultan MA (2022) Modern diagnostic imaging technique applications and risk factors in the medical field: a review. Biomed Res Int 2022:1–19. https://doi.org/10.1155/2022/5164970

Justina M, Virlan R, Miricescu D, Radulescu R, Sabliov CM, Totan A, Calenic B, Greabu M (2016) Molecules organic nanomaterials and their applications in the treatment of oral diseases. https://doi.org/10.3390/molecules21020207

Khalid K, Tan X, Fatimah H, Zaid M, Tao Y, Lye Chew C, Chu D-T, Lam MK, Ho Y-C, Lim JW, Wei LC (2020) Advanced in developmental organic and inorganic nanomaterial: a review. https://doi.org/10.1080/21655979.2020.1736240

Khane Y, Albukhaty S, Sulaiman GM, Fennich F, Bensalah B, Hafsi Z, Aouf M, Amar ZH, Aouf D, Al-kuraishy HM, Saadoun H, Mohammed HA, Mohsin MH, Al-aqbi ZT (2024) Fabrication, characterization and application of biocompatible nanocomposites: a review. Eur Polymer J 214:113187. https://doi.org/10.1016/J.EURPOLYMJ.2024.113187

Kirti K (2017) Quantum dots as biological optical imaging tools. J Appl Biotechnol Bioeng 4(1). https://doi.org/10.15406/jabb.2017.04.00089

Kwong RY, Yucel EK (2003) Computed tomography scan and magnetic resonance imaging. Circulation 108(15). https://doi.org/10.1161/01.CIR.0000086899.32832.EC

Lai J, Luo Z, Chen L, Wu Z (2024) Advances in nanotechnology-based targeted-contrast agents for computed tomography and magnetic resonance. Sci Progress 107(1). https://doi.org/10.1177/00368504241228076

Laurencin CT, Pierre-Jacques HM, Langer R (1990) Toxicology and biocompatibility considerations in the evaluation of polymeric materials for biomedical applications. Clin Lab Med 10(3):549–570

Li X, Wang L, Fan Y, Feng Q, Cui F (2012) Biocompatibility and toxicity of nanoparticles and nanotubes. J Nanomater 2012(1). https://doi.org/10.1155/2012/548389

Manoel PZ, Dike IC, Anis H, Yassin N, Wojtara M, Uwishema O (2024) cardiovascular imaging in the era of precision medicine: insights from advanced technologies–a narrative review. Health Sci Rep 7(11). https://doi.org/10.1002/hsr2.70173

Nicholls M (2019) Sir godfrey newbold hounsfield and Allan M. Cormack. Euro Heart J 40(26):2101–2103. https://doi.org/10.1093/eurheartj/ehz421

Omar R, Khanwalkar G, Singh SK, Saxena S (2022) Biodegradable and biocompatible polymer and their applications as an antimicrobial nano composites. https://doi.org/10.48047/NQ.2022.20.20.NQ109148

Parillo M, Mallio CA, Van der Molen AJ, Rovira À, Dekkers IA, Karst U, Stroomberg G, Clement O, Gianolio E, Nederveen AJ, Radbruch A, Quattrocchi CC (2024) The role of gadolinium-based contrast agents in magnetic resonance imaging structured reporting and data systems (RADS). Mag Reson Mater Phys Biol Med 37(1):15–25. Springer Science and Business Media Deutschland GmbH. https://doi.org/10.1007/s10334-023-01113-y

Peng C, Wu H, Kim S, Dai X, Jiang X (2021) Recent advances in transducers for Intravascular Ultrasound (IVUS) imaging. Sensors 21(10):3540. https://doi.org/10.3390/s21103540

Ranga A, Agarwal Y, Garg K (2017) Gadolinium based contrast agents in current practice: risks of accumulation and toxicity in patients with normal renal function. Indian J Radiol Imaging 27(02):141–147. https://doi.org/10.4103/0971-3026.209212

Roh S, Jang Y, Yoo J, Seong H (2023) Surface modification strategies for biomedical applications: enhancing cell–biomaterial interfaces and biochip performances. BioChip J 17(2):174–191. https://doi.org/10.1007/S13206-023-00104-4

Samir A, Ashour FH, Hakim AAA, Bassyouni M (2022) Recent advances in biodegradable polymers for sustainable applications. NPJ Mater Degrad 6(1). Nature Publishing Group. https://doi.org/10.1038/s41529-022-00277-7

Sánchez-Bodón J, Diaz-Galbarriatu M, Pérez-Álvarez L, Moreno-Benítez I, Vilas-Vilela JL (2023) Strategies to enhance biomedical device performance and safety: a comprehensive review. Coatings 13(12):1981. https://doi.org/10.3390/coatings13121981

Saurwein A, Schaffer A, Wieser C, Rieger B (2021) Synthesis, characterisation and functionalisation of BAB-type dual-responsive nanocarriers for targeted drug delivery: evolution of nanoparticles based on 2-vinylpyridine and diethyl vinylphosphonate. RSC Adv 11(3):1586–1594. https://doi.org/10.1039/D0RA08902H

Shaw LJ, Min JK, Hachamovitch R, Peterson ED, Hendel RC, Woodard PK, Berman DS, Douglas PS (2010) Cardiovascular imaging research at the crossroads. JACC: Cardiovasc Imaging 3(3):316–324. https://doi.org/10.1016/j.jcmg.2009.11.010

Solangi NH, Karri RR, Mubarak NM, Mazari SA (2024) Mechanism of polymer composite-based nanomaterial for biomedical applications. Adv Industr Eng Polym Res 7(1):1–19. https://doi.org/10.1016/j.aiepr.2023.09.002

Suh WM, Seto AH, Margey RJP, Cruz-Gonzalez I, Jang I-K (2011) Intravascular detection of the vulnerable plaque. Circ Cardiovasc Imaging 4(2):169–178. https://doi.org/10.1161/CIRCIMAGING.110.958777

Suhorutsenko J, Oskolkov N, Arukuusk P, Kurrikoff K, Eriste E, Copolovici DM, Langel Ü (2011) Cell-penetrating peptides, PepFects, show no evidence of toxicity and immunogenicity in vitro and in vivo. Bioconjug Chem 22(11):2255–2262. https://doi.org/10.1021/bc200293d

Sun G, Berezin MY, Fan J, Lee H, Ma J, Zhang K, Wooley KL, Achilefu S (2010) Bright fluorescent nanoparticles for developing potential optical imaging contrast agents. Nanoscale 2(4):548. https://doi.org/10.1039/b9nr00304e

Wahajuddin, Arora S (2012) Superparamagnetic iron oxide nanoparticles: Magnetic nanoplatforms as drug carriers. Int J Nanomed 7:3445–3471. https://doi.org/10.2147/IJN.S30320

Wahsner J, Gale EM, Rodríguez-Rodríguez A, Caravan P (2019) Chemistry of MRI contrast agents: current challenges and new frontiers. Chem Rev 119(2):957–1057. American Chemical Society. https://doi.org/10.1021/acs.chemrev.8b00363

Xiao YD, Paudel R, Liu J, Ma C, Zhang ZS, Zhou SK (2016) MRI contrast agents: Classification and application (Review). Int J Mole Med 38(5):1319–1326. Spandidos Publications. https://doi.org/10.3892/ijmm.2016.2744

Xie Y, Jin H, Zeng M, Li D (2017) Coronary artery plaque imaging. Curr Atheroscler Rep 19(9):37. https://doi.org/10.1007/s11883-017-0672-z

Yasuda S, Ide M, Fujii H, Nakahara T, Mochizuki Y, Takahashi W, Shohtsu A (2000) Application of positron emission tomography imaging to cancer screening. Br J Cancer 83(12):1607–1611. https://doi.org/10.1054/bjoc.2000.1496

Yen C, Lin C-L, Chiang M-C (2023) Exploring the frontiers of neuroimaging: a review of recent advances in understanding brain functioning and disorders. Life 13(7):1472. https://doi.org/10.3390/life13071472

Yu H, Guo H, Wang Y, Wang Y, Zhang L (2022) Bismuth nanomaterials as contrast agents for radiography and computed tomography imaging and their quality/safety considerations. WIREs Nanomed Nanobiotechnol 14(6). https://doi.org/10.1002/wnan.1801

Yusefi H, Helfield B (2022a) Ultrasound contrast imaging: fundamentals and emerging technology. Front Phys 10. https://doi.org/10.3389/fphy.2022.791145

Zhu A, Lee D, Shim H (2011) Metabolic positron emission tomography imaging in cancer detection and therapy response. Semin Oncol 38(1):55–69. https://doi.org/10.1053/j.seminoncol.2010.11.012

Zhu G, Jiang B, Tong L, Xie Y, Zaharchuk G, Wintermark M (2019) Applications of deep learning to neuro-imaging techniques. Front Neurol 10. https://doi.org/10.3389/fneur.2019.00869

Chapter 11
Biomedical Nanocomposite Coatings and Their Regulatory Requirements

Neethu Ninan, Rani Shine Raju, Nandhana B. Nair, A. N. Ampadi, Nibu Varghese, and Rhema Grace Edwin

Abstract Biocompatible nanocomposite coatings are a promising field in biomedicine that enhance the performance and functionality of biomedical devices, offering potential for innovative medical therapies and providing solutions to healthcare challenges. These coatings are developed by integrating components such as natural polymers, biodegradable synthetic polymers, and nanomaterials, resulting in increased longevity, efficacy, and safety of biomedical devices in clinical settings. There are various types of nanocomposite coatings, including polymer-based, metal-based and ceramic-based systems, each offering unique characteristics like improved bioactivity, mechanical strength, corrosion resistance, biocompatibility, antibacterial activity, and thermal stability. These properties are vital for increasing cell adhesion, reducing infections, and minimising device-related problems. Their applications vary from therapeutics to diagnostics in medical devices, enhancing performance and increasing device longevity. As research progresses, the potential for further innovation in biocompatible nanocomposites grows, offering promising solutions for the biomedical field and beyond, shaping a more sustainable and technologically advanced future. This review focuses on the properties, types, uses, regulatory standards, and future directions of nanocomposite coatings for biomedical devices,

N. Ninan (✉) · R. S. Raju · N. B. Nair · A. N. Ampadi · N. Varghese
Pushpagiri Research Centre, Pushpagiri Institute of Medical Sciences and Research Centre, Thiruvalla, Kerala, India
e-mail: neethun.ninan@gmail.com

R. S. Raju
e-mail: ranishineraju2001@gmail.com

N. B. Nair
e-mail: nandhana.nair.888@gmail.com

A. N. Ampadi
e-mail: drampadian@pimsrc.edu.in

N. Varghese
e-mail: researchdirector@pushpagiri.in

R. G. Edwin
School of Biosciences, Mar Athanasios College for Advanced Studies, Tiruvalla, Kerala, India
e-mail: gracerhema71@gmail.com

R. Balakrishnan et al. (eds.), *Biocompatible Nanocomposites*, Advanced Structured Materials 239, https://doi.org/10.1007/978-981-96-7450-3_11

highlighting their potential to transform health care by enabling the development of a wide range of devices, from basic bandages to lifesaving implants.

Keywords Biocompatible nanocomposites · Coatings · Medical devices · Nanomaterials · Biodegradable polymers · Biomedicine

11.1 Introduction

Biomedical devices include a wide range of goods from simple bandages to sophisticated life-saving implants and diagnostic devices to treat and monitor health conditions in people (Chen et al. 2024a, b). They can be implanted, externally worn or used in surgical procedures. They include diagnostic devices (electrocardiograms, magnetic resonance imaging systems, or blood glucose monitors); therapeutic devices (pacemakers, dialysis machines, insulin pump); prosthetic devices (cochlear implants, prosthetic eyes, artificial limbs); implantable devices (heart valves, dental implants, or artificial joints); and surgical devices (forceps, robotic surgery systems, scalpels) (Wang et al. 2024; Liu et al. 2024). They often coordinate with the tissues or systems of the body to restore or improve physiological activities. Their development consists of a multifaceted approach integrating biology, medicine, materials science, and engineering.

Medical devices require biocompatible coatings that will enable them to interact with the body without leading to hostile reactions (Li et al. 2009). These coatings can prevent unwanted side effects like rejection or chronic inflammation which will compromise the function of medical devices. They also boost the integration of these devices with nearby tissues, endorsing healing and preventing problems like contamination or clotting (Coombes et al. 2000). Furthermore, biocompatible coatings can increase the durability and efficiency of the device by rendering a defensive barrier against degradation by biological agents or mechanical wear. These coatings are essential for the longevity, efficacy, and safety of biomedical devices in clinical settings (Kravanja and Finšgar 2022; Sahu et al. 2024; Pillai et al. 2025).

Nanocomposite coatings are innovative materials fabricated by integrating nanoparticles into a matrix, enhancing the coating performance. Such coatings are extensively used in the medical device sector due to their improved antibacterial, biocompatible, thermal, and mechanical characteristics (Fig. 11.1). These cutting-edge materials integrate nano-sized fillers in a polymeric or metallic interface to increase the efficiency of devices. Nevertheless, the usage of nanomaterials in biomedical devices presents exclusive regulatory trials.

This chapter delivers in-depth knowledge of the constituent components of nanocomposite, different types of nanocomposite coatings, their applications, and regulatory standards valid for these device coatings based on the frameworks recognized by principal regulatory bodies such as the European Medicines Agency (EMA), United States Food and Drug Administration (FDA), and other internationally approved agencies (Kashoki et al. 2020).

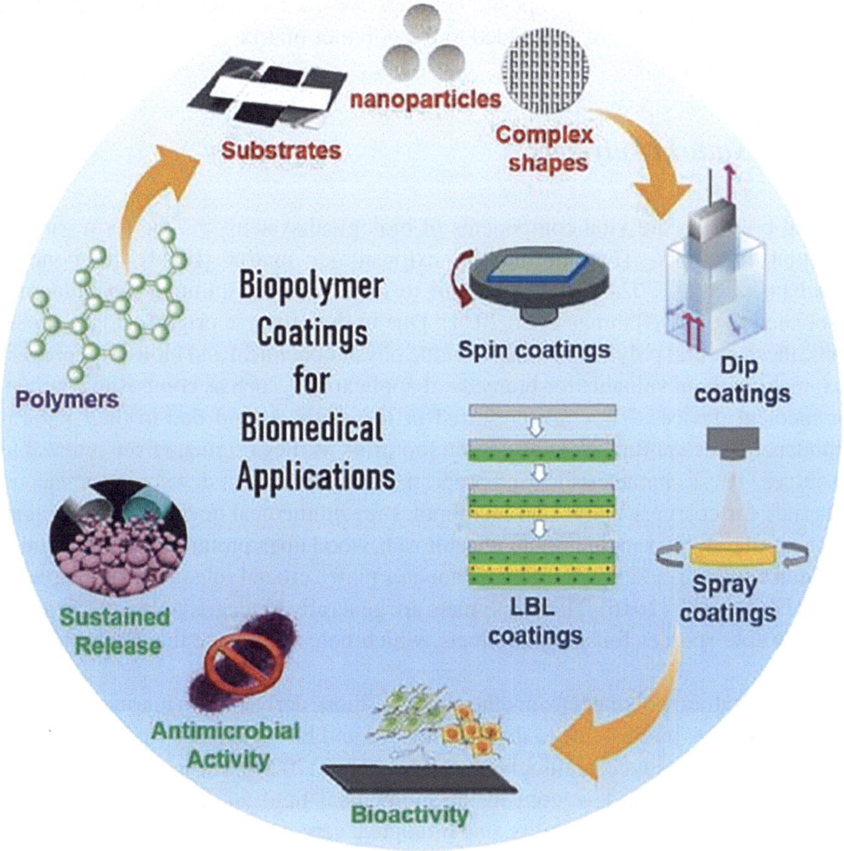

Fig. 11.1 Biopolymer coatings for biomedical applications. Reproduced from [Nathanael and Oh 2020] with copyright permission from MDPI

11.2 Constituents of Nanocomposite Coatings

Natural or artificial biodegradable polymers are combined with nanoscale materials to create nanocomposite coatings. For the development of biomaterials in the biomedical industry (Akintayo et al. 2011), natural and synthetic polymers offer a flexible platform. Natural polymers are naturally found in plants, animals, and microorganisms (Saleh 2021). These are derived from renewable resources like starch, chitosan, wood fiber, protein, cellulose, lignin, pullulan, and alginate. Synthetic biodegradable polymers are artificial polymers that are intended to decompose naturally over time in the environment through natural processes. Among the synthetic biodegradable polymers, some major types of polymers are used, such as biodegradable polyurethanes(PU), polylactic acid (PLA), polyvinyl alcohol (PVA), polybutylene succinate (PBS), poly(glycolic acid) (Haniffa et al. 2016). To improve the desirable characteristics of the nanocomposite coatings for various uses, different types of nanoscale components are also added to the polymer matrix.

11.2.1 Natural Polymers

Natural polymers are vital components of biological systems that perform various essential functions. They mimic the extracellular matrix (ECM) components (Satchanska et al. 2024), enabling them to avoid chronic immune reactions and associated toxicity (Rahman et al. 2024). Due to their diverse origins and compositions, these natural polymers possess unique physicochemical and biological properties, making them valuable for biomedical applications, such as composite coatings for medical devices. They are preferred in the medical field due to their natural abundance, renewability, and low carbon footprint, as they originate from renewable resources. These characteristics and properties are essential for developing advanced materials for coatings, films, and membranes for biomedical devices. Some important natural polymers including starch, chitosan, wood fiber, protein, cellulose, lignin, pullulan, alginate, and polyhydroxyalkanoates play a crucial role in various applications Haniffa et al. (2016). These polymers are generally biodegradable and can serve as renewable sources for many products, which helps to reduce the environmental impact.

Among natural polymers from other sources, those derived from marine resources are of increasing interest due to their abundance and biological activity (e.g., chitin/chitosan, alginate, glycosaminoglycans) (Bose et al. 2023). These marine biopolymers are becoming widely used in the biomedical field due to their availability, biocompatibility, biodegradability, and biological activity (Rodríguez-Jiménez et al. 2025). Some of their properties, such as solubility, viscosity, and surface characteristics, depend on their monosaccharide composition, linkages, chain shapes, and molecular weight (Aravamudhan et al. 2014). Furthermore, their physiological functions make them suitable for the development of composite coatings for biomedical devices. Therefore, the biocompatible bionanocomposite coatings having natural polymers can prevent unwanted side effects which enhances the function of medical devices. They also boost the integration of these devices with nearby tissues by showing similarity to the components of the extracellular matrix, endorsing healing and preventing problems like contamination or clotting (Coombes et al. 2000).

11.2.2 Synthetic Biodegradable Polymers

Synthetic polymers are defined as polymers that are artificially produced in laboratories, commonly referred to as manufactured polymers (Zhang et al. 2022). These man-made compounds are biodegradable synthetic polymers, which can break down spontaneously in biological settings, by providing advantages such as biocompatibility and biodegradability while eliminating the need for surgical removal in the biomedical field. The biomedical industry greatly benefits from these adaptable materials (Suggs et al. 2007). The development of bionanocomposite coatings requires the use of synthetic biodegradable polymers, which offer a flexible platform for application in tissue engineering, implants, and as coatings and films for medical devices. Since the body can naturally break down these polymers, thereby reducing the need for surgical removal, which reduces the risk of long-term complications. Over the last 20 years, synthetic polymer approaches have proven to be an efficient way to improve the structural and functional properties of bionanocomposites (Chiellini et al. 2003).

Synthetic biodegradable polymers are essential in cutting-edge biomedical applications because they enhance mechanical, thermal, and biological properties when paired with nanomaterials. These polymers include polylactic acid(PLA), polyvinyl alcohol (PVA), polybutylene succinate (PBS), poly(glycolic acid), each offering unique advantages tailored to specific medical needs (Haniffa et al. 2016). The most prevalent synthetic polymers utilized in biomedical applications are polylactic acid (PLA), polyglycolic acid (PGA), and their copolymers (Singhvi et al. 2019). Their popularity stems from several properties, such as biodegradability, a highly crystalline structure, hydrophilic polyester nature, outstanding mechanical properties, controlled degradation, high melting point, and low solubility in organic solvents (Khouri et al. 2024). These characteristics make them strong candidates for various biomedical applications.

11.2.3 Nanomaterials

Nanomaterials are defined as materials with a size range between 1 and 100 nm (Mabrouk et al. 2021). These nanoscale materials have emerged as an exciting class of materials that are in high demand for a range of applications, especially in the biomedical field. Nanotechnology involves the arrangement, design, and use of structures in medical devices and systems by controlling the shape and size of nanomaterials (Mabrouk et al. 2021). Due to the unique properties of nanomaterials, different types of biomedical devices can be developed and applied, which predominantly enhance the discovery, therapy, diagnosis, and prevention of diseases while lowering the risk of complications (Prasad 2018).

An important aspect of the biocompatible biomedical device approach is the integration of the nanostructures of materials with biological materials. The unique ability of nanomaterials to alter their structures and properties at the nanoscale makes them ideal for application in biomedical devices. Nanomaterials have significantly altered the properties and applications of nanocomposites due to their unique physical and chemical characteristics. The significance and type of nanoscale material depend on the purpose of the nanocomposite. The properties of the nanocomposite are influenced by both the quantity and shape of these nanomaterials (Haniffa et al. 2016).

Nanomaterials can be classified into two categories based on their origin: natural or synthetic (Khan 2020). Their applications in developing biomedical devices in the human body are governed by their mechanical, bioactivity, degradation, and biocompatibility properties. Natural or synthetic biodegradable nanoplatforms are at the forefront of biomedical device development in medicine. Many nanomaterials show great promise due to their suitable physicochemical properties. Nanomaterials are further divided into organic, inorganic, and carbon-based types, which are used in the fabrication of biomedical devices. Some examples of organic nanomaterials include liposomes, polymersomes, micelles, polymer constructs, and protein complexes. Inorganic nanomaterials such as Au, Ag, Si, TiO2, ZnO, and FeO2 are also commonly used, as well as carbon-based nanomaterials such as fullerenes, quantum dots, and graphene (Mekuye and Abera 2023). Additionally, several nanoparticles, such as natural fiber liquids, nanoscale cellulose crystals, graphene, and carbon nanotubes, are commonly used as fillers in the fabrication of nanocomposite coating films (Arora et al. 2018). Nanomaterials play a crucial role in the development of nanocomposite coatings for medical devices. The process involves several steps, beginning with the selection of nanomaterials based on the specific requirements of the application. Next, uniform dispersion of the nanomaterials within the polymer matrix is achieved through methods like solution blending, melt blending, or emulsion polymerization (Hong and Chen 2014). Surface functionalization is then applied to enhance compatibility and improve dispersion. Finally, the polymer–nanomaterial mixture is processed using techniques such as solution casting, spin coating, or extrusion to form the desired nanocomposite coating. These coatings are then used in biomedical devices to enhance their performance, including properties like durability, biocompatibility, and functionality.

11.3 Bionanocomposite Coatings

One or more discontinuous phases dispersed within a single continuous phase define a bionanocomposite. The discontinuous phase is known as the reinforcing material, whereas the continuous phase is known as the matrix (Haniffa et al. 2016). Biopolymers and nanoparticles serve as the constituent in the fabrication of nanocomposite coatings. Furthermore, bionanocomposites are used for three purposes: chemical functions, self-assembly of copolymer monomers, and controlled incorporation of nanomaterials.

Recently, several conventional and innovative techniques, including pressure extrusion, hot pressing, water evaporation, and suspension casting, have been used to create bionanocomposite coatings. Due to the certain limitations of suspension casting and water evaporation methods, such as excessive water evaporation, time-consuming pressure extrusion is recommended in a recent study (Habibi et al. 2010).

Coating films are divided into two basic categories: monolayer and multilayer films. A monolayer film can be utilized in coating technology to enhance unique features by altering the surface of materials. Multilayer films are created using layer-by-layer technology. In this technology, hydrogen bonds and electrical interactions act as driving forces between two layers that are deposited by solution dipping or spin coating (de Mesquita et al. 2010). However non-electrostatic interactions have an impact on the layer-by-layer assembly of two biomaterials. Examples include the preparation of chitosan- and CNC-based nanocomposite films.

11.4 General Properties of Bionanocomposite Coating

Natural or synthetic biodegradable polymers mixed with nanoparticles improve the barrier properties, mechanical properties (Rhim and Ng 2007), water solubility, functional properties, and thermal stability of bionanocomposite coatings (Wang et al. 2005). The addition of antimicrobial/antioxidant compounds to the biomaterial improves structural stability, microbial resistance, and barrier properties. For example, a chitosan-based nanocomposite with MMT clay particles and rosemary essential oils (as antimicrobial/antioxidant agents) exhibits improved water vaporization, superior mechanical characteristics, and higher water sensitivity (Wang and Wang 2007).

The general properties of nanocomposite coatings mainly depend on the characteristics of the constituent materials. However, the physical and chemical properties of these biomaterials can significantly alter the properties of the nanocomposite coatings. Bionanocomposite coatings are fabricated for their significance in various applications. In addition to biomedical devices, these coatings have a wide range of applications, and the properties may vary depending on the intended use. For biomedical applications, biocompatibility, degradation, and toxicity must be evaluated to ensure the safety and efficacy for patients. Key properties for medical devices include biocompatibility, bioactivity, corrosion resistance, low toxicity, thermal properties (Darder et al. 2007), and both mechanical and nonmechanical properties.

Bionanocomposites are designed with biocompatible materials (e.g., biopolymers and nanoparticles) to interact with biological tissues and minimize the risk of rejection, infection, or toxicity. Some are engineered to promote specific biological responses, such as cell adhesion, proliferation, or differentiation, which is referred to as bioactivity (Saini et al. 2017). Corrosion resistance is another critical property of metal-based nanocomposite coatings, as it helps prevent degradation in the body's harsh environment (e.g., exposure to bodily fluids) and ensures functionality without releasing toxins (Nazari et al. 2022). These coatings must be carefully engineered to ensure both biocompatibility and long-term safety for the success of these biomaterials in clinical applications. These coatings also provide thermal stability, making them resistant to temperature fluctuations caused by the body's natural temperature changes, thus maintaining their integrity and performance (Ge et al. 2025). Biomedical devices often experience temperature fluctuations due to the body's natural temperature variations.

Both mechanical and nonmechanical properties are crucial in the development and performance of bionanocomposite coatings for biomedical devices because they directly influence the device's biological interaction with the surrounding environment. While mechanical properties, such as strength and flexibility, are essential for durability, nonmechanical properties help ensure the device is compatible with human tissues, bioactive, and safe for long-term use (Xing and Tang 2022). Nonmechanical properties (Saini et al. 2017), such as transparency, optical qualities, flexibility, lightweight nature, moisture resistance, adhesion to the substrate, water vapor permeability, thickness, surface hydrophobicity, oil-holding capacity, conductivity, antimicrobial activity, UV-blocking properties, and barrier qualities, play a significant role in these interactions. The interplay of all these properties makes these coatings suitable for a wide range of biomedical applications, particularly in implantable and diagnostic medical devices (Haniffa et al. 2016).

11.5 Types of Nanocomposite Coatings

Biocompatible nanocomposites typically comprise a biocompatible polymer matrix reinforced with nanoscale materials. These materials can be derived from various materials, including metals, ceramics, carbon-based structures, or natural sources like biopolymers and minerals. Some important types such as biomimetic nanocomposite coatings, polymer-matrix nanocomposite coatings, metal-matrix nanocomposites coating, ceramic-matrix nanocomposites coating, hybrid nanocomposites coatings, diamond-like carbon nanocomposites coatings, antibacterial nanocomposite coatings are explained here (Table 11.1).

Table 11.1 Types of bionanocomposite coatings

Biomimetic nanocomposite coatings	Imitate the assembly and function of natural substances, including skin, bone, or teeth
Polymer-matrix nanocomposite coatings	Knee and hip replacements, stents, heart valves, wound dressings, and as tissue engineering scaffolds
Metal-matrix nanocomposites coating	Orthopedic and dental implants
Ceramic-matrix nanocomposites coating	Orthopedic implants, dental implants, surgical tools, and cardiovascular devices
Hybrid nanocomposites coatings	Protective films, medical devices and sensors, hip replacement, spinal cages, catheter coating, biosensors, wound dressing
Diamond-like carbon nanocomposites coatings	Surgical instruments, orthopedic implants, dental implants, and cardiovascular coatings
Antibacterial nanocomposite coatings	Long-lasting antibacterial effect in medical devices

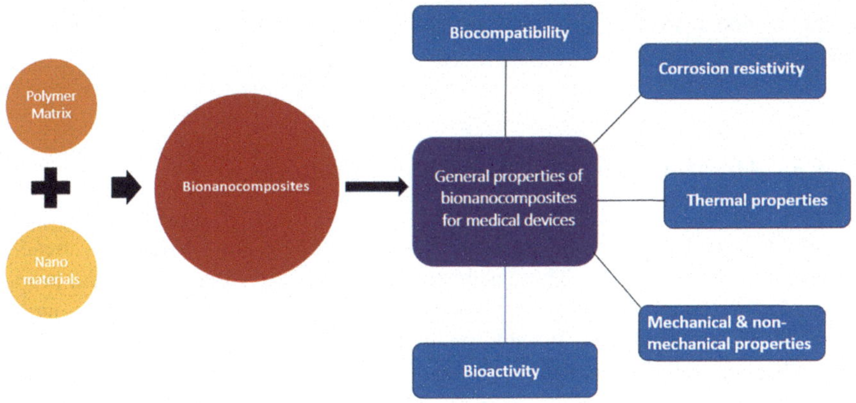

Fig. 11.2 Bionanocomposite and its general properties for medical devices

11.5.1 Biomimetic Nanocomposite Coatings

Biomimetic coatings are intended to imitate the assembly and function of natural substances, including skin, bone, or teeth (Lee et al. 2024). These coatings often embed nanoscale topographies that imitate the graded organization seen in living tissues, increasing the compatibility and efficiency of prosthetics and implants (Woeppel et al. 2023). By mimicking the natural extracellular matrix, these coatings enhance cell attachment, spreading, and differentiation, causing enhanced tissue incorporation and rejuvenation. Wang et al. have fabricated biomimetic nanocomposite coatings using bacteria-like nanoparticles loaded with aggregation-induced emission particles immobilized onto polyoxazoline-based plasma-coated surfaces (Fig. 11.2). Coatings inspired by the design of superhydrophobic lotus leaves are another example employed for antifouling uses.

11.5.2 Polymer-Matrix Nanocomposites Coatings

Polymer–clay nanocomposites are synthesized by dispersing layered silicate clays (montmorillonite) in the matrix of polymer to improve barrier properties, thermal stability, and mechanical stability. The effective intercalation of layers of clay with layers of polymers will lead to good reinforcement, increasing surface area for the transfer of stress (Teepakakorn and Ogawa 2022).

Polymer–carbon-based nanocomposites are formed by dispersing graphene or carbon nanotubes in the polymer matrices, increasing mechanical strength, thermal stability, and electrical conductivity (Zhang et al. 2025). Lightweight components showing high performance can be fabricated due to the high aspect ratio of carbon nanomaterials. They are used for making knee and hip replacements, stents, heart valves, wound dressings, and as tissue engineering scaffolds for applications in regenerative medicine.

11.5.3 Metal-Matrix Nanocomposites Coatings

Metal-matrix nanocomposites employ nano-sized metal grains of titanium, nickel, or copper (Cao et al. 2024).. These nanostructures offer increased mechanical properties like improved strength and hardness owing to the mechanism of strengthening of grain boundaries. In addition to that, the fine grain edifice can yield higher thermal stability and wear resistance, making them ideal for high-stress uses like automotive industries and aerospace. Ceramic nanoparticles including silicon carbide and aluminum oxide are also used as nano-fillers in a metal matrix. Dislocation movements caused by these nanoparticles can increase material strength and increase lifespan in punitive environments. These are mostly used for load-bearing applications like orthopedic and dental implants.

11.5.4 Ceramic-Matrix Nanocomposites Coatings

Ceramic-matrix nanocomposites are of two different types mainly oxide ceramic nanocomposites and non-oxide ceramic nanocomposites (Rashid et al. 2024). In oxide ceramic nanocomposite coatings, the matrix includes oxides such as zirconium dioxide or titanium dioxide in which nanoscale fillers are dispersed to increase thermal and mechanical properties. They are employed for high-temperature uses due to their increased hardness, resistance to oxidation, and thermal stability. Material brittleness is reduced using these nanostructures, which is a common issue faced with ceramics used traditionally.

Silicon nitride or silicon carbide is used in non-oxide ceramic nanocomposite coatings to obtain higher corrosion and wear resistance. Microstructure refinement is achieved by adding nanoparticles, promoting toughness and decreasing the propagation of cracks, which is essential for mechanical applications. Mostly, they are used in orthopedic implants, dental implants, surgical tools, and cardiovascular devices.

11.5.5 Hybrid Nanocomposites Coatings

Organic–inorganic nanocomposite consists of an organic polymer matrix and inorganic nanoparticle-based fillers for fabricating multipurpose coatings for increased optical, thermal, and mechanical properties (Abdel-karim 2023). The interaction between the inorganic and organic phases produces materials that can be custom-made for particular applications like high scratch resistance coatings, optical devices, and flexible electronics. Metal–polymer hybrid nanocomposites are formed when metal nanoparticles are integrated into polymer matrices, enhancing mechanical, antibacterial, and conductivity properties. These are used in protective films, medical devices, and sensors. They are used for hip replacement, spinal cages, catheter coating, biosensors, wound dressing, etc.

11.5.6 Diamond-Like Carbon Nanocomposites Coatings

These are advanced coatings using diamond-like carbon as a matrix and nanoparticles as fillers (Alakoski et al. 2008). Titanium or silicon nanoparticles are added to increase the hardness, wear resistance of coating and reduce friction. They have excellent biocompatibility, corrosion resistance, and tribological properties, promoting durability and efficacy. The incorporated nanoparticles can decrease residual stress and enhance adhesion to different substrates. They are used in surgical instruments, orthopedic implants, dental implants, and cardiovascular coatings.

11.5.7 Antibacterial Nanocomposite Coatings

These coatings contain nanoparticles of copper, zinc, silver, or other agents with known antibacterial properties. These nanoparticles can damage the cell membranes of bacteria, rendering lasting protection against bacterial infection. They can lyse an extensive range of bacteria, along with antibiotic-resistant strains like MRSA. The controlled release of antibacterial agents from their matrices provides them with a long-lasting antibacterial effect.

11.6 Application of Nanocomposite Coatings in Biomedical Devices

Nanocomposite coating films, combining biopolymers and nanomaterials, are increasingly used in biomedical applications to enhance the biocompatibility, durability, and functionality of medical devices (Phadtare et al. 2024). These coatings find applications in diagnostic, therapeutic, prosthetic, implantable, and surgical devices, contributing to advancements in healthcare technology. Applications for biomedical devices have been transformed by bionanocomposite coatings, which improve the devices' biocompatibility, safety, and performance. Their potential in health care is still bright, even though more study is being done to broaden their scope (Thirugnanasambandan 2022).

From simple bandages to complex life-saving implants and diagnostic tools, biomedical devices cover a broad spectrum of goods used to treat and track medical conditions. These devices can be utilized in surgical procedures, worn externally, or can be implanted. They include diagnostic devices (electrocardiograms, magnetic resonance imaging systems, or blood glucose monitors); therapeutic devices (pacemakers, dialysis machines, insulin pump); prosthetic devices (cochlear implants, prosthetic eyes, artificial limbs); implantable devices (heart valves, dental implants, or artificial joints); and surgical devices (Chen et al. 2024a, b) including forceps, robotic surgery systems, scalpels (Liu et al. 2024) (Fig. 11.3). In order to repair or enhance physiological functions, they frequently coordinate with the body's tissues or systems. They are the result of a multipronged approach that combines engineering, materials science, biology, and medicine (Fig. 11.4).

11.6.1 Diagnostic Devices

Bionanocomposite coatings have significant potential to enhance diagnostic devices due to their unique properties, such as biocompatibility, sensitivity, and durability. These coatings improve biomarker detection by enhancing the binding efficiency of specific biomolecules, leading to more accurate diagnostics. They also provide beneficial surface modifications, such as increased surface area and electrical conductivity, which improve the performance of biosensors (Rather 2020). Additionally, bionanocomposite coatings can incorporate antimicrobial nanoparticles, such as Ag, Au, Fe_3O_4, ZnO, and TiO_2, to prevent microbial contamination, ensuring device sterility and reliability (Bardhan 2022). By integrating natural polymers and nanoparticles, these coatings improve biocompatibility, making diagnostic devices safer for prolonged contact with biological tissues. They can also offer self-cleaning properties, reducing the need for maintenance and enhancing the long-term functionality of devices. Wearable diagnostic devices, such as sensors for monitoring vital signs or disease biomarkers, benefit from the flexibility, durability, and enhanced sensing capabilities provided by these coatings (Rebecca et al. 2024). Overall, bionanocomposite coatings are integral to advancing diagnostic technology, offering improved sensitivity, reliability, safety, and user comfort.

Diagnostic devices play a critical role in modern health care, providing essential data for disease detection, monitoring, and management. However, the performance of these devices is often limited by factors such as sensitivity, accuracy, biocompatibility, and durability. Bionanocomposite coatings, which integrate the unique properties of both biological and nanomaterials, have the potential to overcome many of these limitations. These coatings are particularly relevant for devices like electrocardiograms (ECGs), magnetic resonance imaging (MRI) systems, and blood glucose monitors (BGMs), where material performance is critical to accurate diagnosis and patient safety.

Electrocardiograms (ECGs): Electrocardiograms (ECGs) are commonly used to monitor the electrical activity of the heart (Sunwoo et al. 2020). The device relies

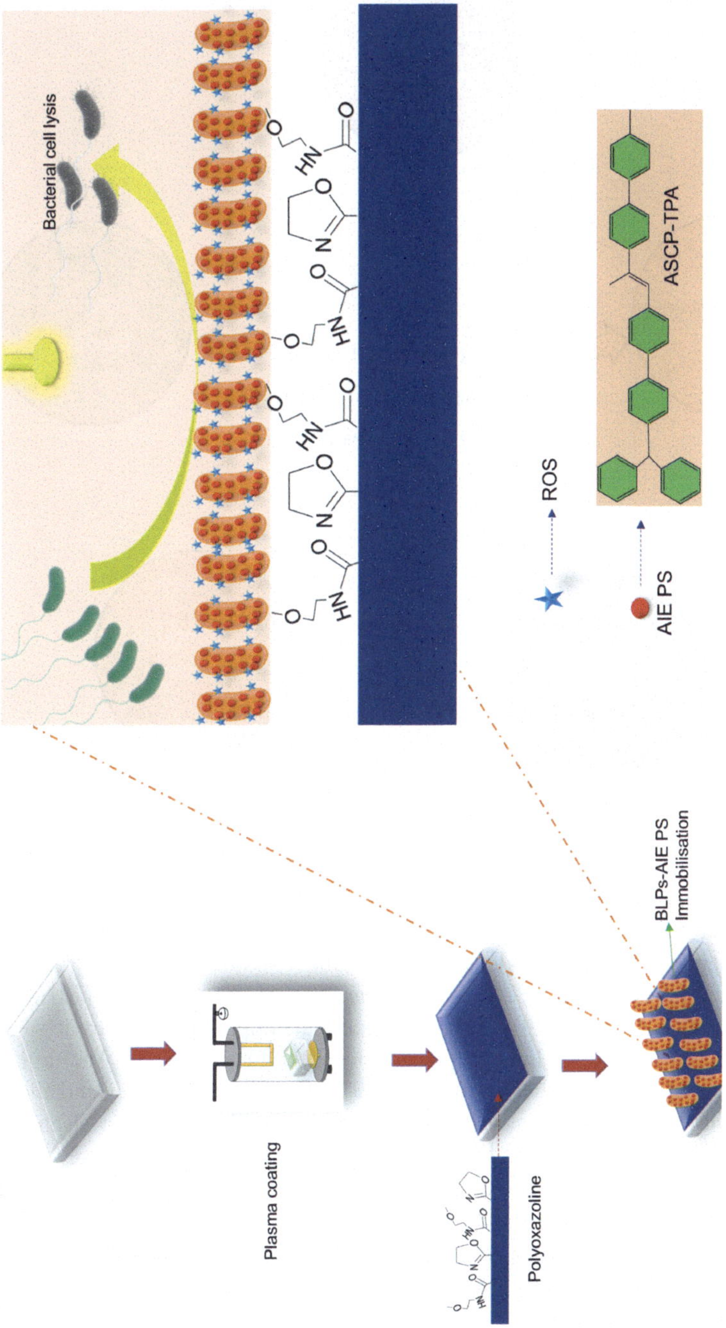

Fig. 11.3 Biomimetic nanocomposite coatings with bacterium-like particles. Reproduced from (Wang et al. 2024) with copyright permission from ACS

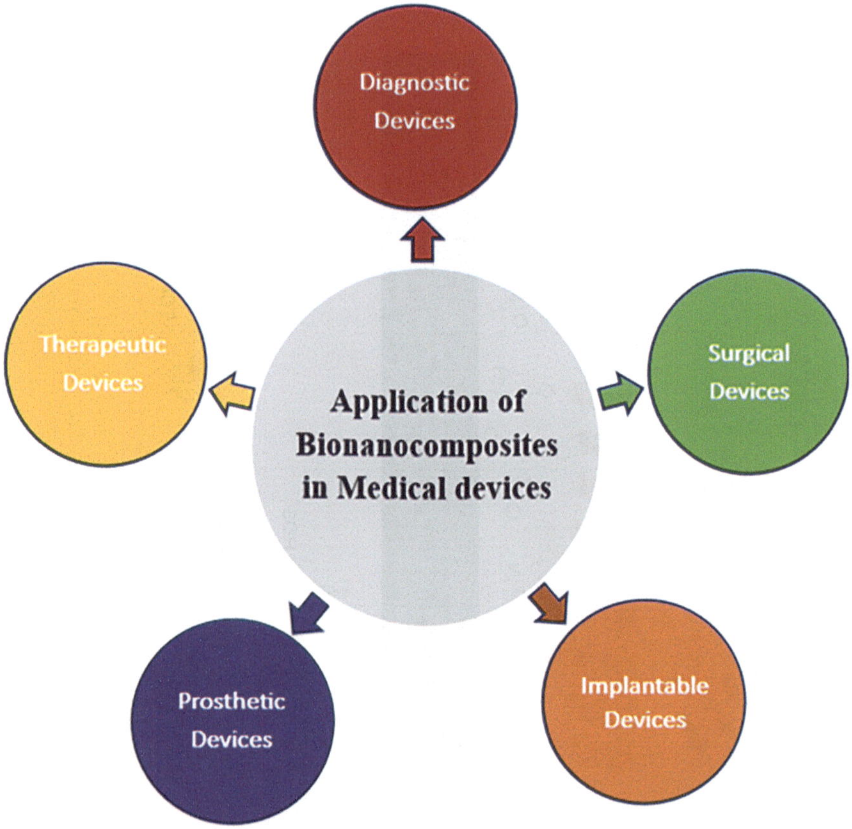

Fig. 11.4 Applications of bionanocomposites in biomedical devices

on electrodes that must exhibit excellent conductivity and biocompatibility to ensure accurate and reliable readings. Bionanocomposite coatings can significantly enhance the performance of ECG electrodes by incorporating conductive nanomaterials such as carbon nanotubes, graphene, or silver nanoparticles into biopolymers like chitosan or alginate (Ashtari et al. 2019). These nanomaterials improve the electrode's electrical conductivity, resulting in more precise electrical measurements. Additionally, the biocompatible nature of these coatings helps minimize skin irritation and allergic reactions, which are common issues during prolonged ECG monitoring (Alshamrani 2022). Furthermore, antimicrobial properties can be incorporated into these coatings to prevent bacterial growth on the electrode surface, improving patient safety and reducing the risk of infections. Nanocomposite coatings increase the conductivity of ECG electrodes, enhance patient comfort by reducing skin irritation, and enable precise cardiac signal recording (Ashtari et al. 2019).

Magnetic Resonance Imaging (MRI) Systems: Magnetic resonance imaging (MRI) is a noninvasive imaging technique used to visualize the body's internal structure. The system utilizes radio frequency (RF) signals and magnetic fields, where high sensitivity and signal fidelity are essential for obtaining high-resolution images. Bionanocomposites can enhance MRI performance through various methods. For example, MRI contrast agents can be coated with magnetic nanoparticles, such as iron oxide, to improve their magnetic properties and the quality of the resulting images.

Additionally, nanomaterials can enhance the amplification of RF signals or minimize signal loss during imaging, leading to clearer and more detailed scans. For MRI-compatible implants, such as pacemakers or neurostimulators, bionanocomposite coatings improve their biocompatibility and mechanical stability, reducing the risk of implant rejection and promoting better interactions with surrounding tissues. Biocompatible polymer-coated nanoparticle-based contrast agents improve imaging stability and specificity. These advancements significantly enhance the diagnostic capabilities of medical imaging (Alshamrani 2022).

Blood Glucose Monitors: Blood glucose monitoring is crucial for individuals with diabetes, as it tracks blood sugar levels and directs treatment (Weinstock et al. 2020). By incorporating conductive nanoparticles, such as gold, silver, or platinum, into biocompatible polymers, bionanocomposite coatings can greatly improve the functionality of glucose sensors. The performance of glucose monitors depends largely on the sensitivity and accuracy of the sensors used to measure glucose concentrations. Bionanocomposite coatings can significantly enhance sensor performance by incorporating conductive nanoparticles, such as gold, silver, or platinum, into biocompatible polymers. These coatings improve the electrochemical properties of the sensors, resulting in glucose readings that are more precise and sensitive. Additionally, the mechanical durability of the sensors is increased by the addition of nanomaterials, which increases their resistance to wear and tear from frequent use. The biocompatible nature of these coatings also reduces the risk of interference or adverse reactions when the sensor interacts with the blood sample, ensuring reliable results for patients. Nanomaterials like carbon nanotubes, when integrated into biopolymers, enhance sensitivity and enzymatic activity in glucose sensors, further increasing the precision of blood glucose monitoring devices (Farajollah et al. 2024).

According to recent research, bionanocomposite coatings have received significant attention in the development of diagnostic devices over the last two decades. These coatings show promise as a subject of study that could lead to the creation of high-performance devices. One prominent area of research in bionanocomposite coatings is their use in biosensors for disease detection, particularly by incorporating various nanomaterials capable of detecting disease-associated biomarkers (Rather 2020). Bionanocomposite coatings are also crucial in the development of portable diagnostic devices (Zarei 2017) and advanced imaging technologies (Alshamrani 2022). There is a growing interest in developing eco-friendly bionanocomposites aimed at improving scalability, biocompatibility, and multifunctionality, paving the way for innovative, environmentally friendly, and affordable healthcare solutions.

11.6.2 Therapeutic Devices

Bionanocomposite coatings have enormous promise for improving therapeutic devices, providing a variety of benefits. One of the key benefits of bionanocomposite coatings is their ability to increase the biocompatibility of therapeutic devices, making them safer for extended interaction with biological tissues, and increasing therapy efficacy while reducing the risk of rejection or inflammation (Haniffa et al. 2016). The use of nanoparticles such as silver, zinc oxide, or titanium dioxide can also provide antibacterial properties, lowering the risk of infection in implanted therapeutic devices. Bionanocomposite coatings have the potential to significantly improve the performance and safety of medical devices such as pacemakers, dialysis machines, and insulin pumps. These coatings can improve biocompatibility while also increasing the overall longevity and functionality of the devices, which is essential for long-term patient use.

In addition to these advantages, bionanocomposite coatings can help preserve the integrity of therapeutic devices over time. For example, if the coating on an insulin pump or pacemaker wears is damaged, a self-healing material can repair itself, increasing the device's lifespan and minimizing the need for maintenance or replacement. It can also increase safety, functioning, and patient comfort, resulting in better therapeutic outcomes and a higher quality of life for patients (Borhani et al. 2018).

Pacemakers: Pacemakers are frequently implanted in the body for long-term use in regulating heart rhythms (Gradaus et al. 2003). Bionanocomposite coatings can increase the biocompatibility, hemocompatibility, and corrosion protection of pacemaker components (Gradaus et al. 2003), reducing the risk of inflammation and tissue rejection. The use of antimicrobial nanoparticles such as silver or zinc oxide can prevent bacterial infections at the implantation site, which is critical for avoiding complications and side effects. Furthermore, these coatings can limit the possibility of electrode corrosion, increasing the device's endurance and performance over time.

Dialysis Machines: Dialysis machines used to treat renal failure must perform precise filtration and interact safely with the patient's blood. Bionanocomposite coatings can be applied to the surfaces of dialysis equipment to increase its biocompatibility and prevent adverse reactions when in contact with blood (Gradaus et al. 2003). Furthermore, these coatings can improve the mechanical properties of the dialysis process, hence increasing its overall reliability and efficiency.

Insulin Pumps: Diabetic people utilize insulin pumps to maintain a constant supply of insulin (Renard et al. 2007). Bionanocomposite coatings can improve insulin pump performance by increasing biocompatibility, ensuring that the device interacts safely with the patient's skin and tissues. Furthermore, these coatings can reduce friction, making them more comfortable to use when worn for extended periods. The coatings can also improve the durability of the pump's components, resulting in consistent insulin delivery over time (Schaepelynck 2019).

Recent investigations have highlighted the use of bionanocomposites in the development of therapeutic applications (Dutt et al. 2023). One of the most important fields is targeted drug delivery systems, where current research has focused on stimuli-responsive bionanocomposites that release therapeutic agents in response to pH, temperature, or specific enzymes, leading to more effective therapy (Xiao et al. 2023). In tissue engineering, bionanocomposites are used to create scaffolds that mimic the extracellular matrix (ECM) (Covarrubias et al. 2018). Another key application is the development of implantable therapeutic devices. Studies have shown that bionanocomposites are effective in the development of durable, biocompatible implants, such as pacemakers, dialysis machines, insulin pumps, and cardiac stents. Additionally, eco-friendly and biodegradable bionanocomposites are being developed to promote sustainability and reduce environmental impact (Haniffa et al. 2016; Song et al. 2018). These materials are particularly beneficial for creating temporary therapeutic devices that degrade safely in the body once they have fulfilled their purpose. Recent advances in bionanocomposites have paved the way for innovative therapeutic devices that are more effective, sustainable, and user-friendly. Ongoing research aims to further enhance their performance, scalability, and multifunctionality for broader therapeutic applications.

11.6.3 Prosthetic Devices

Bionanocomposite coatings help improve the functionality, safety, and comfort of prosthetic devices such as cochlear implants, prosthetic eyes, and artificial limbs (Ngo et al. 2018). They can be designed to reduce friction between the prosthetic and the user's skin, avoiding irritation, pain, and discomfort which is especially significant for people who use prosthetics for extended periods. These coatings may also improve the adhesion between prosthetic components and biological tissues, resulting in a more natural and secure fit.

Cochlear Implants: Cochlear implants, which are used to restore hearing in people with severe hearing loss, require close contact with the fragile structures of the inner ear (Stöver and Lenarz 2009). Bionanocomposite coatings can improve the implant's biocompatibility, lowering the risk of inflammation and tissue rejection. These coatings, which may include antibacterial nanoparticles like silver or zinc oxide, can also help prevent infections in the inner ear or surrounding tissue, which is especially crucial for avoiding implant-related problems. Additionally, coatings that increase the mechanical flexibility and electrical conductivity of the implant's components can improve signal transmission, resulting in better hearing outcomes (Tan et al. 2013).

Prosthetic Eyes: Prosthetic eyes are crucial for restoring an individual's appearance and function after losing an eye (Phadtare et al. 2024). Bionanocomposite coatings can improve the prosthetic eye's functions by creating a long-lasting, scratch-resistant, and realistic surface that mimics the natural appearance of a human eye. In addition to these cosmetic benefits, bionanocomposite coatings can provide antibacterial properties to prevent infections in the eye socket, which is susceptible to bacterial growth. Coatings with increased mechanical strength can further extend the lifespan and resistance of the prosthetic eye to wear and tear, ensuring that the device stays intact and functioning over time (Baino et al. 2014).

Prosthetic Limbs: Artificial limbs, such as prosthetic arms, legs, or hands, must be developed to resist severe mechanical stress while remaining comfortable for the user. Bionanocomposite coatings can increase the wear resistance and durability of prosthetic devices by improving the surface properties of materials such as titanium or polymer-based components. These coatings, which incorporate nanoparticles like carbon nanotubes, graphene, or titanium dioxide, can also improve the flexibility and strength of prosthetic components (Tan et al. 2022). Furthermore, the coatings can have antibacterial qualities, which prevent infections in areas of high friction or where the limb meets the skin. Self-healing coatings could fix slight damage to the prosthetic surface, increasing its life and minimizing the need for frequent maintenance (Yang et al. 2001).

Recent research has highlighted the significance of bionanocomposites in increasing the functionality, durability, and integration of prosthetics with the human body, making them more effective and patient-friendly. One significant area of focus is the creation of lightweight but strong prosthetic materials (Patel and Gohil 2024). Another area of development is the use of conductive bionanocomposites in neural prosthetics and biosensors. Nanomaterials like gold nanoparticles, carbon nanofibers, and conducting polymers are being used to improve electrical conductivity in prosthetic devices. These materials are very beneficial in designing interfaces for prosthetic limbs that are controlled by nerve signals, allowing for more precise and natural movements. Additionally, bionanocomposites are being investigated for prosthetic skin applications. Researchers are working on flexible and biocompatible materials filled with nanomaterials such as silver nanoparticles or zinc oxide to construct antibacterial and self-healing synthetic skin (Liu et al. 2023). In conclusion, new research highlights the revolutionary potential of bionanocomposites in prosthetic devices. These materials improve performance, comfort, and functionality, with continuous research aimed at refining their design and expanding their uses for better prosthetic outcomes.

11.6.4 Implantable Devices

Bionanocomposite coatings improve the functionality, safety, and longevity of implantable devices such as heart valves, dental implants, and artificial joints (Kim et al. 2024). By improving mechanical strength, promoting tissue integration, reducing infection risks, and improving biocompatibility, these coatings ensure that implantable devices perform effectively over time, resulting in better outcomes and a higher quality of life for patients.

Heart Valves: Heart valve implants are subjected to constant mechanical stress and have direct contact with blood (Ciolacu et al. 2022). Bionanocomposite coatings can improve the durability of heart valve materials, reducing the risk of wear and tear. These coatings can also aid in reducing blood clot formation (thrombosis) by providing a smooth surface that discourages clot formation. Incorporating antimicrobial nanoparticles lowers the chance of bacterial infections, which are typical consequences of heart valve implantation. Furthermore, the coatings can improve biocompatibility by stimulating the proliferation of endothelial cells, allowing for better integration with surrounding tissue and lowering the risk of immunological rejection or inflammation. Bionanocomposite coatings improve the performance and longevity of artificial heart valves by reducing calcification and thrombogenicity (Zawidlak-Węgrzyńska et al. 2024).

Dental Implants: Because dental implants are subjected to mechanical pressures during eating, they must be made of strong, durable, and biocompatible materials (Lee et al. 2016). Coatings containing bioactive elements, such as hydroxyapatite, that mirror natural bone structure, can aid in osseointegration, allowing the implant to merge securely with the surrounding bone (Nasar 2019). Chitosan-based coatings reduce the incidence of peri-implantitis by promoting osteointegration, and antibacterial properties embedded in the coating can help prevent infections in the gums and surrounding tissues, resulting in better healing and lowering the risk of irritation or inflammation near the implant site (Xing et al. 2016).

Artificial Joints: Artificial joints, such as those used in hip, knee, and shoulder replacements, must tolerate high mechanical loads over time (Murakami 2023). Bionanocomposite coatings improve the durability and wear resistance of the joint materials, reducing the risk of degradation from friction and ensuring smoother movement. Coatings with antimicrobial nanoparticles lower the incidence of infections in the implant area, which is a major worry for patients having joint replacement surgery. Furthermore, the coatings promote biocompatibility, allowing the prosthetic joint to merge more smoothly with the body. Biopolymer coatings with nanoparticles reduce inflammatory reactions while improving biocompatibility and wear resistance in prosthetic joints (Yousef 2016).

Bionanocomposites have widespread interest in the development of implantable devices. These hybrid materials, which mix biological and nanostructured components, are being researched extensively for use in orthopedics, dental implants, cardiovascular devices, and soft tissue regeneration. Recent advances have centered on increasing the endurance, integration, and performance of implanted devices. Bionanocomposites in orthopedic implants increase osteointegration, lowering implant failure rates and assuring long-term stability (Thirugnanasambandan and Thiagamani 2024). Bionanocomposites have also proven useful in dental implants (Lee 2016). Bionanocomposites have been used in cardiovascular devices such as stents and heart valves to provide biocompatible, antithrombogenic surfaces (Vellayappan et al. 2015). For example, nanomaterials such as silica nanoparticles and graphene are being included into polymeric stent coverings to reduce blood clot formation and improve endothelial cell adhesion. These advances increase the safety and lifespan of cardiovascular implants. Another focus is on soft tissue implantable devices like skin and nerve grafts (Han et al. 2024). Furthermore, researchers are looking into the usage of biodegradable bionanocomposites as temporarily implanted devices (Han et al. 2024). Materials like polylactic acid (PLA) and polycaprolactone (PCL), mixed with bioactive nanoparticles, are being researched to make implants that dissolve safely in the body after serving their purpose, reducing the need for additional procedures (Takagi et al. 2023).

11.6.5 Surgical Devices

Bionanocomposite coatings offer significant advantages in the performance, durability, and safety of surgical devices such as forceps, robotic surgery systems, and scalpels (Abaszadeh et al. 2023). These coatings can improve surgical instrument functionality by increasing biocompatibility, strength, and resistance to microbial contamination, as well as adding to the devices' long-term reliability. Bionanocomposite coatings can increase the biocompatibility of all surgical equipment, ensuring that they do not produce unpleasant reactions when in touch with tissues (Subasri et al. 2023).

Forceps and Scalpels: Forceps are essential surgical equipment for grasping, manipulating, and holding tissues or other objects during procedures. Bionanocomposite coatings can improve forceps grip by creating a nonslip, strong surface that improves surgical control and precision. Additionally, coatings that incorporate antimicrobial nanoparticles, such as silver or copper, can inhibit the growth of bacteria on the forceps surfaces, reducing the risk of surgical site infections. These coatings can also improve the corrosion resistance of the forceps, ensuring that the instruments maintain their integrity over time, even when exposed to human fluids and sterilizing procedures (Aljibori et al. 2023).

Scalpels are vital for making precise incisions during surgery, and their sharpness and durability are critical for surgical success. Bionanocomposite coatings can improve the cutting performance of scalpels by enhancing their sharpness retention and resistance to corrosion. For example, coatings made from materials such as diamond-like carbon (DLC) or titanium dioxide can provide superior hardness, ensuring that the scalpel remains sharp for longer periods and requires less frequent maintenance or replacement (Rajak et al. 2021). Silver nanoparticle-based bionanocomposite coatings provide surgical instruments with antibacterial properties, reducing the frequency of surgical site infections.

Robotic Surgery Systems: Robotic surgery systems, which provide greater precision and minimally invasive procedures, rely on a variety of tools, including robotic arms, instruments, and cameras (Reddy et al. 2023). Bionanocomposite coatings can significantly enhance the performance of these components by improving their durability, wear resistance, and precision. Antimicrobial coatings can also ensure that the components are less prone to bacterial contamination, which is critical in maintaining the sterility of the surgical environment. Furthermore, these coatings can improve the mechanical properties of robotic surgical tools, enhancing their sensitivity and ability to perform delicate tasks with greater accuracy. Robotic surgical systems with biocompatible and friction-reducing coatings reduce tissue injury and improve accuracy during minimally invasive procedures.

Bionanocomposites have emerged as potentially useful materials for surgical instruments (Abaszadeh et al. 2023). These materials, which combine biological components with nanostructured materials, are being thoroughly investigated for use in surgical devices. Recent studies have focused on improving the performance, durability, and antibacterial qualities of surgical instruments to enhance patient outcomes. In sutures, bionanocomposites have been employed to create threads with improved mechanical qualities and regulated biodegradability. These sutures limit postsurgical infections and the requirement for extra medical procedures. Bionanocomposites provide unique options for wound closure devices such as staples and adhesives (Thirugnanasambandan and Thiagamani 2024). Nanoparticles such as silver or zinc oxide are being used in adhesives to generate antibacterial wound closure systems. These materials not only successfully heal wounds, but they also prevent microbial contamination, which improves postoperative recovery. Another area of emphasis is surgical tools and instruments. Researchers are investigating the use of bionanocomposites in surgical tool coatings to improve durability and wear resistance. It also has a significant impact on robotic surgical tools, as biocompatible and friction-reducing coatings prevent tissue harm and enhance precision during minimally invasive procedures (Miranda et al. 2024). Bionanocomposites are driving innovation in surgical equipment by making materials stronger, safer, and more functional. Ongoing research is planned to broaden their uses by concentrating on scalability, cost-effectiveness, and improved patient care. These discoveries have enormous potential for improving surgical results and enhancing healthcare technologies (Thacharodi et al. 2024).

11.7 Regulatory Standards

Nanocomposite coatings are promising materials to promote the efficiency of medical devices, yet they face many regulatory issues. Manufacturers need to comprehend and navigate these complex regulatory procedures to bring these innovative materials to market. By obeying recognized standards and being involved in constant communication with regulatory bodies, manufacturers can safeguard the safety and productive use of nanocomposite coatings for biomedical uses.

11.7.1 Food and Drug Administration (FDA)

FDA is a regulatory body located at the Department of Health and Human Services in the United States which is accountable for protecting the health of the public by maintaining efficacy, security and safety of veterinary and human medicines, medical devices, biological products, substances that discharge radiation, and cosmetics (Schlauderaff and Boyer 2019). Regulation of medical devices is executed by the Center for Devices and Radiological Health of the FDA. Biomedical devices with nanocomposite coatings will be assessed based on their classification, which relies on their application and level of risk. A risk-based approach is followed by FDA, classifying biodevices into three classes (Fig. 11.3):

Class I biomedical devices are low-risk devices that require general controls including registration of facilities, adequate labeling, and implementation of good manufacturing practices (GMP) (Amaral et al. 2024). Devices must have suitable labels with precise evidence about their usage. Facility registration provides a unique facility identifier (UFI) to these facilities which enables easy tracking by FDA monitoring of sites of manufacture. GMP guidelines ensure devices are consistently manufactured and regulated conferring to quality standards. Premarket submission is not needed for Class I biomedical devices. Handheld surgical aids and bandages come under this category.

Class II biomedical devices are moderate-risk medical devices that need special controls as well as general controls. Special controls include precise labeling needs, performance standards, and post-market surveillance. Precise labeling provides warnings and usage instructions in detail. Post-market surveillance inspects the performance of devices once it is marketed to detect and address challenges. Performance standards are explicit criteria which devices must possess to confirm they function properly. Premarket notification (510(k) clearance) is required for these devices to validate considerable correspondence to a marketed device. Powered wheelchairs and infusion pumps come under this category.

Class III biomedical devices are high-risk categories that require premarket approval due to possible risks related to their usage. These devices sustain or support life, deliver a possibly irrational risk of injury or disease, or avoid damage to health. Clinical trials are often required to collect information on efficiency and safety. To confirm whether these devices meet strict regulatory standards, rigorous testing is conducted. Heart valves and pacemakers belong to this category.

Each class demonstrates regulatory control requirements that need to be met to provide reasonable assurance of the device's safety and effectiveness. **Nanocomposite coatings come under Class II or Class III classifications** due to their novelty and possible impact on safety and biocompatibility. Due to these unique features, they need to undergo a thorough evaluation before coming into the market to make sure they abide by safety standards.

The major regulatory pathways for nanocomposite coatings include:

Premarket notification [510(k)] is classically obligatory for Class II devices (Wreh 2023). Manufacturers must validate that their device is significantly comparable to a lawfully promoted device that does not need premarket approval (PMA). By substantial equivalence, it means that the newly invented device has the same envisioned application and technological features. Substantial equivalence means the new device has the same envisioned application and technological features or that any alterations do not increase new queries of security and efficacy. This pathway includes succumbing data and certification that the device achieves as well as the predicate device under similar conditions.

Premarket approval (PMA) is required for Class III devices that carry increased risks and have no predicate device to compare (Van Norman 2016). PMA is a more strict process, necessitating scientific proof to demonstrate the safety and efficacy of the device. Manufacturers must conduct wide-ranging laboratory and clinical tests and submit the results in a comprehensive PMA appeal to the FDA. PMA is essential for nanocomposite coatings that are a portion of life-supporting devices or those that meaningfully influence the functions of the body, like coatings on implantable sensors or pacemakers.

Investigational device exemption (IDE) permits the use of a device in clinical evaluation to collect information on its efficiency and safety, which is required for 510(k) submission or lodging PMA applications. This exclusion is vital for directing clinical experiments on novel or vitally modified devices, such as the one with nanocomposite coatings. The FDA carefully checks these experiments to make sure that the trial is executed ethically and that patients are not facing any perverse risks. These regulatory pathways guarantee that nanocomposite coatings, with their unique characteristics, endure rigorous testing before being applied for medical uses, thereby protecting device reliability and promoting patient health (Fig. 11.5).

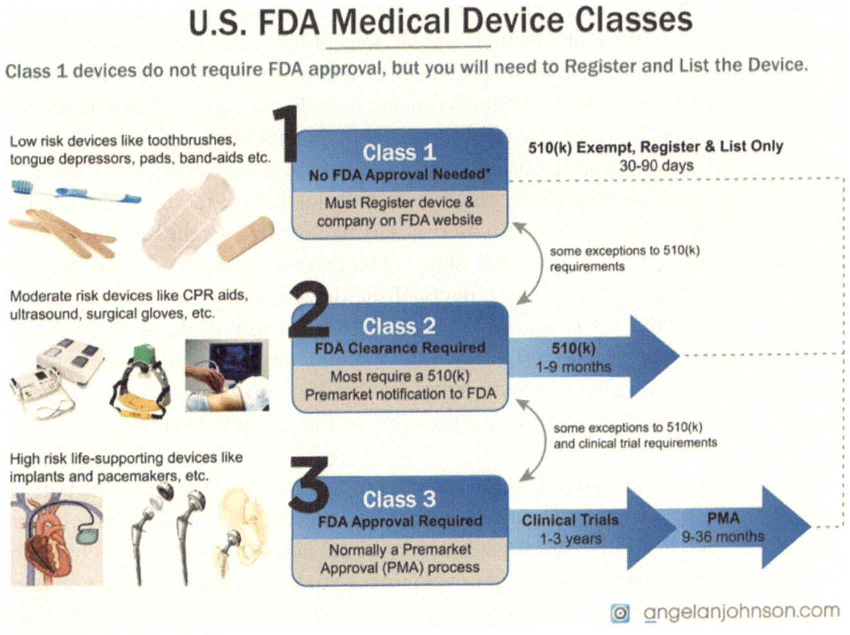

Fig. 11.5 Various classes of medical devices. Reproduced from [Tettey et al. 2024] copyright permission from Springer

11.7.2 European Medicines Agency (EMA)

EMA is a regulatory authority in the European Union accountable for the scientific assessment, management, and safety inspection of medicines (EU) (Greim 2024). Its main role is to confirm that medicines accessible in the EU are harmless, efficient, and of superior quality. Regulation of medical devices is done by Medical Devices Regulation (MDR) (EU 2017/745) of EMU. The MDR presents rigorous necessities for devices containing nanomaterials, highlighting the requirement for a detailed risk valuation due to their exclusive characteristics. The medical devices are categorized into four classes (I, IIa, IIb, III) based on the risk they cause to patients. Nanocomposite coatings fall into Class IIb or III due to their novelty and intricate interactions at the bio–nano interface that impact their safety. The vital regulatory pathways for these coatings include:

Clinical evaluations are important to establish that a medical device with nanocomposite coatings is harmless and functions as envisioned (Kang et al. 2024). This procedure includes a methodical valuation of clinical data, collected from clinical studies or previous literature of identical devices. Novel clinical research especially for the device with nanocomposite coating will be required. This will help to understand the long-term impacts on the body and make sure that they do not cause any negative effects due to their interactions at the nanoscale with biological tissues. For nanocomposite coatings, clinical evaluations are crucial to understand their long-term effects on the body and to ensure they do not cause adverse reactions due to their interaction with biological tissues at the nanoscale.

Conformity assessment is an all-inclusive procedure that confirms a medical device abides by all appropriate regulatory necessities before it can be sold in the EU. For high-risk classes, this valuation is led by Notified Bodies—autonomous organizations chosen by EU member states. The process comprises a comprehensive evaluation of the technical documents of the device, which must establish that the device follows the safety and efficacy necessities of the MDR. A valuation of the quality management system (QMS) of the manufacturer is mainly based on ISO 13485, to guarantee reliable production and control of quality. Nanocomposite coatings need scrupulous certification and tests due to their novelty to confirm they do not present any extra risks.

Post-market surveillance (PMS) is an ongoing process that monitors the performance and safety of a medical device after it has been placed on the market. This comprises the gathering and investigation of data from numerous sources, such as user reactions, contrary event rumors, and novel clinical studies. The intention is to recognize any possible problems that might not have been obvious during the premarket stage and to take corrective actions if essential. For nanocomposite coatings, PMS is primarily significant because the lasting impacts of nano-sized materials may become ostensible after lengthy use. These regulatory needs ensure that medical devices with nanocomposite coatings are meticulously tested and unceasingly checked to protect public health, speaking the exclusive trials modeled by cutting-edge materials and knowledge.

11.7.3 International Organization for Standardization (ISO)

ISO is a nongovernmental independent global body that frames and publishes standards to confirm the safety, interoperability, quality, and efficacy of services and products, across wide industries. International standards are important to harmonize regulatory needs across various regions. The International Electrotechnical Commission (IEC) and ISO have established numerous standards pertinent to nanocomposite coatings for medical devices, such as.

The ISO 10993 series which is a keystone for the biological testing of medical devices, particularly custom-made to discourse the communication between devices and biological systems (Pellevoisin et al. 2022). ISO 10993–1 assists in the basis by framing a risk management process for testing biocompatibility, which includes recognizing biological risks and directing suitable tests to assess these risks. This comprises valuations such as hemocompatibility, cytotoxicity, sensitization, and systemic toxicity. The detailed implementation of the ISO 10993 standards guarantees that nanocomposite coatings meet rigorous safety necessities before approaching the market.

ISO 13485 is a worldwide approved QMS standard explicitly for medical device manufacturing (Pawar and Patel 2024). It establishes criteria for a QMS where an organization is required to validate its capability to deliver medical devices and associated services that constantly meet regulatory and customer needs. This standard needs organizations to provide a complete quality system that incorporates every stage of the lifecycle of a device, from design and fabrication to manufacture, supply, and post-market events. For nanocomposite coatings, obeying ISO 13485 standards confirms that the process of manufacturing is regulated and that the end product is harmless, efficient, and meets regulatory opportunities.

ISO/TR 13,014 guides the physicochemical testing of functionalized nanomaterials, which is vital for toxicological testing. This technical document discourses the exclusive characteristics of nanomaterials, giving an outline for choosing and using suitable methods of characterization. Parameters including surface area, particle size, chemical composition, and shape are important to understand the performance and potential dangers related to nanomaterials. This guidance is chiefly relevant for nanocomposite coatings, where the nanoscale components suggestively impact the complete performance of the material and its safety profile.

11.8 Future Directions

As the arena of nanocomposite coatings evolves, regulatory standards are anticipated to adjust in response to novel scientific insights and technological advancements. Collective efforts among academic researchers, regulatory agencies, and stakeholders from the industry are indispensable in developing comprehensive guidelines that ensure the security and efficacy of these groundbreaking materials. Coordinating global standards will also be essential for enabling the international trade of medical devices with composite coatings. Continuous updates to ISO and IEC standards, informed by pioneering research, will help bridge the gap between scientific potential and regulatory requirements.

The future of biomedical nanocomposite coatings holds significant promise, with numerous exciting advancements on the horizon (Kravanja and Finšgar 2022; Sahu et al. 2024; Pillai et al. 2025). As sustainability becomes more important, the exploration of biodegradable or renewable nanocomposite materials is likely to reduce environmental impact and improve the overall sustainability of medical devices. Future innovations may focus on creating multifunctional coatings that combine antibacterial properties, wear resistance, and enhanced cell adhesion in one solution, minimizing infection risks and improving device performance. The development of personalized medicine is another exciting area where biomedical nanocomposite coatings could be tailored to individual patient needs, improving healing and reducing complications. Smart coatings that dynamically respond to changes in the body's environment, such as pH or temperature, could enhance drug delivery, tissue regeneration, and device monitoring.

As sustainability becomes increasingly important, the development of biodegradable or renewable nanocomposite materials is anticipated to reduce environmental impact and improve the overall sustainability of medical devices. Furthermore, future advancements may focus on generating multifunctional coatings that combine antibacterial qualities, wear resistance, and increased cell adhesion in one solution, lowering infection risks and boosting device performance (Haniffa et al. 2016). Finally, as nanotechnology advances, regulatory agencies will need to enhance their standards and testing techniques to accommodate the unique challenges these materials present, ensuring their safety and efficacy in clinical settings. Overall, the continuous development of biomedical nanocomposite coatings will result in more effective, personalized, and environmentally friendly medical equipment, ultimately benefiting patients (Bharathi and Jayas 2024).

11.9 Conclusion

Biomedical nanocomposite coatings are a promising area in the development of sophisticated biomedical devices, providing enhanced features such as increased biocompatibility, antimicrobial effects, and mechanical strength. These coatings, with their distinct nanostructures, can considerably improve the performance and lifetime of implants, prosthetics, and other medical equipment. The ability to integrate multiple functional properties—such as controlled drug release, wear resistance, and enhanced tissue integration—makes them highly attractive for a wide range of medical applications. However, their implementation into biomedical devices must meet high regulatory standards to ensure patient safety and efficacy. The FDA and other international agencies have built regulatory frameworks that play an important role in the approval and monitoring of nanocomposite-coated devices, addressing concerns about toxicity, long-term stability, and immunological response. These frameworks are constantly changing to keep up with the rapid breakthroughs in nanotechnology, ensuring that materials meet the stringent requirements for premarket approval and post-market supervision. Furthermore, as nanocomposite coatings become more widely employed in sensitive applications like drug delivery systems and implants, there is a rising demand for standardized testing methodologies that can correctly assess their long-term safety and performance in the human body. Collaboration among regulatory authorities, researchers, and manufacturers will be critical in overcoming these obstacles and permitting the safe, effective integration of these new materials into clinical practice.

References

Abaszadeh F, Ashoub MH, Khajouie G et al (2023) Nanotechnology development in surgical applications: recent trends and developments. Eur J Med Res 28:537. https://doi.org/10.1186/s40001-023-01429-4

Abdel-karim AM, Ahmed YM, El-Masry MM (2023) Ag-CuO/epoxy hybrid nanocomposites as anti-corrosive coating and self-cleaning on copper substrate. Sci Rep 13:19248. https://doi.org/10.1038/s41598-023-46533-x

Akintayo CO, Akintayo ET, Ziegler T (2011) Studies on newly developed urethane modified polyetheramide coatings from Albizia benth oil. Prog Org Coat 71(1). https://doi.org/10.1016/j.porgcoat.2011.01.003

Alakoski E, Tiainen VM, Soininen A, Konttinen YT (2008) Load-bearing biomedical applications of diamond-like carbon coatings - current status. Open Orthop J 2:43–50. https://doi.org/10.2174/1874325000802010043

Aljibori H, Al-Amiery A, Kadhum A (2023) Advances in corrosion protection coatings: a comprehensive review. Int J Corros Scale Inhib 12:1476–1520. https://doi.org/10.17675/2305-6894-2023-12-4-6

Alshamrani M (2022) Broad-spectrum theranostics and biomedical application of functionalized nanomaterials. Polymers 14(6):1221. https://doi.org/10.3390/polym14061221

Amaral C, Paiva M, Rodrigues AR, Veiga F, Bell V (2024) Global regulatory challenges for medical devices: impact on innovation and market access. Appl Sci 14(20):9304. https://doi.org/10.3390/app14209304

Aravamudhan A, Ramos DM, Nada AA, Kumbar SG (2014) Chapter 4—natural polymers: polysaccharides and their derivatives for biomedical applications. https://doi.org/10.1016/B978-0-12-396983-5.00004-1

Arora B, Bhatia R, Attri P (2018) 28—Bionanocomposites: green materials for a sustainable future. https://doi.org/10.1016/B978-0-12-811033-1.00027-5

Ashtari K, Nazari H, Ko H, Tebon P, Akhshik M, Akbari M, Naghavi Alhosseini S, Mozafari M, Mehravi B, Soleimani M, Ardehali R, Ebrahimi Warkiani M, Ahadian S, Khademhosseini A (2019) Electrically conductive nanomaterials for cardiac tissue engineering. Adv Drug Deliv Rev 144:59–72. https://doi.org/10.1016/j.addr.2019.06.001

Baino F, Perero S, Ferraris S, Miola M, Balagna C, Verné E, Vitale-Brovarone C, Coggiola A, Dolcino D, Ferraris M (2014) Biomaterials for orbital implants and ocular prostheses: overview and future prospects. Acta Biomater. https://doi.org/10.1016/j.actbio.2013.12.014

Bardhan N (2022) Nanomaterials in diagnostics, imaging and delivery: applications from COVID-19 to cancer. MRS Communications 12:1119–1139. https://doi.org/10.1557/s43579-022-00257-7

Bharathi VSK, Jayas DS (2024) Evolution of bionanocomposites: innovations and applications in food packaging. Foods 13(23):3787. https://doi.org/10.3390/foods13233787

Borhani S, Hassanajili S, Tafti SHA, Rabbani S (2018) Cardiovascular stents: overview, evolution, and next generation. Prog Biomater 7(3):175–205. https://doi.org/10.1007/s40204-018-0097-y

Bose I, Nousheen S, Roy P, Yaduvanshi S, Sharma S, Chandel V, Biswas D (2023) Unveiling the potential of marine biopolymers: sources, classification, and diverse food applications. Materials (Basel, Switzerland) 16(13):4840. https://doi.org/10.3390/ma16134840

Cao C, Killips A, Li X (2024) Advances in the science and engineering of metal matrix nanocomposites: a review. Adv Eng Mater 26:2400217. https://doi.org/10.1002/adem.202400217

Chen Z-Q, Lin B, Liu M-C, Zhang Y, Liu S-J, Li Y, Zhao Q (2024b) Two-photon polymerization-based 3D micro-scaffolds toward biomedical devices. Chem Eng J 493:152469. https://doi.org/10.1016/j.cej.2024.152469

Chen H, Xiang Z, Zhang T, Wang H, Li X, Chen H, Shi Q (2024) Heparinized self-healing polymer coating with inflammation modulation for blood-contacting biomedical devices. Acta Biomater 186. https://doi.org/10.1016/j.actbio.2024.07.010

Chiellini E, Corti A, D'Antone S, Solaro R (2003) Biodegradation of poly (vinyl alcohol) based materials. Prog Polym Sci 28:963–1014. https://doi.org/10.1016/S0079-6700(02)00149-1

Ciolacu DE, Nicu R, Ciolacu F (2022) Natural polymers in heart valve tissue engineering: strategies, advances and challenges. Biomedicines 10(5):1095. https://doi.org/10.3390/biomedicines10051095

Coombes AGA, Breeze V, Lin W, Gray T, Parker KG, Parker T (2000) Lactic acid-stabilised albumin for microsphere formulation and biomedical coatings. Biomaterials 22. https://doi.org/10.1016/S0142-9612(00)00074-0

Covarrubias C, Cádiz M, Maureira Vargas M, Celhay I, Cuadra F, Marttens A (2018) Bionanocomposite scaffolds based on chitosan–gelatin and nanodimensional bioactive glass particles: in vitro properties and in vivo bone regeneration. J Biomater Appl 32. https://doi.org/10.1177/0885328218759042

Darder M, Aranda P, Ruiz-Hitzky E (2007) Bionanocomposites: a new concept of ecological, bioinspired, and functional hybrid materials. Adv Mater 19:1309–1319. https://doi.org/10.1002/adma.200602328

de Mesquita JP, Donnici CL, Pereira FV (2010) Biobased nanocomposites from layer-by-layer assembly of cellulose nanowhiskers with chitosan. Biomacromol 11(2):473–480. https://doi.org/10.1021/bm9011985

Dutt Y, Pandey RP, Dutt M, Gupta A, Vibhuti A, Vidic J, Raj VS, Chang CM, Priyadarshini A (2023) Therapeutic applications of nanobiotechnology. J Nanobiotechnology 21(1):148. https://doi.org/10.1186/s12951-023-01909-z

Farajollah R, Alavi Nikje MM, Abedin Dorkoosh F (2024) Hierarchical porous bio polyurethane film as an implantable delivery system for sustained release of insulin: synthesis, characterization, and in-vitro evaluation. React Funct Polym 197. https://doi.org/10.1016/j.reactfunctpolym.2024.105860

Ge P, Li L, Jiang M, Wang G, Wen F, Gao X (2025) A polymer nanocomposite for high-temperature energy storage with thermal stability. Cell Rep Phys Sci 6. https://doi.org/10.1016/j.xcrp.2024.102361

Gradaus R, Breithardt G, Böcker D (2003) ICD leads: Design and chronic dysfunctions. Pacing Clin Electrophysiol: PACE 26(2 Pt 1):649–657. https://doi.org/10.1046/j.1460-9592.2003.00112.x

Greim H (2024) European regulations for compounds and products. In: Encyclopedia of toxicology, 4th ed. https://doi.org/10.1016/B978-0-12-824315-2.00568-6

Habibi Y, Hoeger I, Kelley SS, Rojas OJ (2010) Development of Langmuir-Schaeffer cellulose nanocrystal monolayers and their interfacial behaviors. Langmuir: ACS J Surf Colloids 26(2):990–1001. https://doi.org/10.1021/la902444x

Han WB, Jang T-M, Shin B, Naganaboina VR, Yeo W-H, Hwang S-W (2024) Recent advances in soft, implantable electronics for dynamic organs. Biosens Bioelectron 168:116472. https://doi.org/10.1016/j.bios.2024.116472

Haniffa MAC, Ching YC, Abdullah LC, Poh SC, Chuah CH (2016) Review of bionanocomposite coating films and their applications. Polymers 8(7):246. https://doi.org/10.3390/polym8070246

Hong RY, Chen Q (2014) Dispersion of inorganic nanoparticles in polymer matrices: challenges and solutions. In: Kalia S, Haldorai Y (eds) Organic-Inorganic Hybrid Nanomaterials, vol 267. Springer, Cham. https://doi.org/10.1007/12_2014_286

Kang SL, Kwon JY, Kim SM (2024) Insights into post-marketing clinical validation of companion diagnostics with reference to the FDA, EMA, PMDA, and MFDS. Mol Therapy—Methods Clin Dev 24:101346. https://doi.org/10.1016/j.omtm.2024.101346

Kashoki M, Hanaizi Z, Yordanova S, Veselý R, Bouygues C, Llinares J, Kweder SL (2020) A comparison of EMA and FDA decisions for new drug marketing applications 2014–2016: concordance, discordance, and why. Clin Pharmacol Ther 107(1). https://doi.org/10.1002/cpt.1565

Khan FA (2020) Nanomaterials: types, classifications, and sources. In: Khan F (ed) Applications of nanomaterials in human health. Springer, Singapore, pp 1–26. https://doi.org/10.1007/978-981-15-4802-4_1

Khouri NG, Bahú JO, Blanco-Llamero C, Severino P, Concha VOC, Souto EB (2024) Polylactic acid (PLA): properties, synthesis, and biomedical applications—A review of the literature. Mol Struct. https://doi.org/10.1016/j.molstruc.2024.138243

Kim HJ, Choi H, Kim DH, Son D (2024) Stretchable functional nanocomposites for soft implantable bioelectronics. Nano Lett 24(28):8453–8464. https://doi.org/10.1021/acs.nanolett.4c01163

Kravanja KA, Finšgar M (2022) A review of techniques for the application of bioactive coatings on metal-based implants to achieve controlled release of active ingredients. Mater Des 217:110653. https://doi.org/10.1016/j.matdes.2022.110653

Lee SG, Carrow JK, Fraser LA, Yan J, Jeyamogan S, Sambandam Y, Clemons TD, Kolberg-Edelbrock AN, He J, Mathew J, Zhang ZJ, Leventhal JP, Gallon L, Palmer LC, Stupp SI (2024) Single-cell coating with biomimetic extracellular nanofiber matrices. Acta Biomater 177:244–256. https://doi.org/10.1016/j.actbio.2024.02.002

Lee J-H, Kim H-W, Seo S-J (2016) Polymer-ceramic bionanocomposites for dental application. J Nanomater 2016(3795976):8. https://doi.org/10.1155/2016/3795976

Li B, Jiang B, Boyce BM, Lindsey BA (2009) Multilayer polypeptide nanoscale coatings incorporating IL-12 for the prevention of biomedical device-associated infections. Biomaterials 30. https://doi.org/10.1016/j.biomaterials.2009.01.042

Liu L, Lei J, Liu T, Shang B, Zhao X, He J, Pan F, Li L (2023) Artificial skin with fast and robust self-healing ability for durable intelligent protection of magnesium alloys. Chem Eng J. https://doi.org/10.1016/j.cej.2023.141482

Liu T, Feng H, Jin C, Pawlak M, Saeb MR, Kuang T (2024) Green segregated honeycomb biopolymer composites for electromagnetic interference shielding biomedical devices. Chem Eng J 493:152438. https://doi.org/10.1016/j.cej.2024.152438

Mabrouk M, Das DB, Salem ZA, Beherei HH (2021) Nanomaterials for biomedical applications: production, characterizations, recent trends and difficulties. Molecules (Basel, Switzerland) 26(4):1077. https://doi.org/10.3390/molecules26041077

Mekuye B, Abera B (2023) Nano Select 4:486. https://doi.org/10.1002/nano.202300038

Miranda MX, Fiocco D, Caneva T, Yiapanis P, Orgill DP (2024) Current and future applications of artificial intelligence in surgery: Implications for clinical practice and research. Front Surg. https://doi.org/10.3389/fsurg.2024.1393898

Murakami T (2023) Biotribology of artificial joints. In: Biotribology of natural and artificial joints. Elsevier, pp 1–17. https://doi.org/10.1016/B978-0-12-823669-7.00001-6

Nasar A (2019) 8—Hydroxyapatite and its coatings in dental implants. In: Applications of nanocomposite materials in dentistry. https://doi.org/10.1016/B978-0-12-813742-0.00008-0

Nathanael AJ, Oh TH (2020) Biopolymer coatings for biomedical applications. Polymers 12(12):3061. https://doi.org/10.3390/polym12123061
Nazari MH, Zhang Y, Mahmoodi A, Xu G, Yu J, Wu J, Shi X (2022) Nanocomposite organic coatings for corrosion protection of metals: a review of recent advances. Prog Org Coat 162. https://doi.org/10.1016/j.porgcoat.2021.106573
Ngo TD, Kashani A, Imbalzano G, Nguyen KTQ, Hui D (2018) Additive manufacturing (3D printing): a review of materials, methods, applications, and challenges. Compos B Eng 143:172–196. https://doi.org/10.1016/j.compositesb.2018.02.012
Van Norman GA (2016) Drugs, devices, and the FDA: Part 2: an overview of approval processes: FDA approval of medical devices. JACC: Basic Transl Sci 1(4):277–287. https://doi.org/10.1016/j.jacbts.2016.03.009
Patel P, Gohil P (2024) Design, analysis, and development of prosthetic and orthotic elements by additive manufacturing process. Int J Light Mater Manuf. https://doi.org/10.1016/j.ijlmm.2024.09.004
Pawar H, Patel M (2024) Harmonization of regulatory frameworks for medical devices in BRICS countries: a path to enhanced trade and investment. Ann Pharm Fr. https://doi.org/10.1016/j.pharma.2024.08.007
Pellevoisin C, Coleman KP, Hoffmann S (2022) ISO 10993–23 In vitro irritation testing for medical devices: Substantiating applicability to mild irritants and non-extractables. Toxicol in Vitro. https://doi.org/10.1016/j.tiv.2022.105371
Phadtare P, Viswapriya V, ShindeV, Mathure D, Salunke MR (2024) Hybrid bionanocomposites as advancements in biomedical utility. Hybrid Adv 8. https://doi.org/10.1016/j.hybadv.2024.100365
Pillai VV, Ramasubramanian B, Sequerth O, Pilla S, Wang, T, Mohanty AK, Govindaraj P, Alhassan SM, Salim N, Kingshott P, Fuss FK (2025) Nanomaterial advanced smart coatings: emerging trends shaping the future. Appl Mater Today 42:102574. https://doi.org/10.1016/j.apmt.2024.102574
Prasad M, Lambe UP, Brar B, Shah I, Ranjan K, Rao R, Kumar S, Mahant S, Khurana SK, Iqbal HMN, Dhama K, Misri J, Prasad G (2018) Nanotherapeutics: an insight into healthcare and multi-dimensional applications in the medical sector of the modern world. Biomed Pharmacother 97:1521–1537. https://doi.org/10.1016/j.biopha.2017.11.026
Rahman MT, Hismat U, Tripura R, Choudhury PD, Debnath B, Das S (2024) Overview on natural polymers: a promising pharmaceutical excipient in mucoadhesive drug delivery system. Res J Pharm Technol 17(2):915–919. https://doi.org/10.52711/0974-360X.2024.00142
Rajak DK, Kumar A, Behera A, Menezes PL (2021) Diamond-like carbon (DLC) coatings: classification, properties, and applications. Appl Sci 11(10):4445. https://doi.org/10.3390/app11104445
Rashid AB, Haque M, Islam SMM, Uddin Labib KMR, Chowdhury P (2024) Breaking boundaries with ceramic matrix composites: a comprehensive overview of materials, manufacturing techniques, transformative applications, recent advancements, and future prospects. Adv Mater Sci Eng. https://doi.org/10.1155/2024/2112358
Rather SU (2020) Chapter 20—Bionanocomposites in sensor technology. In Zia KM, Jabeen F, Anjum MN, Ikram S (eds) Micro and nano technologies, bionanocomposites, pp 475–510. https://doi.org/10.1016/B978-0-12-816751-9.00020-9
Rebecca BPN, Durgalakshmi D, Balakumar S, Rakkesh A (2024) Carbon-based bionanocomposites for biosensors. In: Venkatesan J (ed) Biocomposite nanomaterials and their applications, nanostructure science and technology. Springer, Cham. https://doi.org/10.1007/978-3-031-74356-6_9
Reddy K, Gharde P, Tayade H, Patil M, Reddy LS, Surya D (2023) Advancements in robotic surgery: a comprehensive overview of current utilizations and upcoming frontiers. Cureus 15(12):e50415. https://doi.org/10.7759/cureus.50415
Renard E, Schaepelynck-Bélicar P (2007) Implantable insulin pumps. A position statement about their clinical use. Diabetes Metablsm 33. https://doi.org/10.1016/j.diabet.2006.10.004

Rhim JW, Ng PK (2007) Natural biopolymer-based nanocomposite films for packaging applications. Crit Rev Food Sci Nutr 47(4):411–433. https://doi.org/10.1080/10408390600846366

Rodríguez-Jiménez PE, Oropeza García de Quevedo P, Rodríguez Rodríguez R (2025) Chapter 22—Marine biopolymers in drug delivery applications. In: Marine biopolymers, pp 1–15. https://doi.org/10.1016/B978-0-443-15606-9.00022-X

Sahu R, Ninan N, Nguyen NH, Wang J, Nguyen M, Vasilev K, Tang Y (2024) Antibacterial plasma coating with aggregation-induced emission photosensitizers to prevent surgical site infections (Adv. Mater. Interfaces 15/2024). Adv Mater Interfaces 11:202470039. https://doi.org/10.1002/admi.202470039

Saini R, Bajpai AK, Jain E (2017) Fundamentals of bionanocomposites. In: Handbook of biopolymer-based nanocomposites: processing, performance, and applications, pp 285–312. https://doi.org/10.1016/B978-0-08-100970-3.00012-2

Saleh TA (2021) Polymer science and polymerization methods toward hybrid materials. In: Polymer hybrid materials and nanocomposites, pp 49–77. https://doi.org/10.1016/B978-0-12-813294-4.00004-2

Satchanska G, Davidova S, Petrov PD (2024) Natural and synthetic polymers for biomedical and environmental applications. Polymers 16(8):1159. https://doi.org/10.3390/polym16081159

Schaepelynck P (2019) The implantable insulin pump. In: Reznik Y (ed) Handbook of diabetes technology, pp 65–78. https://doi.org/10.1007/978-3-319-98119-2_5

Schlauderaff A, Boyer KC (2019) An overview of food and drug administration medical device legislation and interplay with current medical practices. Cureus 11(5):e4627. https://doi.org/10.7759/cureus.4627

Singhvi MS, Zinjarde SS, Gokhale DV (2019) Polylactic acid: synthesis and biomedical applications. J Appl Microbiol 127:1612–1626. https://doi.org/10.1111/jam.14290

Song R, Murphy M, Li C, Ting K, Soo C, Zheng Z (2018) Current development of biodegradable polymeric materials for biomedical applications. Drug des Dev Ther 12:3117–3145. https://doi.org/10.2147/DDDT.S165440

Stöver T, Lenarz T (2009) Biomaterials in cochlear implants. GMS Curr Top Otorhinolaryngol Head Neck Surg 8:Doc10. https://doi.org/10.3205/cto000062

Subasri R, Patra R, Yadav M et al (2023) Biofilm-inhibiting nanocomposite coatings on surgical sutures: durability and mechanistic insights. J Coat Technol Res 20:377–392. https://doi.org/10.1007/s11998-022-00678-y

Suggs LJ, Moore SA, Mikos AG (2007) Synthetic biodegradable polymers for medical applications. In Mark JE (ed) Physical properties of polymers handbook. Springer, New York, pp 1053–1076. https://doi.org/10.1007/978-0-387-69002-5_55

Sunwoo S-H, Han SI, Kang H, Cho YS, Jung D, Lim C, Cha M-J, Lee S-P, Hyeon T, Kim D-H (2020) Epicardial mesh electronics: Stretchable low-impedance nanocomposite comprised of Ag–Au core–shell nanowires and Pt black for epicardial recording and stimulation. Adv Mater Technol 5:2070014. https://doi.org/10.1002/admt.202070014

Takagi A, Hsu Y-I, Uyama H (2023) Biodegradable poly(lactic acid) and polycaprolactone alternating multiblock copolymers with controllable mechanical properties. Polym Degrad Stab. https://doi.org/10.1016/j.polymdegradstab.2023.110564

Tan F, Walshe P, Viani L, Al-Rubeai M (2013) Surface biotechnology for refining cochlear implants. Trends Biotechnol 31(12). https://doi.org/10.1016/j.tibtech.2013.09.001

Tan Q, Wu C, Li L, Shao W, Luo M (2022) Nanomaterial-based prosthetic limbs for disability mobility assistance: a review of recent advances. J Nanomater 2022(1):3425297. https://doi.org/10.1155/2022/3425297

Teepakakorn A, Ogawa M (2022) Interactions of layered clay minerals with water-soluble polymers; structural design and functions. Appl Clay Sci 222. https://doi.org/10.1016/j.clay.2022.106487

Tettey F, Parupelli SK, Desai S (2024) A review of biomedical devices: classification, regulatory guidelines, human factors, software as a medical device, and cybersecurity. Biomed Mater Devices 2:316–341. https://doi.org/10.1007/s44174-023-00113-9

Thacharodi A, Singh P, Meenatchi R, Ahmed ZH, Kumar RRS, V N, Kavish S, Maqbool M, Hassan S (2024) Revolutionizing healthcare and medicine: the impact of modern technologies for a healthier future—a comprehensive review. Healthc Sci 3(5):329–349. https://doi.org/10.1002/hcs2.115

Thirugnanasambandan T, Thiagamani SMK (2024) Cytotoxicity and biocompatibility of the bionanocomposites in orthopedic implants and wound dressing applications. In: Biocomposites for industrial applications. Elsevier, pp 205–221. https://doi.org/10.1016/B978-0-323-91866-4.00009-3

Thirugnanasambandan T (2022) Bio-nanocomposites in biomedical application. In: Muthukumar C, Thiagamani SMK, Krishnasamy S, Nagarajan R, Siengchin S (eds) Polymer based bio-nanocomposites. Springer, Singapore, pp 413–436. https://doi.org/10.1007/978-981-16-8578-1_15

Vellayappan MV, Balaji A, Subramanian AP, John AA, Jaganathan SK, Murugesan S, Supriyanto E, Yusof M (2015) Multifaceted prospects of nanocomposites for cardiovascular grafts and stents. Int J Nanomed 10:2785–2803. https://doi.org/10.2147/IJN.S80121

Wang L, Wang A (2007) Adsorption characteristics of Congo Red onto the chitosan/montmorillonite nanocomposite. J Hazard Mater 147(3):979–985. https://doi.org/10.1016/j.jhazmat.2007.01.145

Wang SF, Shen L, Tong YJ, Chen L, Phang IY, Lim PQ, Liu TX (2005) Biopolymer chitosan/montmorillonite nanocomposites: Preparation and characterization. Polym Degrad Stab 90:10–16. https://doi.org/10.1016/j.polymdegradstab.2005.03.001

Wang J, Ninan N, Nguyen NH, Nguyen MT, Sahu R, Nguyen TT, Mierczynska-Vasilev A, Vasilev K, Truong VK, Tang Y (2024) Biomimetic bacterium-like particles loaded with aggregation-induced emission photosensitizers as plasma coatings for implant-associated infections. ACS Appl Mater Interfaces 16(15):18449–18458. https://doi.org/10.1021/acsami.3c19484

Weinstock RS, Aleppo G, Bailey TS (2020) The role of blood glucose monitoring in diabetes management. Arlington (VA): American Diabetes Association. https://www.ncbi.nlm.nih.gov/books/NBK566165/

Woeppel K, Dhawan V, Shi D, Cui XT (2023) Nanotopography-enhanced biomimetic coating maintains bioactivity after weeks of dry storage and improves chronic neural recording. Biomaterials 302. https://doi.org/10.1016/j.biomaterials.2023.122326

Wreh E (2023) Chapter 3—Premarket notification [510(k)]. In: Medical device regulation. https://doi.org/10.1016/B978-0-323-95354-2.00006-2

Xiao R, Zhou G, Wen Y, Ye J, Li X, Wang X (2023) Recent advances on stimuli-responsive biopolymer-based nanocomposites for drug delivery. Compos Part B: Eng 266. https://doi.org/10.1016/j.compositesb.2023.111018

Xing Y, Xu Q, Li X, Chen C, Ma L, Li S, Che Z, Lin H (2016) Chitosan-based coating with antimicrobial agents: preparation, property, mechanism, and application effectiveness on fruits and vegetables. Int J Polym Sci 4851730. https://doi.org/10.1155/2016/4851730

Xing W, Tang Y (2022) On mechanical properties of nanocomposite hydrogels: searching for superior properties. Nano Mater Sci 4:1–16. https://doi.org/10.1016/j.nanoms.2021.07.004

Yang S, Leong KF, Du Z, Chua CK (2001) The design of scaffolds for use in tissue engineering Part I. Traditional factors. Tissue Eng 7(6):679–689. https://doi.org/10.1089/107632701753337645

Yousef S (2016) Polymer nanocomposite artificial joints. INTECH. https://doi.org/10.5772/62269

Zarei M (2017) Portable biosensing devices for point-of-care diagnostics: recent developments and applications. TrAC Trends Anal Chem 91. https://doi.org/10.1016/j.trac.2017.04.001h

Zawidlak-Węgrzyńska B, Rydz J, Musioł M, Radziwon-Balicka A (2024) Polymer-drug anti-thrombogenic and hemocompatible coatings as surface modifications. Pharmaceutics 16(3):432. https://doi.org/10.3390/pharmaceutics16030432

Zhang Y, Yu X, Cheng Z (2022) Research on the application of synthetic polymer materials in contemporary public art. Polymers 14(6):1208. https://doi.org/10.3390/polym14061208

Zhang Z, Wang J, Hou L, Zhu D, Xiao H-J, Wang K (2025) Graphene/carbohydrate polymer composites as emerging hybrid materials in tumor therapy and diagnosis. Int J Biol Macromol 287:138621. https://doi.org/10.1016/j.ijbiomac.2024.138621

The manufacturer's authorised representative in the EU is Springer Nature Customer Service Centre GmbH, Europaplatz 3, 69115 Heidelberg, Germany. If you have any concerns regarding our products, please contact ProductSafety@springernature.com

Printed and bound by CPI Group (UK) Ltd, Croydon, CR0 4YY
01/10/2026
02225921-0013